I0069849

HANDBOOK

OF

VACUUM TUBES

AND

TECHNIQUES

IN

MILITARY ELECTRONICS

edited by

Benjamin M. Dodd

Electronic Engineering Series

Wexford Press
2008

TABLE OF CONTENTS

1. INTRODUCTION

This publication is Appendix-A to Military Handbook No. 211, Techniques for Application of Electron Tubes in Military Equipment, which was published by the Government Printing Office in December, 1958.

Handbook No. 211 represents the first attempt to present within one cover all available application data pertaining to the receiving tube types designated for use in military electronic equipments (i.e., those receiving types listed in MIL-STD-200). The material includes not only general instructions in the interpretation of specifications and in the use of statistical design techniques, but also detailed data derived from MIL-E-1 specifications and from manufacturers' life-test records.

This Appendix goes beyond the handbook in dealing with various application problems which are commonly encountered in complex equipments. Part I discusses the design of circuits. Part II provides -- for receiving tube types which have been added to MIL-STD-200 -- the same type of application data which are given in Parts III and IV of the handbook.

2. APPLICATION NOTES

This part of the **Appendix** is a compilation of application notes based on experience in solving part-reliability problems encountered in the design of electronic circuits. The discussion describes the measures taken to prevent or overcome the most commonly observed tube-reliability problems.

2.1 Long-Life Assurance for Electron Tubes

This section discusses two important approaches to the improvement of tube life: (1) reduction of cathode operating temperature to the minimum value, and (2) reduction of bulb temperature.

2.1.1 The Effect of Cathode Operating Temperature on Tube Life

The effect of cathode temperature, and therefore of heater voltage, on the life expectancy of tubes has frequently been observed,* and the observations agree with what is known about the physical mechanisms involved in the relationship.

In studies of the phenomena involved in the deterioration of tubes, ARINC has cited both theoretical concepts and experimental data which show that the exponential relationship is very frequently observed in association with thermal effects. Figures 1, 2, 3, and 4 show, respectively, the effects of temperature on the formation of interface resistance and the decay of pulse emission, on heater burnout, on heater-cathode leakage, and on insulation leakage in general. For the sake of simplicity, it might be said that all these phenomena, including emission decay, follow Arrhenius' Law, which states that the reaction rate (r) is an exponential function of temperature. Thus

$$r = A\epsilon^{-E/kT}$$

If the constant $\hat{E}$ (the activation energy of the phenomenon) is expressed in electron volts, the activation energy corresponds roughly to the work function of the Dushman equation controlling the emission of electrons.

In all the deterioration studies made by ARINC, activation energy has been estimated to have values between 1.5 and 3 electron volts. The estimates have a tendency to be lower than the true value because

---

* See, for example, Thomas H. Briggs, Electron Tube Operation as Influenced by Temperature and Voltage (Technical Report 56-53), Wright Air Development Center, January 1956.

3

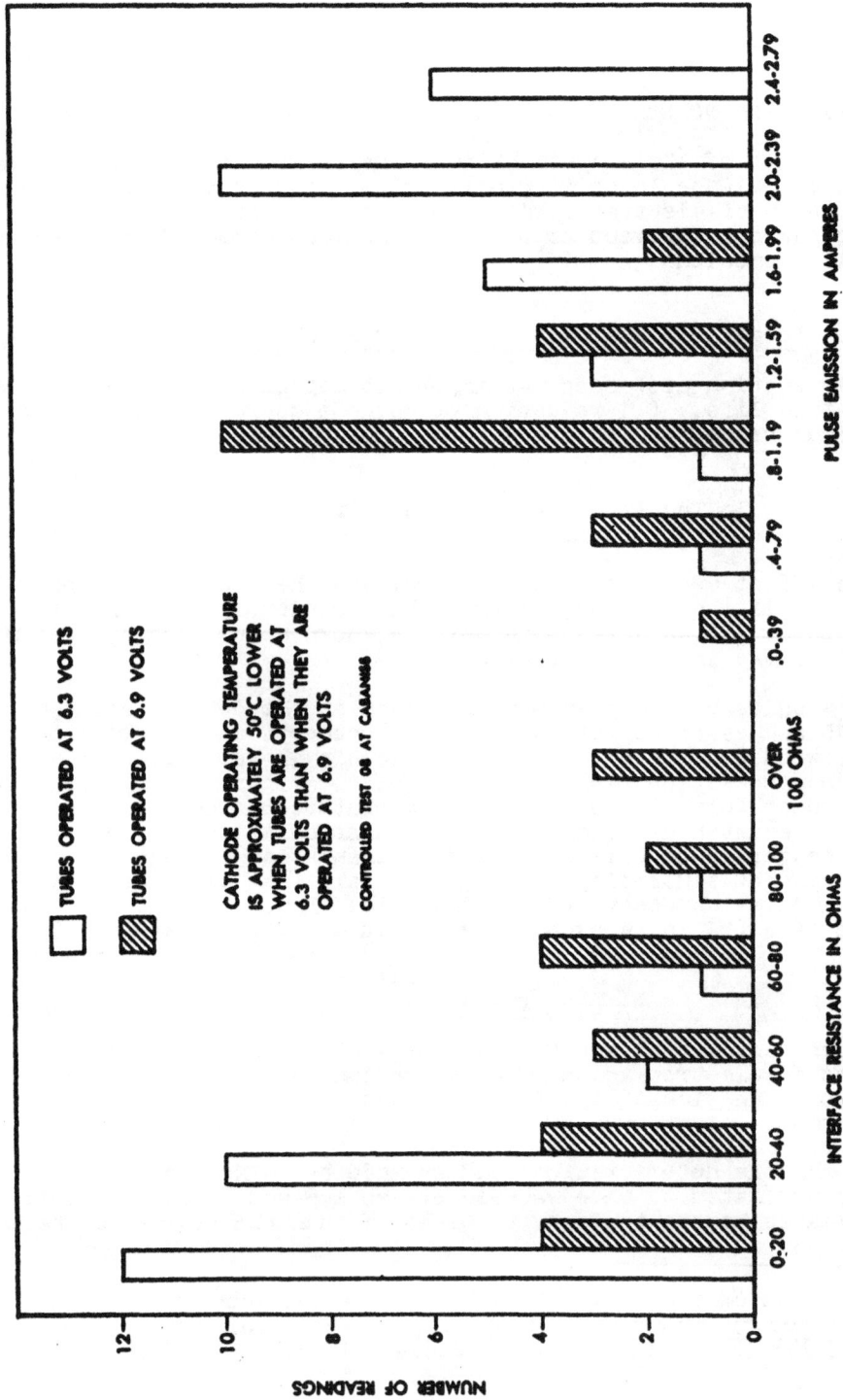

FIGURE 1

TUBE TYPE 12AT7: INCREASE IN INTERFACE RESISTANCE AND DECREASE IN PULSE
EMISSION, AS FUNCTIONS OF CATHODE OPERATING TEMPERATURE

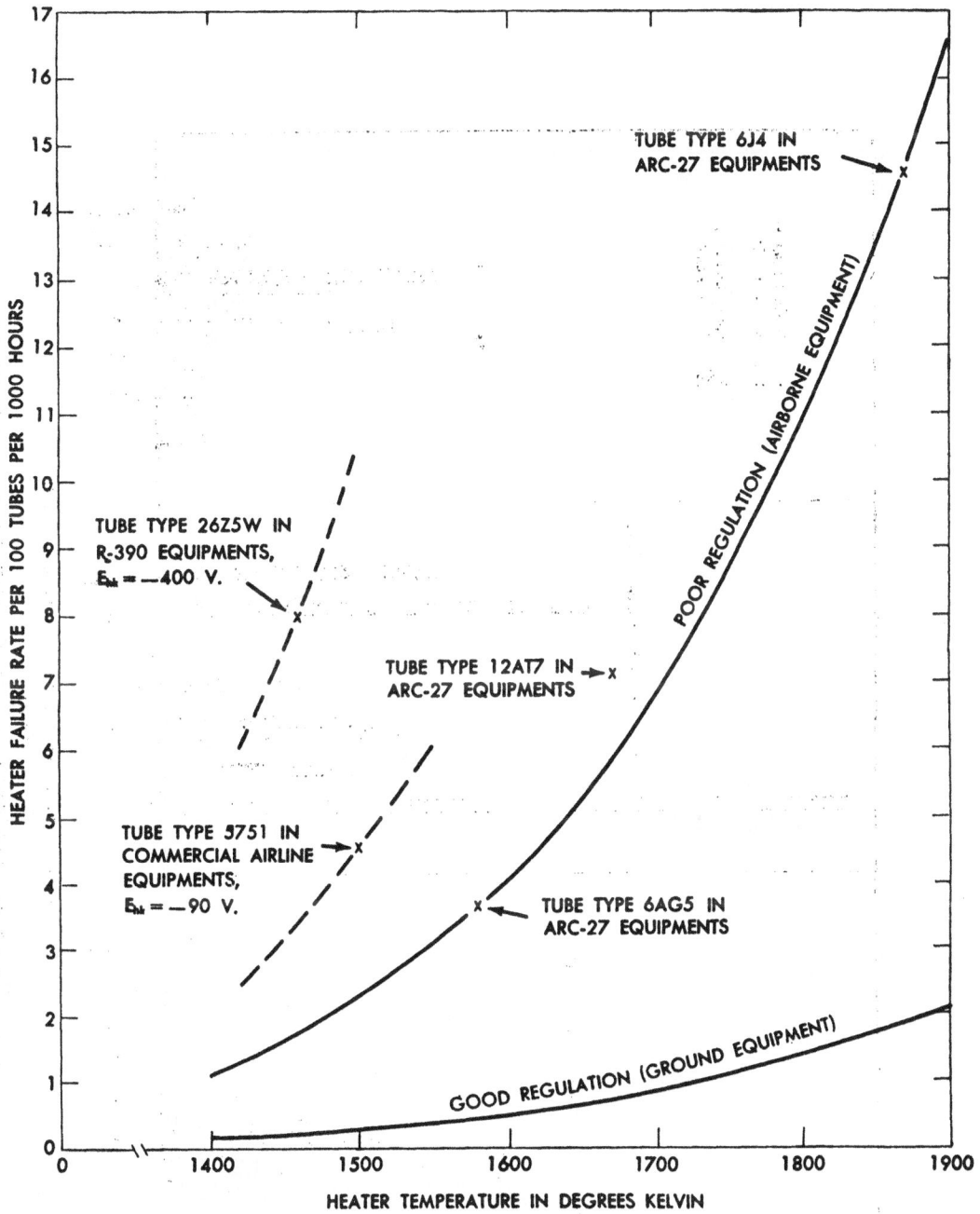

FIGURE 2
THEORETICAL MODEL FOR HEATER BURN-OUT FAILURE AS A FUNCTION
OF HEATER TEMPERATURE
(SHOWING EFFECT OF VOLTAGE REGULATION ON BURN-OUT FAILURES)

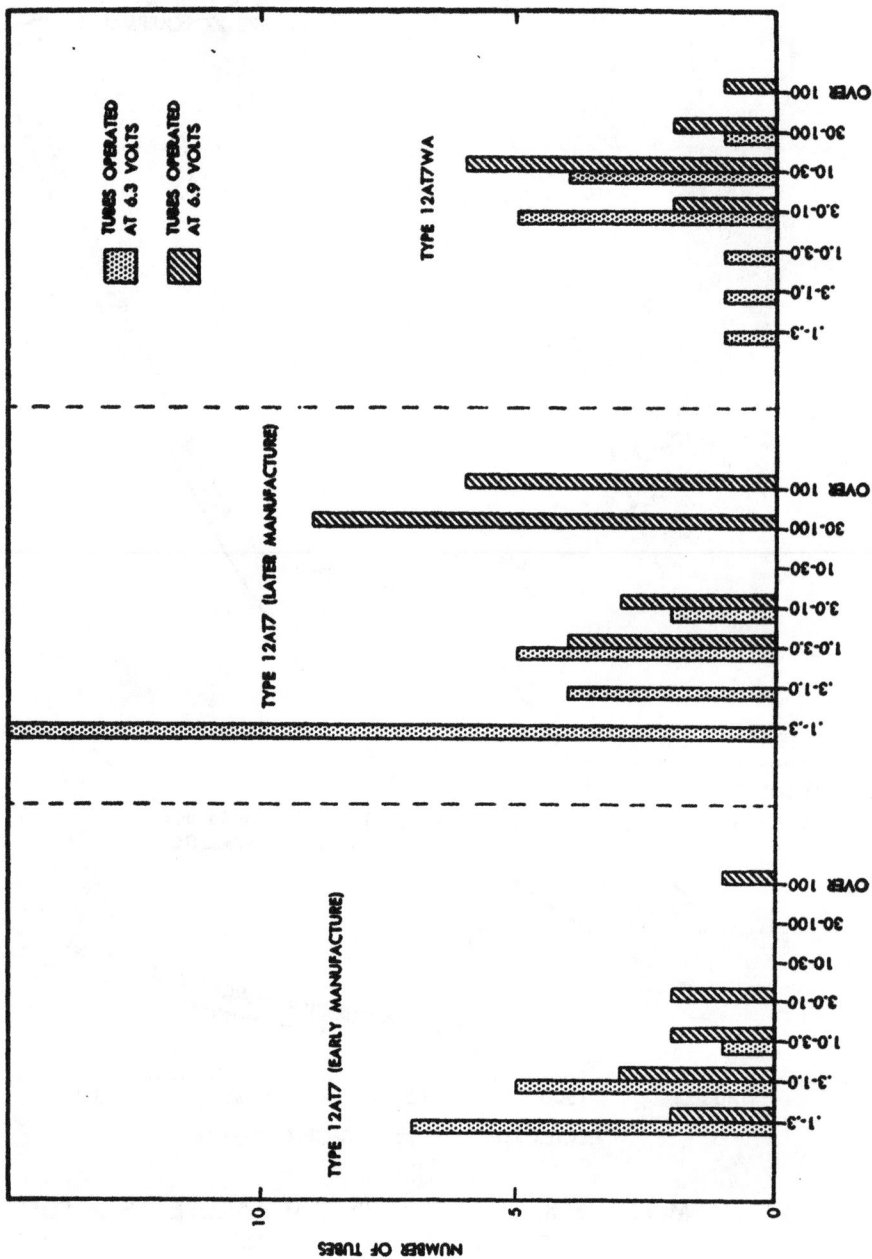

POSITIVE HEATER-CATHODE LEAKAGE, IN MICROAMPERES

FIGURE 3

DISTRIBUTION OF HEATER-CATHODE LEAKAGE UNDER TWO CONDITIONS OF
HEATER-SUPPLY VOLTAGE

THREE VERSIONS OF TUBE TYPE 12AT7—MEASUREMENT AT APPROXIMATELY 700 HOURS

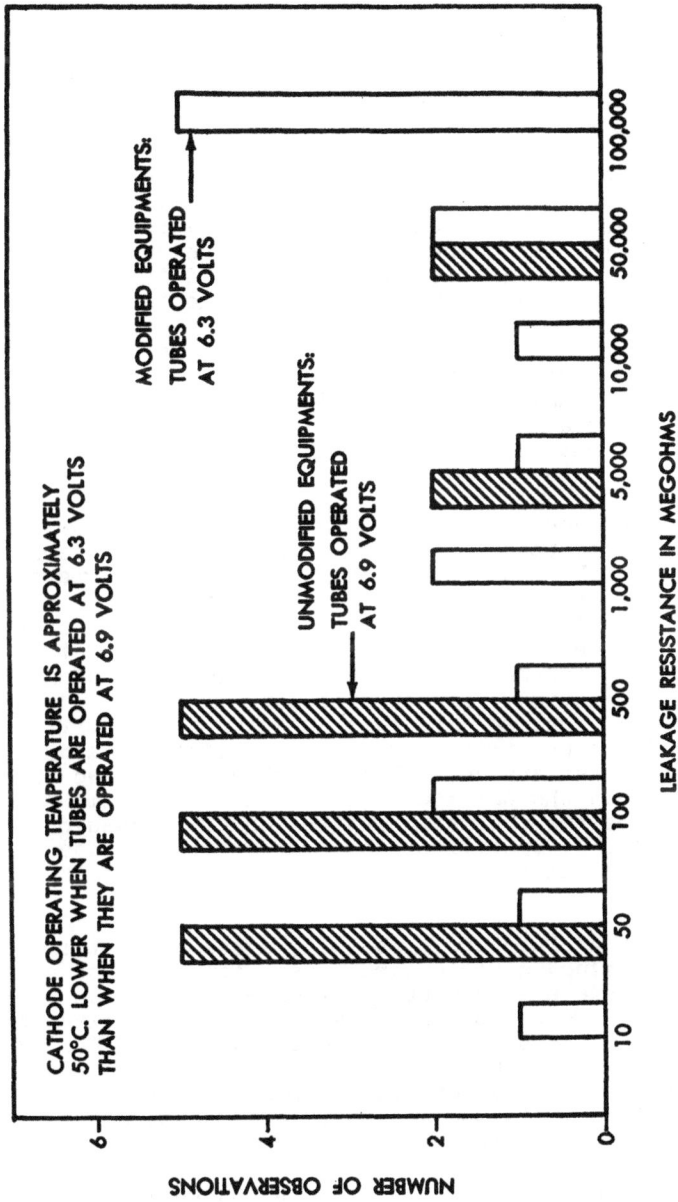

FIGURE 4

VALUE OF LEAKAGE RESISTANCE BETWEEN GRID AND OTHER ELECTRODES AS A FUNCTION OF TWO LEVELS OF CATHODE TEMPERATURE, AFTER APPROXIMATELY 700 OPERATING HOURS

7

of effects due to phenomena other than those studied. This situation
implies that deterioration phenomena, in general, vary with temperature
at a faster rate than emission does. The inescapable conclusion to be
drawn is that if an attempt is made to improve emission by increasing
the temperature of operation, deterioration effects will be encountered
at a much earlier point in the life of the tubes. Therefore, the best
way to obtain long life is to reduce the operating temperature to the
minimum value compatible with emission requirements.

The major obstacle to the successful application of the principle
stated above is the complete lack of control of the emission-
temperature characteristics of present-day receiving tubes. Usually,
tubes are operated at the point where the saturation current available
at the operating temperature is between 100 and 1000 times larger than
current density. It is possible, therefore, to make tubes passing
usual specifications for most characteristics, but exhibiting differ-
ences as great as 10-to-1 in saturation-emission characteristics --
either at the operating temperature corresponding to the center value
of heater voltage, or at any other temperature. It is also possible
to make tubes whose saturation-emission characteristics are very similar at one temperature but significantly different at every other
temperature.

To illustrate the type of information required by the designer,
and usually not available to him, Figures 5 and 6 show the variation
in transconductance which occurs with time when cathode temperature
is varied by variation of heater voltage. The curves for zero hours
show the previously mentioned effect of saturation emission on the
value of transconductance. The curves for various periods of operation show the effect of interface resistance superimposed upon the
effect of saturation emission. If the operating condition is that of
a high-gain, resistance-coupled amplifier in which plate current is
very near the cut-off point, transconductance is limited mainly by the
formation of interface resistance. (Here it should be noted that the
curves probably fit only the particular production lot from which the
tubes were taken, since the operating regions under discussion are
completely uncontrolled by specifications. However, in spite of wide
variations among individual readings, the general shape of the curves
is typical.)

The curves show that for tube type 5814, a heater voltage of 12.6
volts (the specified value) gives optimum operation only for a period
of 500 hours. To achieve a life expectancy greater than 500 hours
under the conditions of operation of the amplifier under discussion,
it is necessary to decrease the value of heater voltage. For example,
operation for 8000 hours at a heater voltage of 12.6 volts would result in a transconductance value of 1200 micromhos at the end of life;
however, operation at 10 volts would result in a transconductance value
of 1700 micromhos after the same period of time.

It should also be noted here that the data shown in Figures 5
and 6 were measured under the operating conditions given in military
specifications. In the curves for zero hours, the difference between
the transconductance values measured at 12.6 volts and 10 volts is not
an indication of the difference which would occur under the operating
conditions of the amplifier circuit. Under MIL-E-1 conditions, plate
current is well above the cut-off value and saturation emission has a
strong effect on transconductance. At conditions near cut-off,

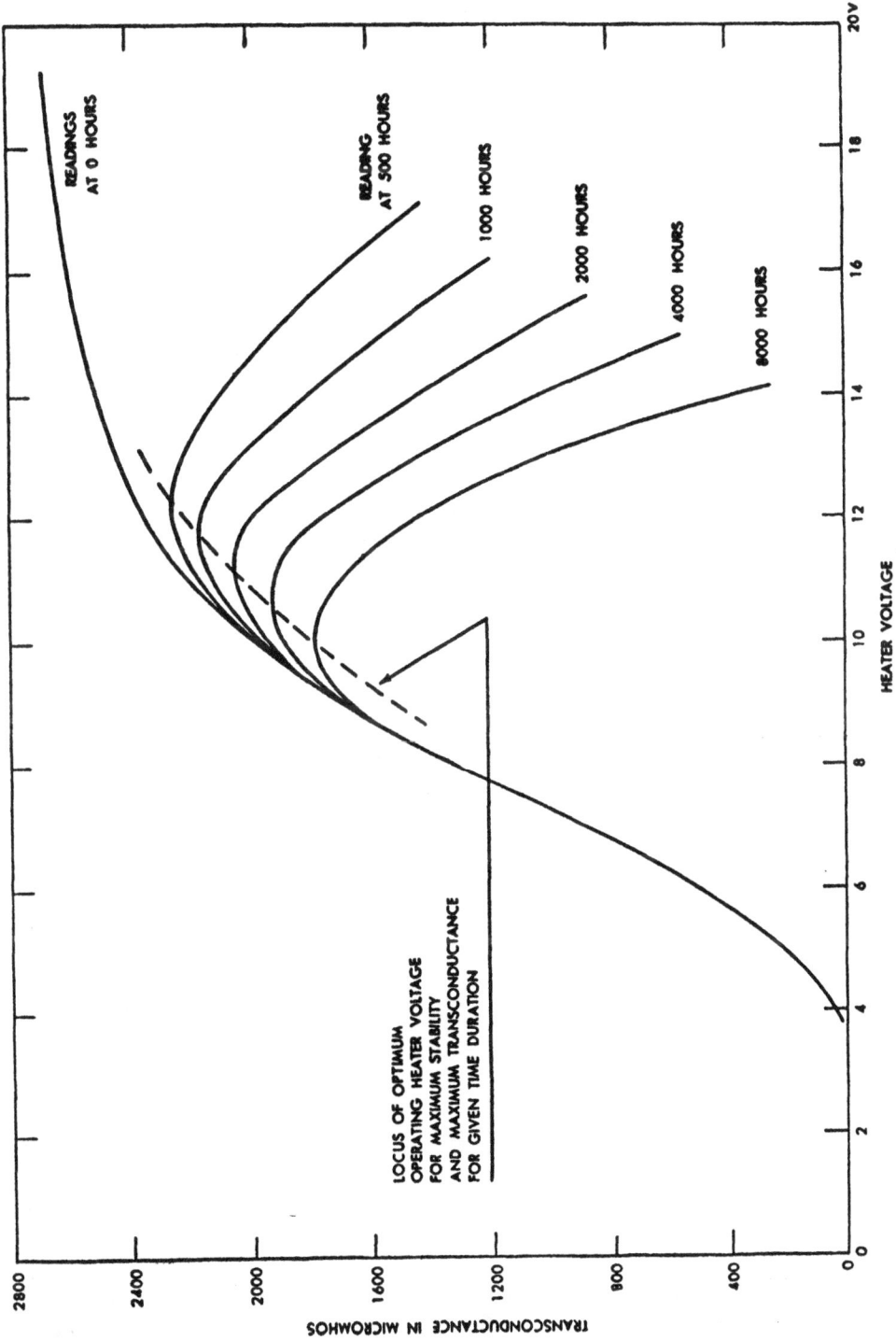

HEATER VOLTAGE
FIGURE 5

VARIATION IN TRANSCONDUCTANCE WITH TIME AND OPERATING TEMPERATURE
OF THE CATHODE (HEATER VOLTAGE) TUBE TYPE 5814 OPERATED NEAR CUT-OFF

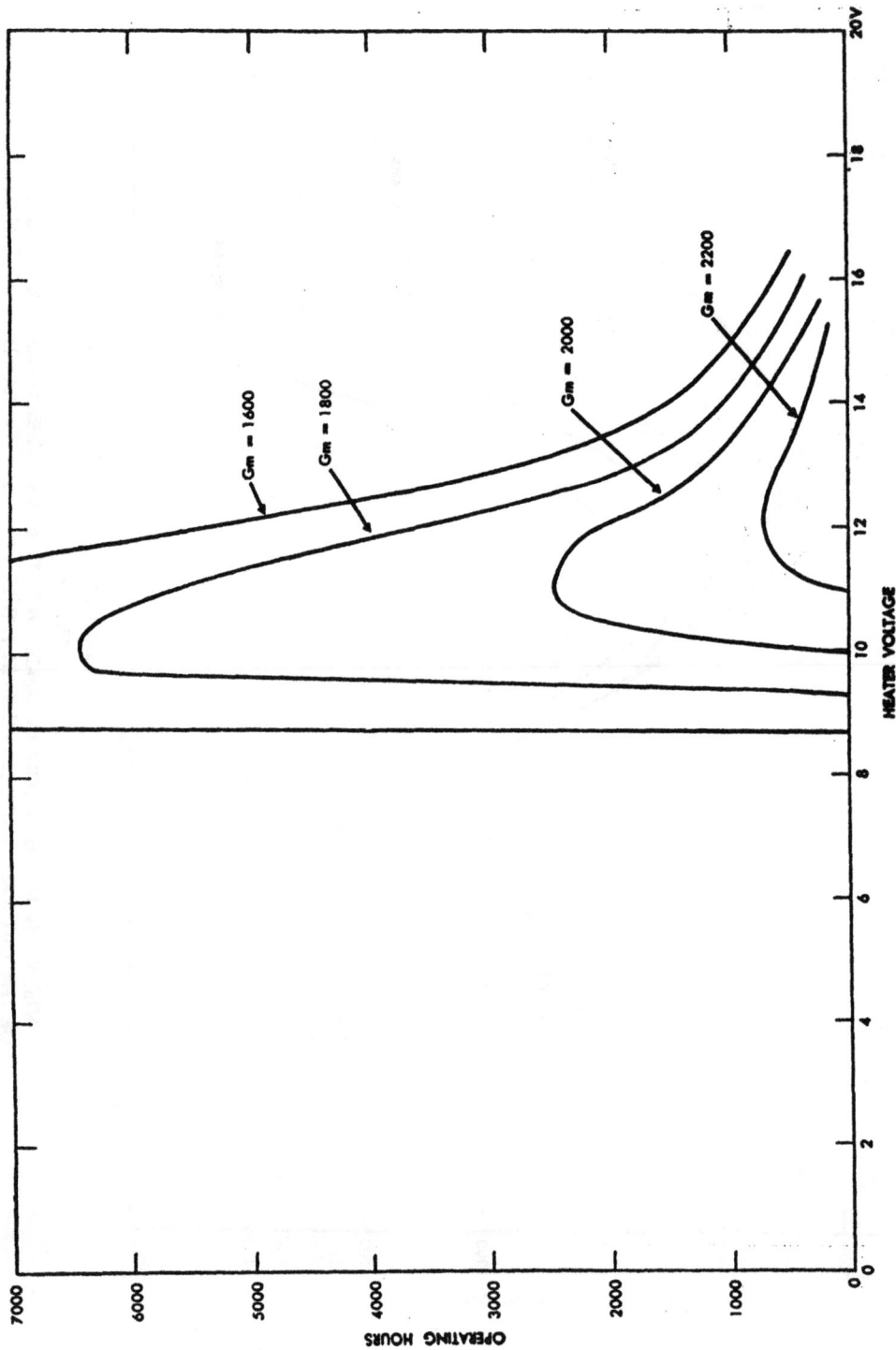

CONTOURS OF CONSTANT TRANSCONDUCTANCE IN THE: OPERATING TIME/OPERATING TEMPERATURE (HEATER VOLTAGE) PLANE
TUBE TYPE 5814 OPERATED NEAR CUT-OFF

saturation current at low heater voltages is plentiful enough to
produce full space-charge-limited operation, and the gain of the amplifier will not vary with heater voltage. However, transconductance will
decrease with time as a function of the formation of interface resistance, and therefore will vary with heater voltage as shown in the
figures.

The effect of variations in voltage supply requires that great
care be taken in the design of electronic equipment in which long life
is desired. It has been the practice of designers of military electronic equipment to assure performance even under the extreme conditions imposed by the use of a minimum voltage supply. This practice
has generally produced equipments in which deterioration is rapid, but
in which performance is practically unaffected by voltages 5 or even
10 percent lower than the center value.

Having observed such conditions of operation during field investigations, ARINC has performed several experiments in which the voltage
applied to equipments was reduced in order to obtain a lower rate of
tube removal.

In a recent test involving 16 communication receivers operated
between 6000 and 8000 hours, half of the receivers were operated with
a supply voltage purposely reduced 8 percent below the center value;
the rest were operated at a center value of supply voltage. The mean
time-between-malfunctions for the receivers operated at reduced supply voltage was almost three times as long as that for the equipments
operated at center-value supply voltage.

In all the experiments performed so far, the expected improvement
has been observed. The amount of improvement under each condition of
operation has not always been consistent -- usually because concomitant
phenomena mask the effect under investigation -- but the experimental
results have agreed so well with expectations that it now seems safe to
to formulate a working hypothesis* which can be applied to nearly all
electronic equipment:

(1) Determine the average and expected extreme variations
of supply voltage used for the equipment.

(2) Analyze the probability of occurrence of the lowest
range of voltage, and the effect of this range on the
performance characteristics of the equipment.

(3) If the above analysis shows no dangerous shortcomings,
reduce the center value of the voltage by no more than
one-third of the extreme deviation from the original
center value.

----

* Regulation of electron-tube heater supplies in the past has been
limited to constant-voltage transformers, complex regulating circuits or ballast tubes. The use of the recently developed power-
zener regulator diodes now appears to be feasible, offering the
advantages of small size and weight.

Several methods of reducing supply voltage are possible. The simplest one is readily applicable to equipments operated by a-c current, 60 or 400 cycles per second. In these equipment, the primary winding of the power transformer is frequently supplied with three separate taps: one for the center value, one for high voltage, and one for low voltage. Each tap causes a 5- or 10-percent change in secondary voltage.

For equipments not using primary taps, or for equipments using the highest tap, a separate auto-transformer may be satisfactory. For equipments operated by d-c current, a dropping resistor provided with proper short-circuiting switch may be used. In the latter case, the dropping resistor can be inserted in the tube heater strings only. Most high-voltage power supplies are regulated, and therefore would not be affected by the lower voltage, except insofar as the regulator tube is concerned.

The procedure described here assures flexibility of choice in case previous analysis shows that some tubes, such as transmitting tubes, are sensitive to variations in heater voltage. Usually heater voltage produces the major effect in retarding deterioration phenomena, because the cathode in electron tubes is the element most subject to variations in deterioration with temperature.

Whenever the supply voltage of an in-service equipment is decreased, as suggested above, it must be realized that some part which operated satisfactorily at higher voltages may become marginal at the lower voltage. This is particularly true of electron tubes which have given long service under high-voltage conditions. When voltage is reduced, cathode temperature decreases, and tube performance will rapidly reach an unsatisfactory level. The resulting increase in tube removals may appear to disprove the theory that reduction of supply voltage has beneficial effects. Actually, the increase in removals is due to other causes -- principally, the high rate of deterioration which takes place when tubes are operated at high voltages. To avoid multiple tube failures after a reduction in supply voltage, it is recommended that all tubes be tested and replaced if found to be near the end of life. Time spent in replacing tubes at the time of voltage change-over will be amply compensated by the lower rate of removal in subsequent operation.

2.1.2   The Effect of Bulb Temperatures on Tube Life*

The influence of temperature upon the reliability of electronic equipment in general has been recognized only in the last few years. It is most apparent in aircraft, where space is at a premium, because airborne equipments have a much higher "packing factor" than do ground-operating equipments. Recent investigations have thrown some light on the way bulb temperatures affect electron-tube reliability.

---

* The material in Sections 2.1.2 and 2.1.2.1 is taken from ARINC's General Report No. 2, pp. 75 and 78 - 85.

There have been many studies on the effect of temperatures on tube life. One in particular dealt with the effect of plate temperatures.* The hypothesis underlying this research was that the gases absorbed by the plate during fabrication are diffused in the tube at a rate proportional to the plate current, and that these gases cause poisoning of the cathode. The study resulted in the development of de-rating curves for various tube types: for each condition of ambient temperature, a reduced plate-dissipation value was assigned, at which a constant plate temperature would be maintained.

The study just described was based on the assumption that plate temperature is the only important consideration in determining power rating. However, research by a manufacturer has shown that contaminating gas is released not only by the plate current, but also -- and in larger quantities -- by the bulb.** Therefore, deterioration effects in tubes would be accelerated by high bulb temperatures, even if the plate temperature were held constant. This fact is so well recognized by tube engineers that extensive operations are performed in order to outgas the bulbs for electron tubes. However, recent research has revealed the complexity of the phenomena involved and the limitations on operating temperatures.

Gas is adsorbed by the surface of the glass bulb and absorbed in the glass itself. The adsorbed gas can be practically eliminated by the usual vacuum-heating process during the exhaust cycle, but the absorbed gas presents an entirely different problem. Dr. B. J. Todd has published a report on studies of the outgassing of glass, at different temperatures, as a function of time.† In this report, he shows that the gas diffused from the interior of the glass is composed principally of water vapor (98 to 100 percent of the total gas depending upon temperature). He points out that the rate of diffusion of water vapor from glass is proportional to the square root of the time involved for an almost infinite duration; therefore, it would be practically impossible to heat the glass long enough, or hot enough, to draw all the water vapor out of it. For example, he indicates that it would take 37 years of heating glass of one-millimeter thickness at 430°C to produce a depletion effect that would result in a deviation in the volume of gas diffusion from the square-root law quoted above.

---

* Schmidt, B.M., Temperature-Pressure Derating of Electron Tubes, Phase I, University of Dayton, Dayton, Ohio; Contract AF 33 (616)-113, Test Report WCLCV - 6, May 25, 1953. Also, Schmidt, B.M., A Study of Environmental Temperature and Pressure Effects on the Plate Dissipation Rating of Receiving Tubes, WADC Technical Report 53-433, University of Dayton, Dayton, Ohio, December 1957.

** Bowie, W.S., Study of Electron Tube Bulb Temperature Ratings, General Electric Company, Reliability Analysis Laboratory, Owensboro, Ky., Contract AF 33(038)28636, Proposal I and Quarterly Progress Reports.

† Todd, B.J., "Outgassing of Glass", Journal of Applied Physics. Vol. 26, No. 10, October 1955; American Institute of Physics, New York City.

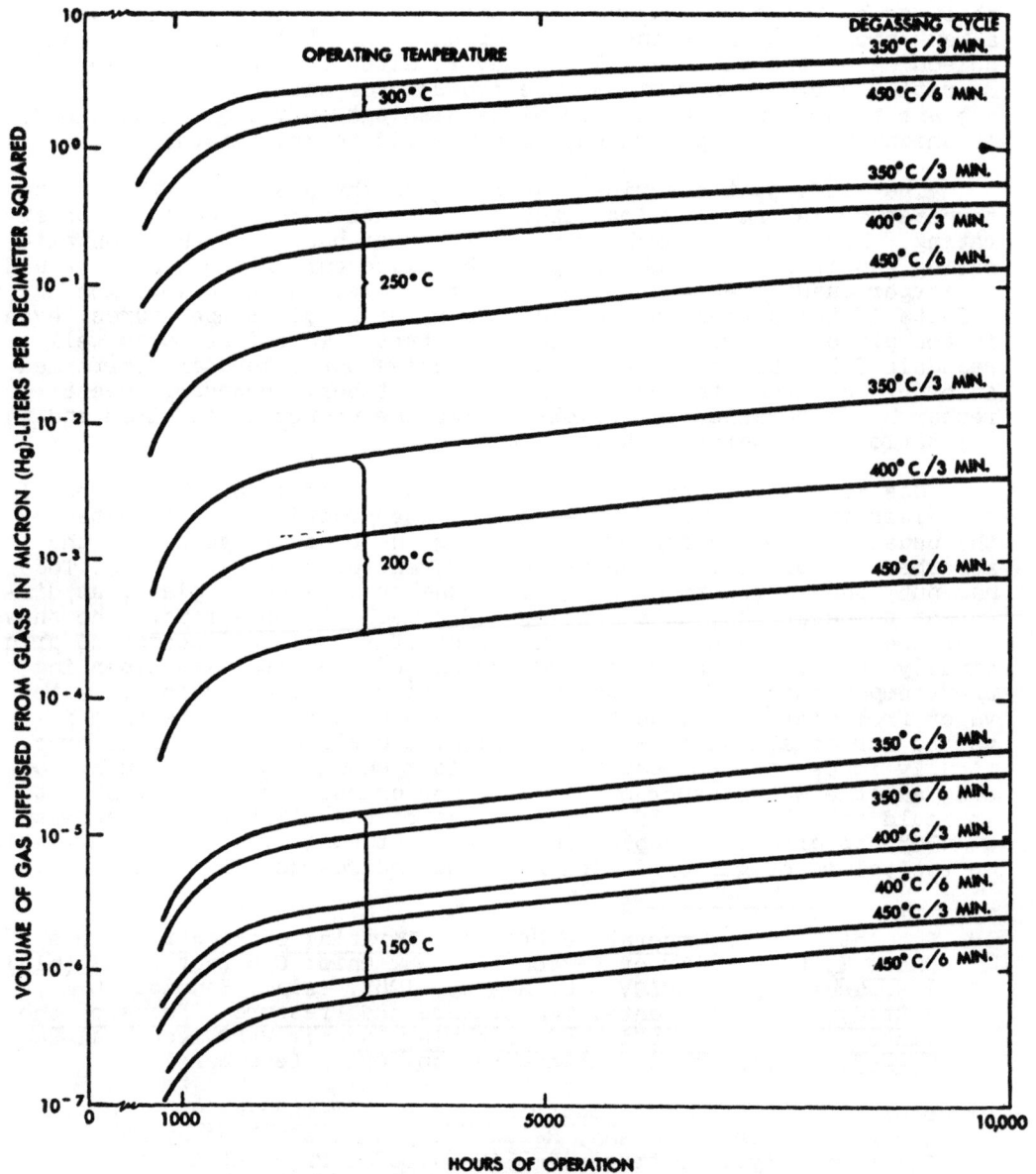

FIGURE 7
EFFECT OF DEGASSING CYCLE ON GAS DIFFUSION
(VARIATION IN GAS DIFFUSION FROM LIME-GLASS BULB (008),
SHOWN AS A FUNCTION OF OPERATING TIME UP TO 10,000 HOURS)

Figure 7 was derived from the data in Dr. Todd's report. It shows the volume of gas diffused from a bulb made of lime glass (008) at four different temperatures, as a function of operating time up to 10,000 hours. For each of the four temperatures, several degassing cycles are shown. Thus, the processes usually taking place in a tube during exhaust can be simulated.

This figure suggests several interesting conclusions. The first is that, after approximately 1000 hours of operation at a bulb temperature as high as $300^\circ C$, there is little difference between the amount of gas evolved from the bulb of a thoroughly degassed tube and the amount from the bulb of a tube that was not so completely degassed during fabrication. This observation contradicts the wide-spread belief that a tube can be made to operate for as long a period at high temperatures as it does at lower temperatures, provided that it is more thoroughly degassed during exhaust.

On the other hand, another observation to be made from Figure 7 is that the difference between thoroughly degassed tubes and tubes less completely degassed is quite significant when the operating temperature of the bulb is $200^\circ C$ or lower, and that this difference is maintained for long periods of time. Still another remarkable fact is that the 50-degree difference between an operating temperature of $150^\circ C$ and one of $200^\circ C$ results in a difference of almost 1000-to-1 in the volume of gas diffused. Evidently, the tube-degassing process is much more effective in minimizing the amount of gas evolved during operation if the tubes are operated at the lower temperatures than it is if they are operated at temperatures above $200^\circ C$.

It is evident from Figure 7 that the amount of water vapor diffused from glass bulbs at bulb temperatures over $150^\circ C$ is by no means insignificant. If no gas were absorbed by the getter or the cathode, a tube in a miniature-medium bulb that had been processed at $400^\circ C$ for three minutes would build up approximately 10^{-3} micron (Hg)-liters of water vapor in 10,000 hours of operation at $200^\circ C$. This volume of gas in a tube of that size would alone be enough to produce a pressure from 10 to 100 times as great as is expected in a well evacuated tube. Actually, it is unlikely that so much pressure would develop, because the total volume of gas evolved would be influenced to some extent by the effect of various absorbing and adsorbing materials, the most active of which is the getter.

Depletion of the metallic barium deposit on the getter of tubes operating for extended periods of time at high bulb temperatures has been observed repeatedly by ARINC and many other investigators. Such depletion and other phenomena just described point to water vapor from the bulb as the dominant phenomenon causing deterioration in tubes operated at high bulb temperatures. Other observations confirm this theory. The metal parts in tubes are usually degassed during exhaust at temperatures much higher than the tube operating temperature, and the gas desorbed from such parts is principally hydrogen carried over from the firing process. Mica, the only material not subjected to high-temperature degassing, is known to produce large quantities of gas when its temperature is sufficiently high to cause release of its waters of crystallization, or when it is stressed mechanically in a way that causes separation of the crystals at cleavage planes. Water vapor emanating from the mica would add to the water vapor from the bulb, and would accentua⁺ the effect. Thus, it seems logical to accept water vapor from the bulb as the dominant phenomenon causing deterioration in tubes operated at high temperatures. Just how this gas produces deterioration and reduces tube life remains to be explained.

One of the direct and pronounced effects of gas is well recognized. An ion current, proportional to the gas pressure and to the total current flowing in the tube, is collected by the control grid. Flowing through the external grid-circuit resistance, it produces a positive grid voltage that increases the total current in the tube. If the increase in current is sufficiently large to cause a substantial increase in power dissipation at the surface of the bulb, the phenomenon may become self-sustaining, and may result in arc-over and complete destruction of the tube.

A less violent but more important effect of gas is gradual poisoning of the emitting surface. The work-function value required for good emission is obtained by an excess of donor centers in the barium-oxide lattice or by a vacancy of oxygen ions. It is logical to assume that the presence of oxygen gas tends to destroy the donor centers by oxidation, a process which is proportional to the amount of oxygen available. If the effect of oxidation is greater than the total effect of reduction due to other causes, the lattice balance will be restored and emission will drop to a low value.

The effect of oxygen poisoning on cathode emission has been studied by many experimenters. Herrmann and Wagener* have reported several experiments which indicate that, when the cathode is operating at 1000°K, an oxygen pressure of the order of one micron (Hg) can reduce the emission by several orders of magnitude in less than a minute. As was mentioned earlier, the effect of oxidation depends upon the temperature of the cathode, and is balanced out by reducing factors which are dependent upon the temperature and condition of the activating materials.

2.1.2.1 Heat-Transfer Mechanisms Affecting Bulb Temperature

The preceding discussion of bulb temperatures has been presented in order to convey a general idea of the order of magnitude of the phenomenon. An attempt has also been made to indicate the limits beyond which the phenomenon becomes overpowering, but much more investigation is needed to determine exactly what quantity of water vapor from the bulb can be tolerated in a tube. As has been shown, this depends upon the amount of gas absorbed by other tube elements; the amount of oxygen that finds its way to the emitting surface; the effect of other gases on cathode poisoning; and the interactions among gas poisoning, cathode temperature, and activation of the barium-oxide layer.

A prerequisite to investigations of these factors is adequate information on the bulb temperatures to be expected in various equipments in the field. Such information can be obtained only by a careful analysis of the heat-transfer mechanism in electron tubes.

---

* Herrmann, G., and Wagener, S., The Oxide-Coated Cathode, Vol. II, 1951; Chapman and Hall, London, England.

The elements in a tube must dissipate their power chiefly by radiation; but, contrary to a general belief, the glass bulb is practically opaque to infrared radiation of the frequency encountered in most receiving tubes. Therefore, the internal power, dissipated by the elements through radiation to the bulb, must be conducted through the thickness of the glass and then dissipated by radiation, conduction or convection. The heat conductivity of a glass bulb is low (approximately 1/500 that of the conductivity of copper). As a result, the distribution of heat on the surface of the bulb is far from uniform -- a fact to be kept in mind when bulb-temperature measurements are reported. The method used in obtaining bulb-temperature measurements given in this report is to select the hottest spot on the bulb and consider it an index of the temperature of the bulb.

The mechanisms through which a bulb is cooled are conduction through the pins and base of the tube, and convection and radiation through the surrounding atmosphere. If the atmosphere is still, as it is in many electronic equipments where no provision has been made for ventilation, there is little transfer of heat. Consequently, the ambient temperature increases, causing the bulb temperature to increase still further.

Circulation of air improves heat-transfer, if the air is cooled by an effective heat-sink. In this event, the heat-transfer is proportional to the square root of the air velocity and to the difference in temperature between the bulb and the cooling air.

Heat-conduction through the base of the bulb can be effective only if the chassis on which the base is secured is maintained at a temperature lower than that of the surrounding air by some artificial means.

Many studies have been conducted recently* in an attempt to obtain a correlation between internal power dissipation and bulb temperature

* (1) Klass, P., "New Shield Insert Reduces Tube Heat," Aviation Week, Vol. 64, No. 8, February 20, 1956; McGraw-Hill Publishing Co., New York City; (2) Woods, L., An Evaluation of Shields for Subminiature Electron Tubes, International Electronic Research Corp., Burbank, Cal.; (3) Woods, L., Effect of Tube Shields on Miniature Electron Tubes, International Electronic Research Corp., Burbank, Cal.; (4) Sylvania Electric Products, Inc., High Temperature and Altitude Life Evaluation of Subminiature Tubes, Engineering Report on Supplemental Agreement No. 8 to Contract AF 33(038)-9853, December 1,1955; Radio Tube Division, Emporium, Pa.; (5) Mills, B.D., A Method of Determining Vacuum Tube Rating At Very High Altitudes, Report No. 55, February 6, 1952; Brimar Valve Works, Kent, England; (6) Mills, B.D., and Wright, W., "The Rating of Thermionic Valves for Use Under Abnormal Conditions," Journal of Electronics, Vol. 1, November 1955; Taylor and Francis, Ltd., London, England; (Brimar Valve Works Report No. 82-G); (7) Wallin, D.R., Vacuum Tube Envelope Temperature Measurements, Research Report No. 594, March 1955; Naval Electronics Laboratory, San Diego, Cal.; (8) Weeks, P.T., "Reliability in Miniature and Subminiature Tubes", Proceedings of the IRE, Vol. 39, No. 5., p. 499 (May 1951); Institute of Radio Engineers, New York City. (9) Passman, H.M., Thermal Evaluation of Various Miniature Tube Shields -- II, Report No. CTR-156, Collins Radio Company, Cedar Rapids, Iowa.

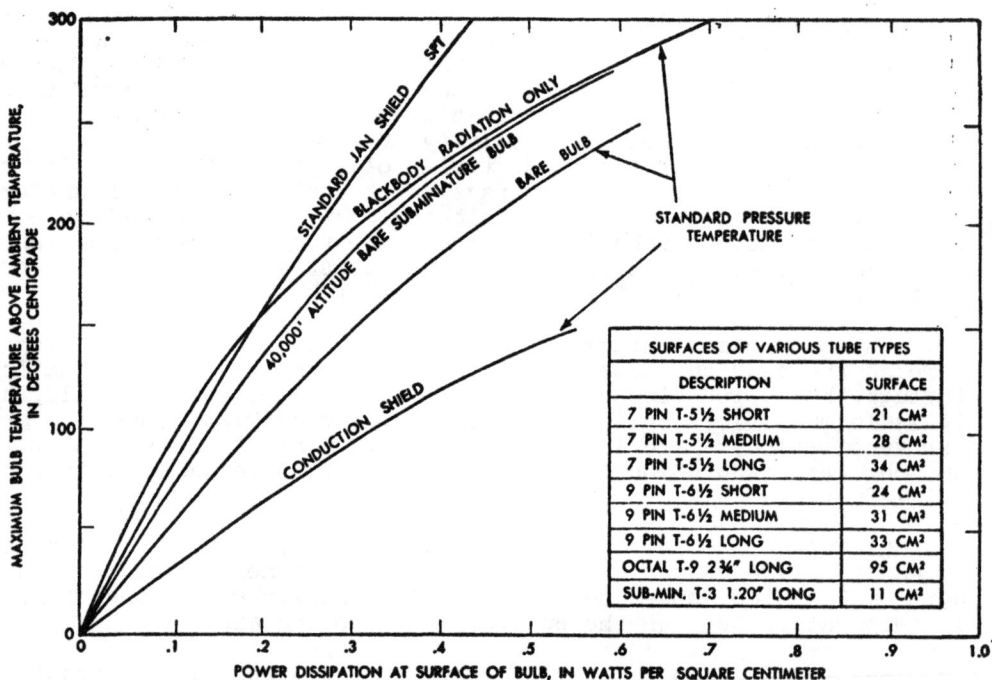

FIGURE 8

INCREASE IN BULB TEMPERATURE AS A FUNCTION OF POWER DISSIPATION AT SURFACE OF BULB

for given values of ambient temperature and pressure. The results are in fairly good agreement, considering the fact that bulb temperature is non-uniform, and, therefore, results obtained by one worker may differ from those obtained by others, merely because of the size or position of the thermocouple used for temperature measurement. All of the measurements that were considered to be in reasonable agreement were coordinated and condensed into graphical form. The result is Figure 8, in which the difference between bulb temperature and ambient temperature for an ambient temperature of 25°C is plotted against the power dissipation per unit surface of the bulb. The table on the figure lists the bulb surfaces of the most commonly used tubes, in order to permit direct determination of the temperature of a bulb, if the total dissipation of the electrodes is known.

In Figure 8, the curve of fundamental importance is the one for a bare bulb in a still atmosphere at standard pressure and temperature (atmospheric pressure at 25°C). At ambient temperatures higher than 25°C, the increase in bulb temperature is slightly less than the figure indicates; but the difference is so small that it can be disregarded for most applications, and the curves can be used for all ambient temperatures that are commonly encountered.

For purposes of comparison, the theoretical curve due to radiation of a black body in vacuum is also shown. Comparison of the shape of this curve with that of the curve for a bare bulb at standard pressure and temperature leads to the conclusion that conduction and

18

convection are the dominant phenomena in the lower region of dissipation for the bare bulb. In the higher region, the curvature of the bare-bulb curve approaches that of the black body, indicating that radiation has become the dominant phenomenon.

Another curve of interest is that for a bare bulb at the pressure corresponding to an altitude of 40,000 feet. The shape of this curve closely approximates that of the curve for the black body, indicating that conduction and convection are rather low at the pressure associated with the selected altitide. For higher altitudes (or lower pressures), the bare-bulb curve would attain higher temperature values than the black body curve, thus reflecting the absence of convection cooling and the effect of the relatively low coefficient of radiation for glass.

However, electron tubes are not ordinarily used as bare bulbs; instead, they are inserted in sockets with shields intended to provide electrostatic shielding and also -- particularly in airborne equipments -- to hold the bulb in place under conditions of shock or vibration. Use of the standard JAN shield is almost universal in military equipments -- in fact, until a few years ago, it was the only approved shield. This shield is made of bright metal and, as Figure 8 shows, its use produces a marked increase in the temperature of the bulb. This standard shield is made in the form of a cylinder which completely surrounds the bulb and is clamped to the socket. The air enclosed between the shield and the bulb cannot escape, and so reaches a high temperature. The effect is aggravated because the bright inner surface of the shield reflects heat back to the bulb.

Many attempts have been made to correct this situation. A logical step is to blacken the surface of the shield, thus increasing the absorptive property of the inner surface and the radiating property of the outer surface. Another method is to put vents in the shield to permit circulation of air, and thus increase the amount of convection. Used in combination, these devices are successful in reducing the bulb temperature, but not quite to the level of bare-bulb temperature.

A more drastic method of cooling the bulb is to provide for direct metallic conduction from the surface of the bulb to the surface of the shield. This has been done in several laboratories* by inserting a thin strip of corrugated metal between the bulb and the shield, so positioned that it is in good contact with both of them. The great advantage of this insert is that it is directly adaptable for use with the standard JAN shield, and so can be utilized to effect an immediate improvement in existing equipments.

A special shield now generally available may provide an even better solution of the problem. This shield consists of a black cylinder with many springlike members that make tight contact with the bulb. As the shield also makes good contact with the base of the socket, thermal conduction through the chassis aids in dissipating the heat from the bulb. This type of shield appears quite effective in decreasing bulb temperatures. However, because of the variation in methods

* Passman, H. M., Thermal Evaluation of Various Miniature Tube Shields -- II, Report No. CTR-156, Collins Radio Company, Cedar Rapids, Iowa.

of measurement used by different investigators, it was not possible
to derive a curve indicating the difference in bulb temperature obtained through use of this shield rather than the JAN shield with the
metal nsert described above. The curve in Figure 8 that is labeled
"conduction shield" is applicable to all shields mentioned that use
the conduction principle; therefore, it should be interpreted with
some degree of approximation. The temperature of the chassis is a
consideration in the evaluation of a shield with a good property of
heat-transfer to the chassis, but it is unimportant in evaluation of
other shields.

The temperature of the heat-sink is extremely important in the
final determination of bulb temperature. This consideration should
be kept in mind, particularly with reference to subminiature bulb
shields. Some of these shields are designed to hold a tube so that,
along its entire length, it is in contact with the chassis. Under
conditions of such close proximity to the chassis, and good heat-
transfer through the shield, the bulb temperature of subminiature
tubes is practically the same as the temperature of the chassis. Even
when subminiature tubes are used bare, conduction of heat through the
stem-leads accounts for a larger percentage of the total heat dissipation than it does in other tubes. This is why a curve for pressure
corresponding to a 40,000 foot altitude has been presented only for
subminiature tubes. None of the data on miniature tubes in this kind
of operation were found to be in sufficient agreement for use with the
other curves.

In view of the considerations discussed above, it is obvious that
the curves in Figure 8 should be used only as a guide in a first ap-
praisal of tube operation. Figures estimated from theoretical consideration should always be checked experimentally, as soon as a first
model having the same configuration as the final product is built.

2.2 Tube Characteristic Spreads and Cathode Bias*

One of the problems confronting the equipment designer using
electron tubes is the spread in electrical characteristics, both from
tube to tube and for any given tube under various operating conditions.
Variations in characteristics will also occur in a particular tube
during its operating life. Unless properly allowed for, such spreads
and variations may cause circuit performance to vary to such an extent
as to render the equipment virtually useless.

For designers of military equipment, an important source of information concerning spreads in characteristics may be the MIL-E-1
specification. Military specifications are quality-control documents,
but they can provide a wealth of information to a designer. For example, the specification sheet for type 5670 shows that the plate current for an individual tube should not differ from design center by
more than ± 28 percent; for type 5814A these limits are ± 39 percent.
Superficially it would seem that tighter controls have been imposed
on type 5670. However, a glance at the test conditions near the top
of the specification sheet shows that bias for the 5670 is derived
from a cathode resistor of 240 ohms, whereas the 5814A is tested with

* Reproduced with permission, A.G.E.T. News Bulletin, January 1, 1958.

fixed bias. As the specifications are written, the limits on plate current for type 5670 with fixed bias would be about ± 75 percent of design center, while with cathode bias the spread for type 5814A would be narrowed to ± 12 percent. Figures 9 and 10 present this information graphically, together with plate-current distributions based on measurements of typical lots of tubes.

One point that is immediately apparent from these figures is that the specified test conditions exert an enormous influence on the characteristic spreads that a circuit designer will encounter in a lot of tubes. Obviously, the test conditions must be taken into account when selecting a tube type or when determining the spread

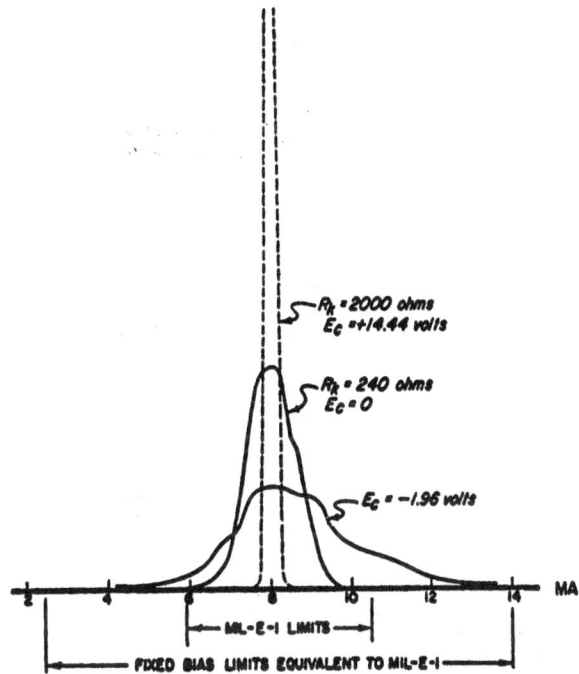

FIGURE 9
PLATE-CURRENT DISTRIBUTION
FOR TUBE TYPE 5670

that the proposed circuit will be required to tolerate. Incidentally, this spread should always be determined from the specification, not from measurements made on a few tubes procured for design purposes. The economics of tube manufacture require that most of the product be somewhere near design center, or the risk of rejecting a lot will be too high. A few tubes chosen at random might all be fairly close to design center, and since both cathode bias and fixed bias yield virtually identical measurements on a design-center tube, the results of measurements made on the small sample might be very misleading. On the other hand, when tubes are procured in production quantities, there is a high probability that a non-trivial number of tubes will approach the specification limits.

2.2.1 Advantages and Disadvantages of Cathode Bias

Referring again to Figures 9 and 10, it is apparent that cathode bias is an effective means of narrowing the spread in plate current that would exist with fixed bias. This is only one of the advantages of employing cathode bias, which are summarized below:

1. Narrowing of spread in characteristics from tube to tube.

2. Reduction of changes in characteristics caused by variations of electrode potentials.

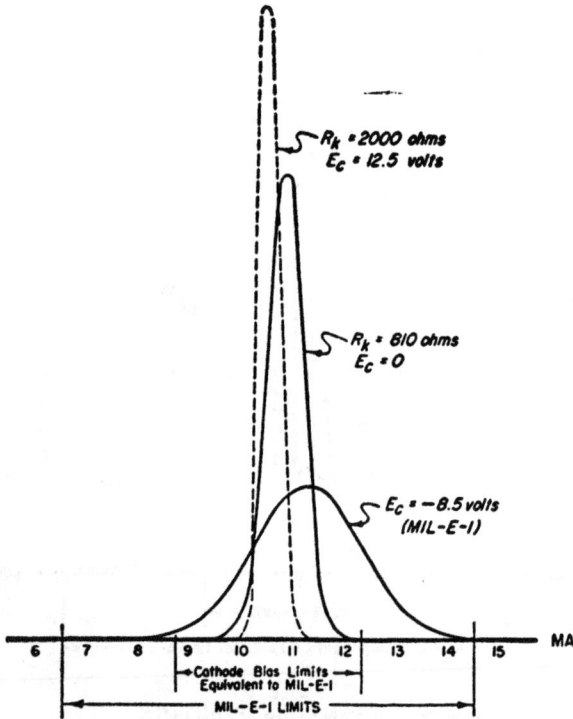

FIGURE 10
PLATE-CURRENT DISTRIBUTION
FOR TUBE TYPE 5814A

3. Reduction of changes in characteristics during operating life caused by emission deterioration, development of interface resistance, etc.

It should be noted that these advantages are obtained even when the cathode resistor is bypassed, as they are the result of d-c negative feedback. The improvements in signal amplification at the expense of gain resulting from unbypassed operation are well known and will not be discussed here.

One disadvantage of cathode bias is that the circuit becomes increasingly sensitive to heater-cathode leakage* as the cathode resistance is increased. Other disadvantages include (a) the need for a bypass capacitor where gain is to be maintained, and (b) complications arising in the design of certain high-frequency circuits and certain television amplifier circuits.

2.2.2 Bias Circuits

Figure 12 illustrates the three grid-bias circuits used in obtaining the curves presented in Section 2.2. In Figure 12B, the full amount of the bias of Figure 12A is obtained from a cathode resistor; the effect of this arrangement on restricting operating point to a small area is illustrated in Figure 11. The circuit in Figure 12C maintains the same operating point as those in Figures 12A and 12B, while enabling the use of greater cathode resistance; the dotted curves in Figures 9 and 10 illustrate the effect of this method on the spread of plate current from tube to tube.

---

*   See Section 2.6; also see "Heater-Cathode Leakage", by the Application Engineers of the Advisory Group on Electron Tubes, Tele-Tech and Electronic Industries, January 1956.

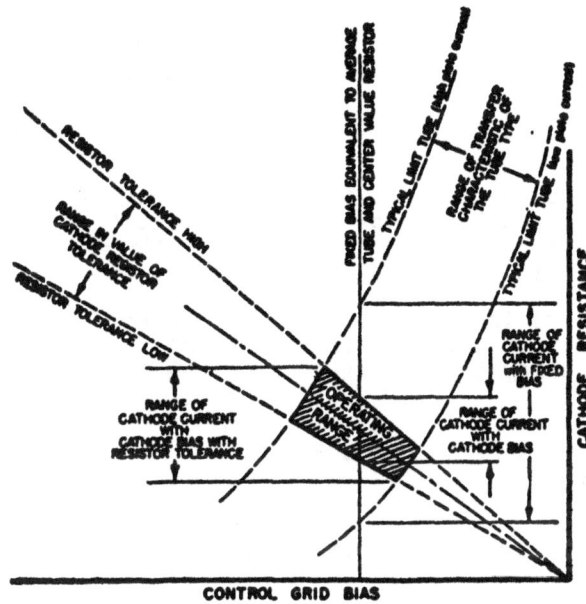

FIGURE 11
INFLUENCE OF BIASING METHOD ON
TRANSFER CHARACTERISTIC

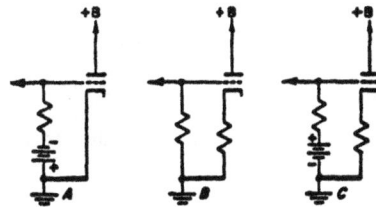

FIGURE 12
COMMON BIAS CIRCUITS

23

2.2.3 Cathode-Current Stability and Maximum Permissible Grid Resistance

Obviously, the cathode resistors in Figures 12B and 12C cause these circuits to be degenerative with respect to anything that tends to change the cathode current. To illustrate the resulting increase in stability, suppose that a 5670 has Sm = 550 µmhos, Rk = 240 ohms, Rg = 1.0 megohm, RL = 0, and that the grid current changed from 0.0 to 0.3 µA because of gas or leakage or grid emission, etc. Employing the formula

$$\Delta Ik = \Delta Ec \frac{Sm}{1 + SmRk}$$

it is readily calculated that the cathode current would increase by only 0.71 mA with cathode bias compared to 1.65 mA with fixed bias. If the circuit of Figure 12C is used, with Rk = 200 ohms and positive bias of 14.44 V, the increase in cathode current would be only 0.14 mA.

This increase in stability can be exchanged for an increase in grid circuit resistance. The maximum permissible grid-circuit resistance for triodes may be calculated from the following formula:

$$(1) \quad Rg = \frac{\Delta Ik}{\Delta Ic} \left[\frac{1}{Sm} + Rk \left(1 + \frac{1}{\mu} \right) + \frac{1}{\mu} RL \right] *$$

where

ΔIk = change in cathode current permitted by desired circuit stability or maximum ratings

ΔIc = change in control-grid current permitted by specification limits

Rg = maximum permissible control-grid circuit resistance in ohms

Rk = cathode circuit resistance in ohms

RL = plate-load resistance in ohms

μ = amplification factor of triode

Sm = grid-plate transconductance in mhos at operating point.

For pentodes or tetrodes with no series screen resistance, the formula is:

$$(2) \quad Rg = \frac{\Delta Ik}{\Delta Ic} \left[\frac{1}{Sm} \frac{Ib}{Ik} + Rk \left(1 + \frac{1}{\mu t} \right) \right] *$$

where
μt = amplification factor of the tube connected as a triode
Ib = plate current
Ik = cathode current

---

* F. Langford Smith, Radiotron Designer's Handbook (Sydney, Australia: Wireless Press, 1953), reproduced and distributed by R.C.A, Harrison, N.J., p. 82.

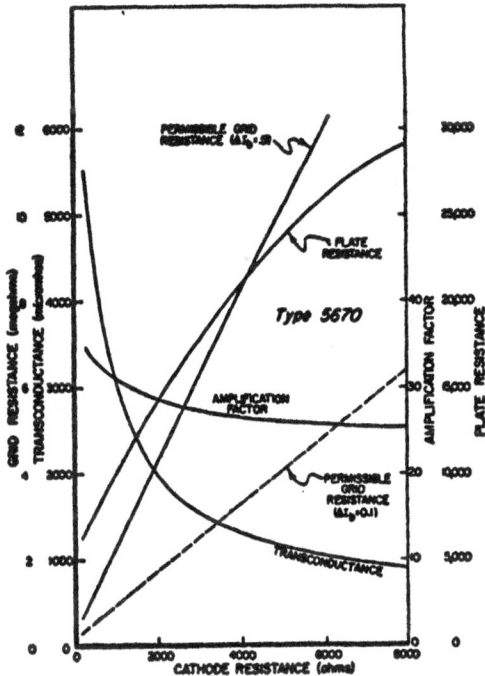

FIGURE 13
TUBE-CHARACTERISTICS VARIABILITY
AS A FUNCTION OF CATHODE RESISTANCE

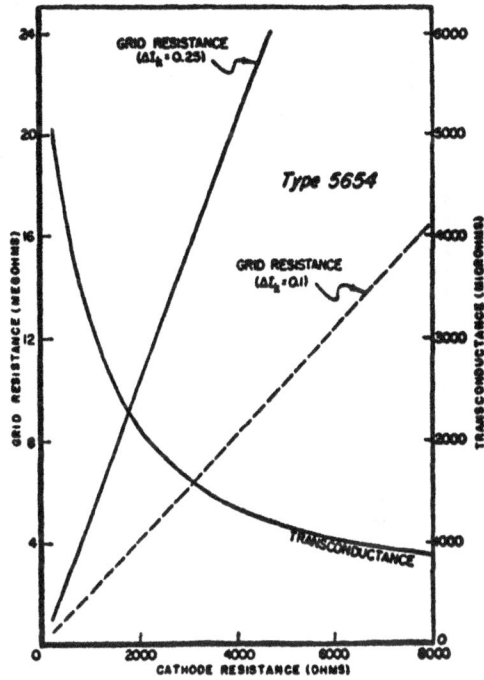

FIGURE 14
GRID RESISTANCE AND TRANSCONDUCTANCE AS A
FUNCTION OF CATHODE RESISTANCE

When applying these formulas, the value of ΔIc employed should always be the acceptance-test limit given in the MIL specification or other procurement specification. For reasons already discussed, permissible ΔIc should never be determined from measurements on individual tubes.

When the value of ΔIk is to be that permitted by tube ratings (the most liberal value), it can be estimated from the maximum value of Rg allowed in the MIL specification for fixed bias. By putting Rk = RL = 0 in (1), we obtain

(3) $\Delta Ik = Sm \Delta IcRg(0)$,

where $\Delta Rg(0)$ = maximum rated grid resistance at zero bias.

Where the maximum permissible grid-circuit resistance permitted by MIL specification limits and ratings is desired, the following forms of (1) and (2) are more convenient:

(1a) $Rg = Rg(0) \left[1 + SmRk + \dfrac{RL + Rk}{rp} \right]$

       $Rg = Rg(0) \left[1 + SmRk \dfrac{Ik}{Ib} \left(1 + \dfrac{1}{\mu t} \right) \right]$

25

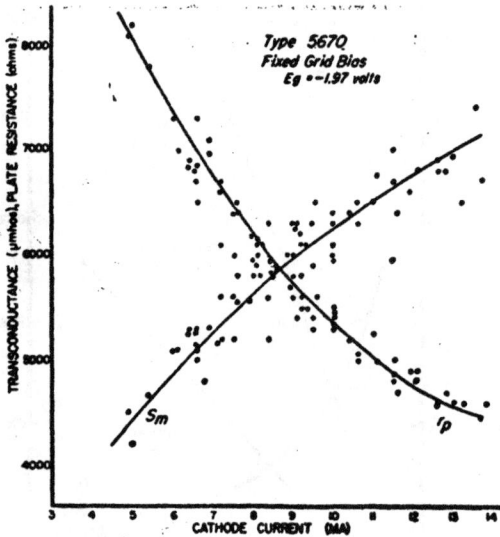

FIGURE 15
TRANSCONDUCTANCE AND PLATE—RESISTANCE
VARIABILITY AS A FUNCTION OF CATHODE CURRENT
UNDER FIXED—BIAS CONDITIONS

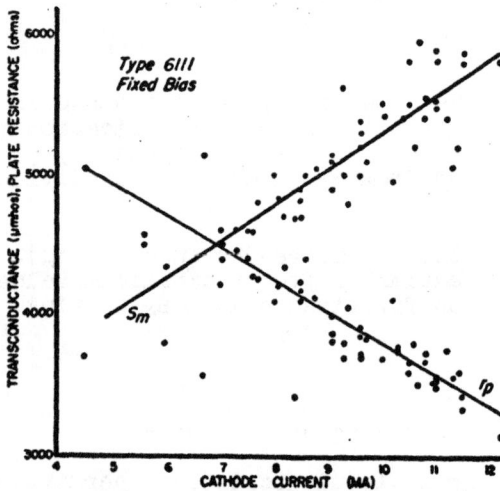

FIGURE 16
TRANSCONDUCTANCE AND PLATE—RESISTANCE
VARIABILITY AS A FUNCTION OF CATHODE CURRENT
UNDER FIXED—BIAS CONDITIONS

Figures 13 and 14 illustrate the application of (1) and (2) to types 5670 and 5654, respectively. In both figures, ΔIc is the MIL specification limit and RL = 0.

2.2.4 Correlation Between Cathode Current and Other Tube Properties

Thus far only the effect of cathode bias on cathode-current spreads and stability has been discussed. Any tube property having a high degree of correlation with cathode current will be similarly affected. Among the tube properties that have such correlation are Sm, μ, and rp in triodes and Sm, Ip, and Ic_2 in pentodes. Figures 15, 16, and 17 illustrate correlations observed in actual lots of tubes; the curves shown are the statistical mean relationship between Ik and other tube properties for the particular lots of tubes.

Figure 18 is a replot of Figure 15, showing the effect of cathode bias on the Sm vs. Ik relationship. It is seen that the degenerative action of the cathode resistance has narrowed the spread in Sm as well as that in Ik; this narrowing in Sm spread has occurred because Sm behaves to a considerable extent as a function of Ik. However, it is also seen that the deviations of the points from the mean curve are about the same with both fixed bias and cathode bias. These deviations represent the extent to which Sm is independent

26

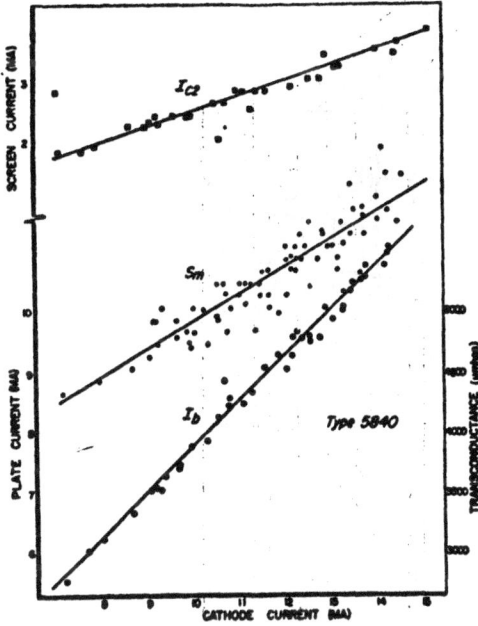

FIGURE 17
TUBE-CHARACTERISTICS VARIABILITY
AS A FUNCTION OF CATHODE RESISTANCE

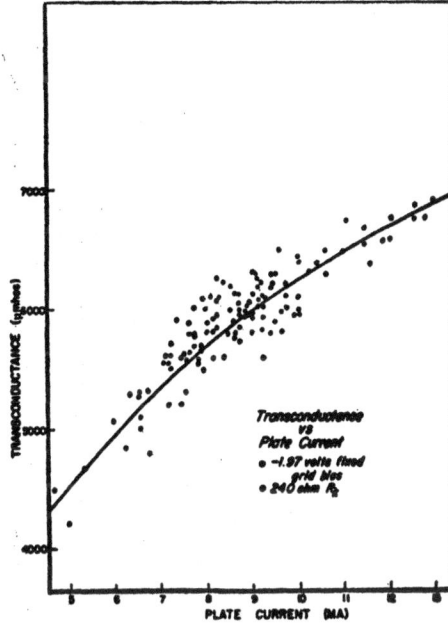

FIGURE 18
TRANSCONDUCTANCE VERSUS
PLATE CURRENT FOR TUBE TYPE 5670

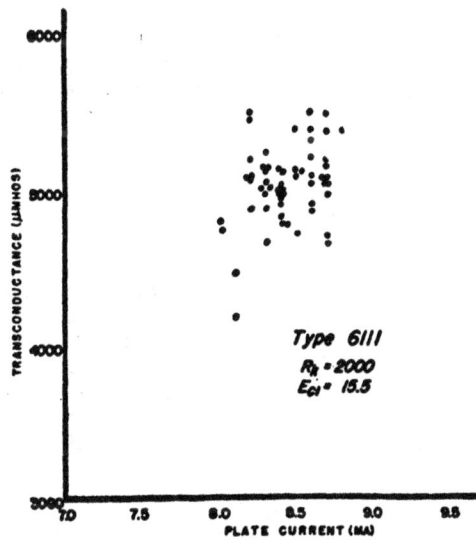

FIGURE 19
TRANSCONDUCTANCE VERSUS
PLATE CURRENT FOR TUBE TYPE 6111

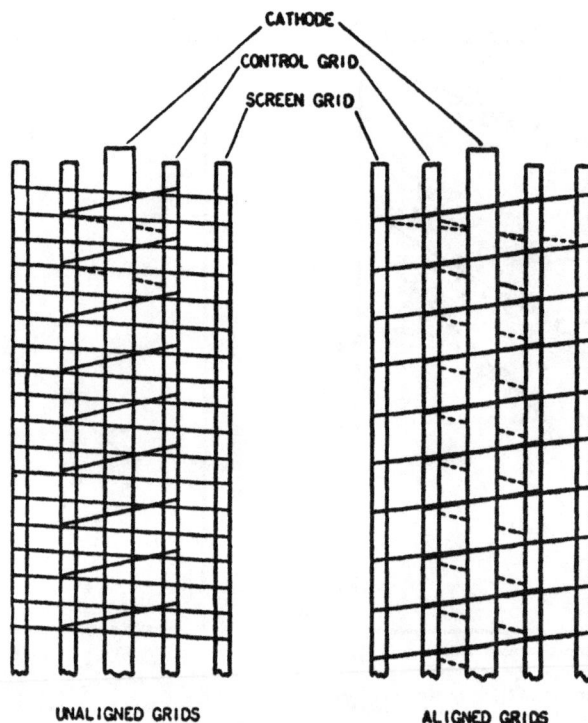

FIGURE 20
TYPICAL CONTROL-GRID AND
SCREEN-GRID CONFIGURATIONS

of Ik; this component of Sm spread will not be reduced by any amount of current feedback. Figure 19 illustrates how a large amount of cathode resistance can be so effective in removing that part of the spread in Sm due to spread in Ik that it is no longer possible to fit·a meaningful average curve to the points.

2.2.5 Summary

Cathode bias provides current feedback, which is effective in reducing the spread of cathode current from tube to tube and in stabilizing the cathode current in a particular tube. The improved stability can be exchanged for an increase in maximum permissible grid circuit resistance. Because other important tube properties correlate well with cathode current, these properties will be improved in the same way as cathode current, but to a somewhat smaller degree.

2.3 Screen-Grid, Voltage-Dropping Resistors*

The equipment design-engineer using electron tubes is always seeking methods of reducing the effects of the wide variations of properties which seem to be inherent in tubes. These variations must be

* Reproduced with permission, A.G.E.T. News Bulletin, October 1, 1957.

given consideration when characteristics such as transconductance (Sm)
and plate current (Ib) are involved. Not only do these characteristics vary among tubes, but in a given tube they vary with changes in
the voltage supply and also with the operational age of the tube.
These variations may cause excessively wide spreads in equipment performance characteristics.

In some respects, the effects of variations are more easily min-
imized in a pentode or a tetrode than in a triode, the screen or accel-
erator grid making possible some compensation for plate-characteristic
variations. A resistor in series with the screen tends to maintain
the screen current at a selected value and thus offers several advantages; however, under certain conditions, the resistor may actually
accentuate variations in some characteristics. Some of these advantages and disadvantages are:

1. The use of a series resistor reduces the variation in screen
current, which: (a) reduces the variation of plate characteristics
for most pentodes and tetrodes; (b) usually expands the variation of
plate characteristics for beam tubes.

2. The maximum dissipation of the screen grid can be limited to
a safe value.

3. The maximum value of control-grid resistance can be increased
as the value of screen-grid resistance is increased.

2.3.1 Effects of Structure on Pentodes, Tetrodes, and Beam-Power Tubes

There are two configurations of control grid and screen grid
commonly used in the construction of electron tubes that greatly affect
the relation of screen current to plate current. The most common is
shown in Figure 20A. In this type of construction there is usually an
attempt to prevent the alignment of opposing turns of the grids. Here
the screen grid has maximum exposure to the cathode and any lateral
shift does not materially affect this exposure. A result of this condition is that the number of electrons intercepted usually bears a
definite ratio to the plate current. (See Figure 21.) The screen
grid in this case acts almost as a plate so that the screen current
increases with the applied voltage as in a triode. These conditions
cause a definite compression of the plate-current range when the
screen-current range is limited by use of a series resistor. (See
Figures 22 and 23.)

The second type of structure is shown in Figure 20B. This
arrangement is used to form the plate current into a high-intensity
beam or to obtain a high ratio of plate current to screen-grid current. Theoretically, this type of structure behaves in a manner
similar to that described above with regard to voltage-current relations. Unfortunately, in this design deviations from perfect alignment cause difficulties. When the grid turns are mechanically distorted or the two grids move laterally relative to each other, the
screen-grid turns may no longer be in the shadow of the control-grid
turns. As the turns emerge from the control-grid shadow, they in-
tercept a number of electrons at the expense of plate current without
materially changing the cathode current. Not only would measurements

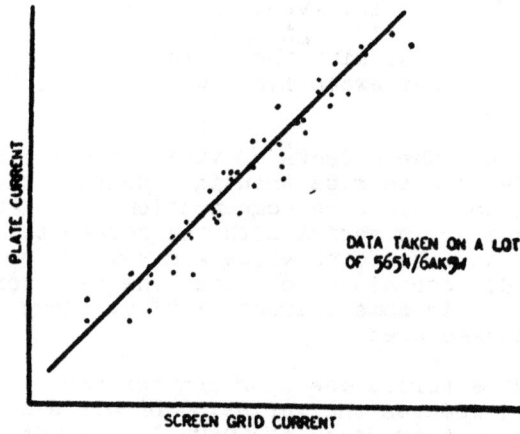

FIGURE 21
PLATE CURRENT VERSUS
SCREEN—GRID CURRENTS

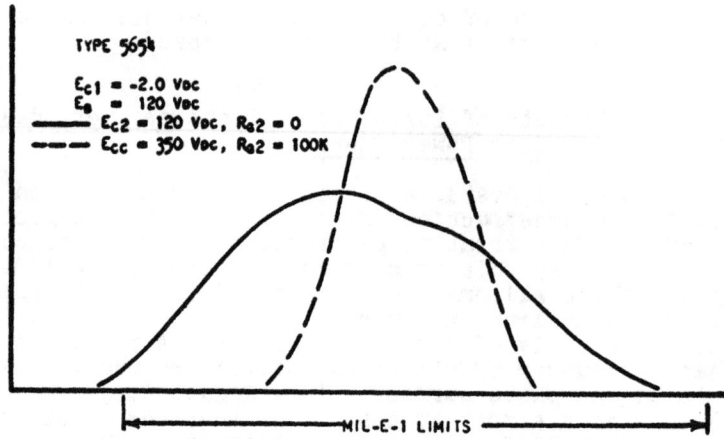

FIGURE 22
PLATE-CURRENT DISTRIBUTIONS

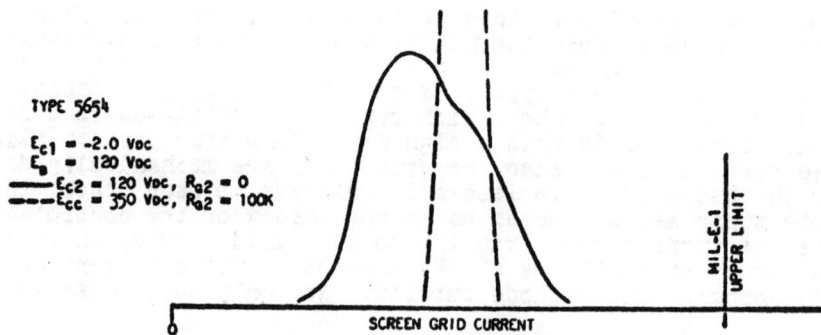

FIGURE 23
SCREEN—GRID CURRENT DISTRIBUTIONS

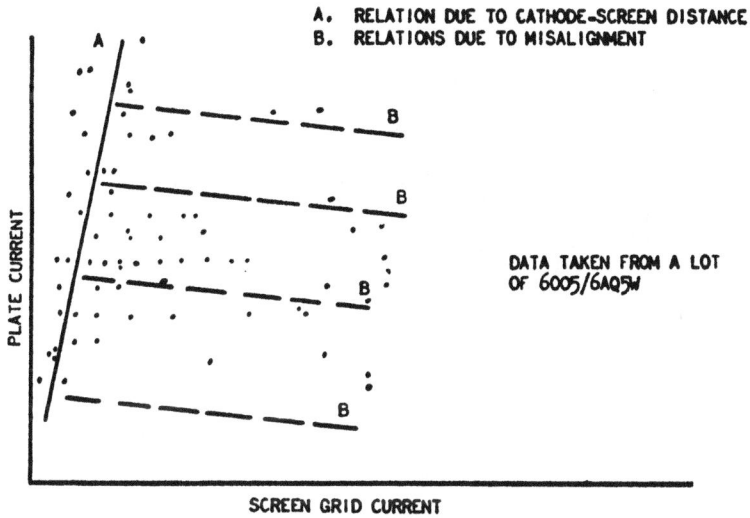

A. RELATION DUE TO CATHODE-SCREEN DISTANCE
B. RELATIONS DUE TO MISALIGNMENT

DATA TAKEN FROM A LOT
OF 6005/6AQ5W

FIGURE 24
EFFECTS OF STRUCTURAL SPACING ON
PLATE CURRENT VERSUS SCREEN—GRID CURRENT

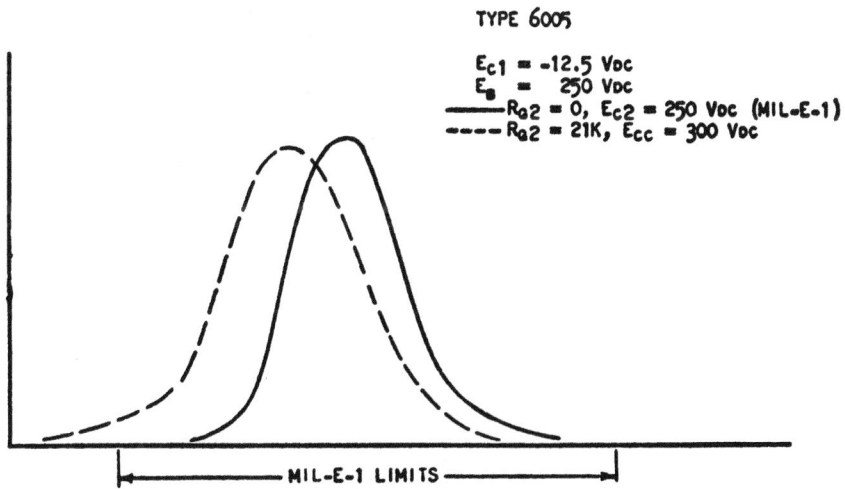

TYPE 6005

E_{c1} = -12.5 Vdc
E_a = 250 Vdc
——— R_{a2} = 0, E_{c2} = 250 Vdc (MIL-E-1)
– – – R_{a2} = 21K, E_{cc} = 300 Vdc

MIL-E-1 LIMITS

FIGURE 25
DISTRIBUTION OF PLATE CURRENT

TYPE 6005

E_{c1} = -12.5 Vdc
E_b = 250 Vdc
—— E_{c2} = 250 Vdc, R_K = 0 (MIL-E-1)
----E_{c2} = 300 Vdc, R_K = 21K

MIL-E-1 LIMIT

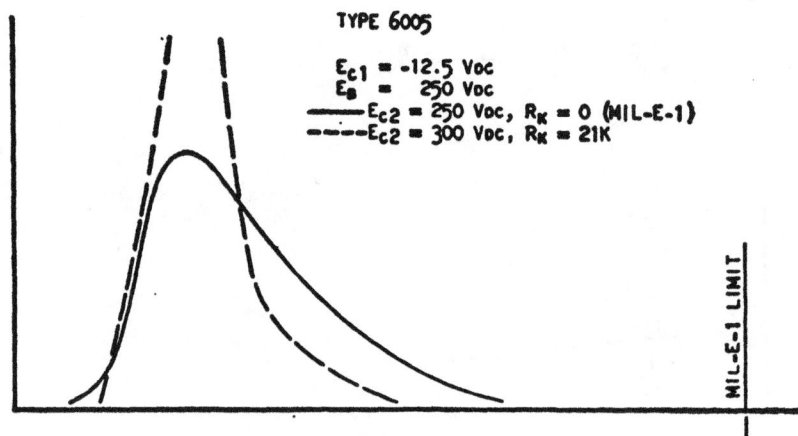

FIGURE 26
DISTRIBUTION OF SCREEN-GRID CURRENT

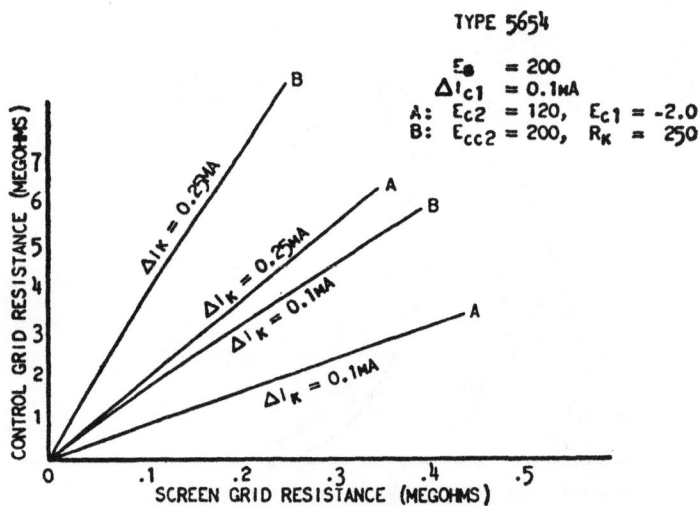

TYPE 5654

E_b = 200
ΔI_{c1} = 0.1mA
A: E_{c2} = 120, E_{c1} = -2.0
B: E_{cc2} = 200, R_K = 250

B

ΔI_K = 0.25mA

ΔI_K = 0.25mA

ΔI_K = 0.1mA

ΔI_K = 0.1mA

A

B

A

CONTROL GRID RESISTANCE (MEGOHMS)

SCREEN GRID RESISTANCE (MEGOHMS)

0   .1   .2   .3   .4   .5

FIGURE 27
PERMISSABLE CONTROL-GRID RESISTANCE
VERSUS SCREEN-GRID SERIES RESISTANCE

32

made on such tubes show poor correlation between screen-grid current, plate current, and cathode current, but also the slope of the line would, by and large, be negative. (See Figure 24.)

Various mechanical distortions, such as distorted individual grid turns and relative movement of grid-structures, can cause such poor correlation. Because the introduction of a screen-grid voltage-dropping resistor tends to maintain a constant screen-grid current at the expense of the accelerating potential, the result is compression of screen-grid current (Figure 25) and an increase in the range of plate current (Figure 26).

2.3.2 Limitation of Screen-Grid Dissipation by Series Resistor

The maximum allowable screen dissipation can be exceeded and permanent damage done to the tube and the circuit if the plate supply fails and the screen voltage is maintained. Similar damage will result when the screen-grid voltage is applied before the plate voltage. Where other considerations permit, a simple remedy is to supply both the plate and screen from the same source, using sufficient resistance to limit the screen dissipation to the permissible maximum.

2.3.3 Use of a Screen Resistor Permitting Higher Values of Control-Grid Resistance

In many applications a cathode resistor as well as a screen-dropping resistor is used. The combination has a decided effect on the permissible resistance in the control-grid circuit. A formula that may be used for tetrodes and pentodes for the determination of these resistances is:

$$ Rg = \frac{\Delta Ik}{\Delta Ic} \left[\frac{Ib}{SmIk} + Rk \left(1 + \frac{1}{\mu t}\right) + \left(\frac{Ic_2}{Ik} \cdot \frac{Rg_2}{\mu t} \right) \right] \quad * $$

where

Rg = grid resistance in ohms

ΔIk = permissible change in cathode current, im amperes, determined by desired stability or limited by ratings

ΔIc = specified maximum grid current in amperes (MIL-E-1)

Ib = plate current in amperes

Sm = transconductance in mhos

Rk = cathode resistance in ohms

μt = amplification factor, control grid to screen grid

Ic_2 = screen-grid current in amperes

Rg_2 = screen-grid series resistance in ohms

---

* A.G.E.T. News Bulletin, October 1, 1957.

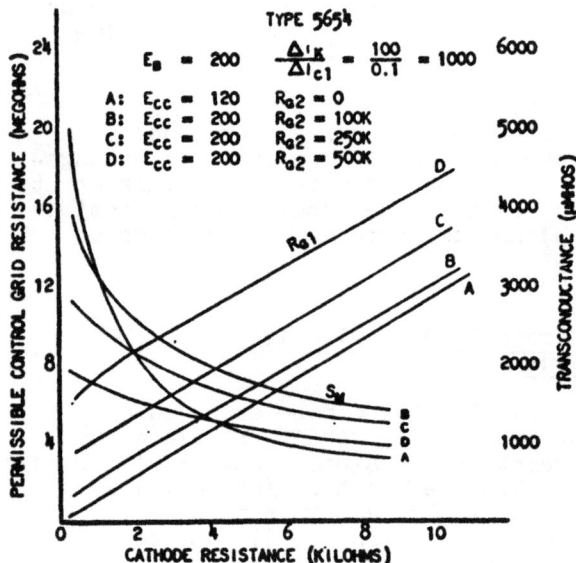

TYPE 5654

$E_B = 200$ $\quad \dfrac{\Delta i_K}{\Delta i_{c1}} = \dfrac{100}{0.1} = 1000$

A: $E_{cc} = 120$ $\quad R_{a2} = 0$
B: $E_{cc} = 200$ $\quad R_{a2} = 100K$
C: $E_{cc} = 200$ $\quad R_{a2} = 250K$
D: $E_{cc} = 200$ $\quad R_{a2} = 500K$

FIGURE 28
CONTROL—GRID RESISTANCE AND
TRANSCONDUCTANCE VERSUS
CATHODE RESISTANCE FOR FOUR VALUES
OF SCREEN-GRID SERIES RESISTANCE

Figure 27 shows the curve for permissible control-grid resistance plotted against screen-grid series resistance. The screen supply voltage is adjusted to give proper screen voltage. Figure 28 is a plot of the control-grid resistance and transconductance against cathode resistance for various values of screen-grid resistance.

2.3.3.1 Stability

To say that the maximum permissible grid resistance is increased by the introduction of screen resistance is another way of saying that a tube is more stable when a screen resistor is employed. This is true for individual tubes of either aligned or unaligned grid structure.

In the case of aligned grids, the use of a screen-grid voltage-dropping resistor would aggravate the effects of mechanical deformation, such as might be caused by shock or severe vibration. Therefore, under certain circumstances, the use of a screen resistor might be undesirable.

In pentodes and tetrodes, the cathode current remains essentially constant when the plate voltage is varied. When large signals are applied to the control grid, the screen-grid current will rise rapidly as the input signal approaches the positive region. The decrease during the negative excursion of the signal is quite small, resulting in an increase in the average screen-grid current. When a dropping resistor is used, a bypass capacitor to the cathode is necessary to prevent degeneration and distortion. When the resistor is adequately bypassed, the non-linear current results in an increase in the voltage drop across it. Complete compensation is not possible, but the use of cathode resistance for obtaining control-grid bias minimizes this effect.

2.3.3.2 Determination of Type of Grid Structures

Tube data sheets state the type of grid structure employed only in the case of beam-power tubes. It is therefore impossible to determine with certainty the type of grid structure used in other types. However, in modern tube types having effective suppression of secondary emission from the plate, the grids are very probably aligned if the

34

screen current is less than 10 percent of the cathode current under typical operating conditions. In old tube types having non-negligible secondary emission or in tubes where the screen current is more than 10 percent of the cathode current, the only sure way of determining the type of grid structure is by direct-inquiry of the manufacturer.

When it is known that a tube has aligned grids, 'screen voltage-dropping resistors should be used cautiously, giving due weight to the considerations outlined in Section 2.3.

2.4 Series Operation of Heater Strings*

One of the major causes of catastrophic failures in electronic equipments has been heater burn-outs due to stresses brought about by operation of heaters in series strings.

Although great progress has been made in heater construction and design -- so that, at present, heaters can operate without failure throughout a very large number of cycles at higher-than-bogie voltage-- the operation of heaters in series is still a source of excessive stress during the warm-up period. If high reliability is of primary importance, equipment circuitry must be carefully designed to minimize the stresses imposed on heaters.

The reasons for using series strings are different in military equipments and in equipments intended for entertainment purposes. In the latter equipments, series heater strings are used in order to permit direct application of the 115-volt supply, and thus save a transformer. In military equipments, specifically those used in aircraft, the major reason for series strings is to permit utilization of the primary electric-power supply, the only supply available when the engines are not running. This supply is a 12-cell storage battery of the lead-acid type. Under discharge conditions, the battery delivers a nominal voltage of 25.2 volts. However, since the capacity of the battery is relatively low, considering the number of equipments which must be supplied in modern aircraft, the voltage is kept constant at a nominal value of 27.5 volts.

The necessity for keeping the battery under charge has been recognized by the agencies concerned, and the voltage supply has been standardized at a center value of 27.5 volts, with a minimum of 25 volts and a maximum of 29 volts. This condition of operation -- known as a "service condition" -- actually corresponds to a voltage of 27.0 volts ± 7.5 percent.

To assure operation of the electronic equipment which may be needed most urgently in case of engine failure, a "special service condition" with voltage limits from 21 to 29 volts was established. Under these conditions, the most vital electronic equipments will

---

* See also: Application Engineers of the Advisory Group on Electron Tubes, "Series Heater and Filament Strings in Military Equipment", Tele-Tech and Electronic Industries, January 1955. Material appearing in Sections 2.4.1 through 2.4.5 is from the Sylvania Subminiature Tube Manual, Sylvania Electric Products, Inc., and is reproduced with permission.

continue to operate, but with somewhat reduced performance. Because of the reduction in performance, the special service condition should not be considered the controlling condition of operation, and the design should not call for operation at a center voltage of 25 volts, with ± 16 percent tolerance limits. If this were done, the tubes would operate at over-voltage most of the time -- for the distribution of supply voltage is skewed to the right, the mode occurring in the region near 27.0 volts.

The situation described above is exactly analogous to that found in automobile radios. An automobile battery has a nominal voltage of 6.3 volts, but has a charge of 7 volts under usual driving conditions. If the radio were designed to operate at a center voltage of 7 volts, it would fail if turned on when the battery was not under charge. The radio industry has compromised by using a center value of 6.3 volts, but the compromise was made at the expense of reliability, since deterioration is accelerated if a radio designed for 6.3 volts is operated at 7 volts most of the time.

In the case of a car radio using five or six tubes, deterioration is an unimportant consideration. However, in the case of aircraft equipment or military electronic equipment in general, the loss of reliability would be intolerable. Therefore, as a first step in the design of series heater strings, it is important to provide for a dropping resistor whenever four tubes having a 6.3-volt heater are connected in series across a 27.5-volt battery. This has been standard practice in commercial airline equipment for many years, but has been introduced in military aircraft equipment only in the last few years.

When the heaters of tubes are connected in an arrangement such that the rated supply voltage is equal to the design-center heater voltage, proper cathode temperature is assured -- provided the heater supply does not vary beyond the limits specified.

Under the constant-current conditions of a series string, heater current becomes quite critical, since the steady-state distribution of heater voltage is determined by the hot-heater resistance of the individual tubes. For example, a tube having lower than normal hot-heater resistance (high heater current at rated heater voltage) will receive less than its rated heater voltage when placed in a series string. Likewise, a tube having a high hot-heater resistance (low current at rated heater voltage) will receive more than its share of heater voltage. The situation is further aggravated by the fact that string current is dependent upon the spread of the average heater current of the tubes composing the string. As a result of these factors, values of heater power and cathode temperature which are beyond recommended operating limits are possible. Variations in heater power will naturally occur with constant voltage circuitry. However, the range values normally encountered are less than those found in circuits using series strings.

2.4.1 Development of Composite Operating Curves

Figure 29 is a composite heater voltage-current characteristic curve as shown in the Sylvania Technical Manual for 450-mA subminiature tubes and will aid in the analysis of series string condition

36

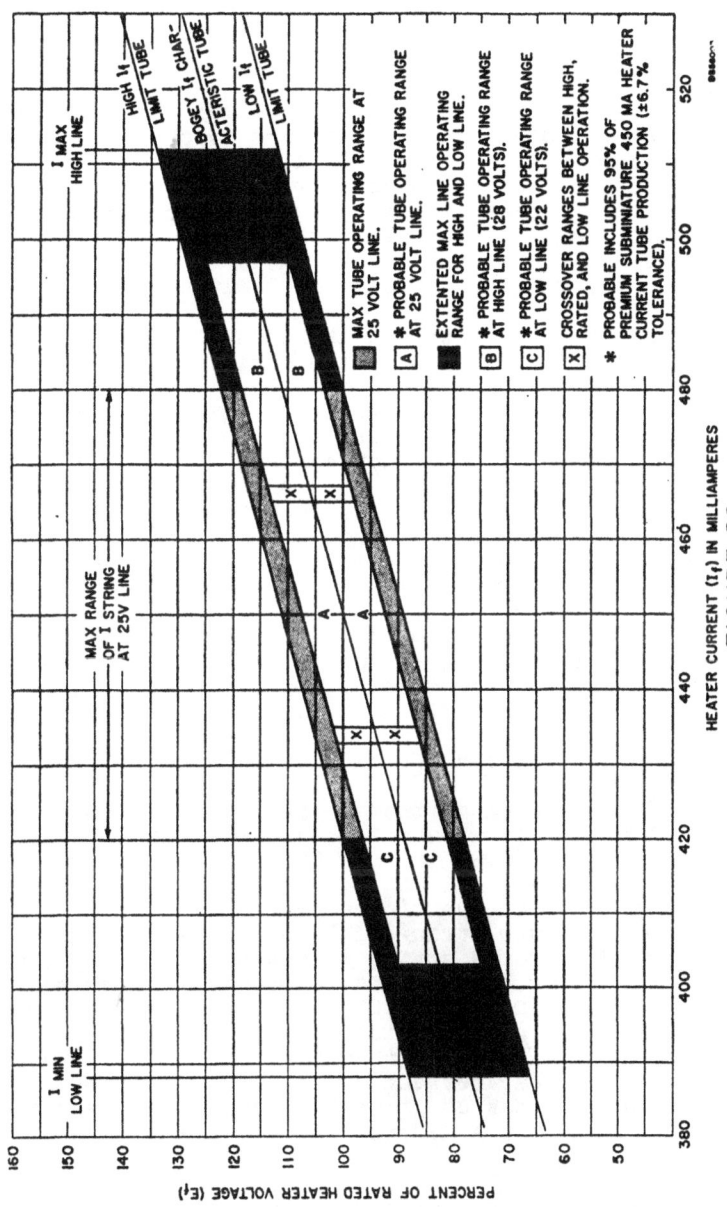

FIGURE 29
COMPOSITE OPERATING CURVE FOR A FOUR TUBE
STRING OF 450 Ma PREMIUM SUBMINIATURE TUBES

and enable prediction of approximate operating points or power inputs
of individual tubes in a four-tube string for line voltages between
22 and 28 volts. The center value of 25 V and the range of 22 to
28 V used in this figure are shown to indicate the wide range of variations possible. The same reasoning and technique used here can be
applied to a center value of 27 V with a range of 25 V to 29 V.

Referring to Figure 29, the heater voltage-current curve for a
bogie tube, by nature of its designation, passes through 450 mA at
100 percent rated heater voltage. The curves for low and high current
limit tubes are parallel to, but displaced from, the bogie curve as
determined by the heater current limits, 420 mA and 480 mA, at 100
percent rated heater voltage. The characteristics of other than bogie
tubes will, therefore, be represented by a family of curves, which
parallel the bogie curve, ranging between the limit curves.

The maximum range of string current at rated line voltage is
dependent upon the heater current production limits, since the string
current range cannot be greater than the heater-current spread of the
tubes composing the string. String currents at high and low line conditions, Figure 29, are based on predetermined supply-voltage limits
and strings composed of only high- or low-current limit tubes.

The total enclosed area of Figure 29, therefore, represents the
maximum possible theoretical operating conditions between low and high
line, divided as shown. A tube may possibly operate from 73-130
percent rated heater power at rated line conditions, or 57-153 percent
at low and high line. These conditions, however, are extremes, since
the current of the tube in question must be at one limit and the string
current at the other limit; and the string current could not be at the
top limit if one of the four tubes were at the opposite limit.

2.4.2 Probable Operating Conditions of a Four-Tube String

Statistical analysis indicates that 95 percent of the typical
subminiature tube production is within ± 4.45 percent of the design
center heater current rating, ± 20 mA for 450 mA types. (Top and
bottom boundary lines of areas A, B and C, Figure 29). As a result,
the string current for 99.7 percent of the probable combinations of
tubes of an arbitrary four-tube string will be within ± 3.33 percent
of the rated value, ± 15 mA for a string composed for 450 mA types.
This probable range of string currents is represented by the vertical
boundaries of areas A, B and C, Figure 29. Areas A, B and C, then,
represent the probable areas of operation at rated (25 V), low (22 V),
and high (28 V) line conditions. It should be noted that areas of
crossover (X, Figure 29) appear between areas A, B and C. These areas
must also be considered as areas of probable operation since areas
B and C will slide through area A as the line voltage changes.

2.4.3 Comparison with Constant Voltage Operation

It was previously stated that the steady-state heater power variations in a series string may be in excess of those encountered in a
constant-voltage arrangement. Let us consider the probable maximum
heater power variation of 450 mA types at heater current limit conditions of ± 20 mA (represents at least 95 percent of production),

when operated from a 6.3-volt line. Assume also, that the line voltage variation is comparable to that shown for the four-tube string, ±12 percent. Reference to Figure 29 indicates that the low limit operating curve (430 mA) corresponds to the upper boundary of areas A, B and C. Therefore, at low line conditions, 88 percent rated voltage, the heater current will be approximately 397 mA and the resultant heater power approximately 78 percent rated value. Substitution of the upper current limit and high line conditions indicates a probable operating range of 78 to 125 percent rated heater power. The corresponding range for a four-tube string is 67 to 138 percent rated value, or 24 percent greater.

2.4.4 Application of Composite Heater Voltage-Current Curves

To determine the operating point of a tube on the composite curves, Figure 29, the string current at rated line voltage and the heater current of the particular tube at rated heater voltage must be known. This heater current value will determine which curve of the family of heater voltage-current curves, will be used. The operating point of this tube is then determined by the intersection of its curve with the vertical line representing the string current. Assuming linearity of the heater voltage-current characteristics, operation along the curves is described by the equation:

$$I = I_f - KE_f + KE_t \tag{1}$$

or

$$E_t = \frac{I - I_f + KE_f}{K} \tag{2}$$

where

I = heater string current in mA

I_f = heater current at rated E_f in mA

E_f = rated heater voltage in Volts

E_t = applied heater voltage in Volts

K = 42.9 mA/V. for 450 mA types

27.8 mA/V. for 300 mA types

13.6 mA/V. for 150 mA types.

The above treatment has considered only the steady state condition. If the transient conditions (when the heater power is turned on) were considered, a much more complex situation would appear, in which the heating time of the individual heaters would have to be considered. In fact, since all electron tube heaters are made of pure tungsten wire and are operated at a temperature in the neighborhood of $1500°K$, the temperature coefficient of resistance of the tungsten wire is such that the resistance of the heater when cold is approximately seven times smaller than the resistance of operating temperature. The conditions obtained in case of different heating time can be visualized by making the simplifying assumption that one tube is already

39

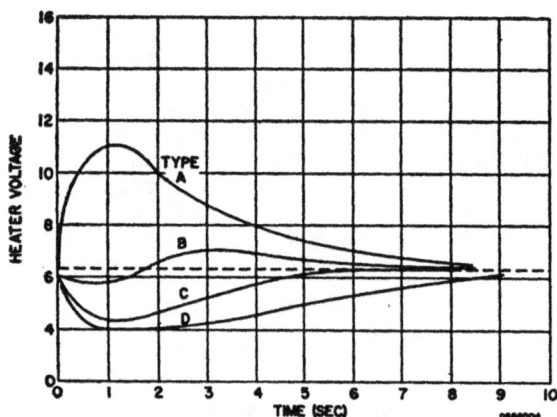

FIGURE 30
HEATER VOLTAGE-TIME
CHARACTERISTICS OF INDIVIDUAL TUBES
IN A FOUR TUBE STRING

fully heated while the other in a two-tube string is still completely cold. In such a case, the voltage in the series string would be distributed in proportion to the resistance of the heaters; that is, only approximately one-seventh would be applied across the cold tube while six-sevenths would be applied across the hot tube. For a 12.6 V string this would mean 10.8 V applied to the hot tube instead of 6.3 V. Also the current will be one and three-quarters times the normal operating current. The power applied to the warm tube while the cold tube is warming up will be three times more than the rated power.

2.4.5 Transient Conditions

The transient condition is shown in Figure 30 for a four-tube string. It is possible to avoid this transient phenomenon by using in the series only tubes that have either identical or very similar heaters. The commercial applications have developed heaters that are controlled within narrow margins for warm-up time characteristics. These have not been made available as yet in military preferred tube specifications, but there is no reason why they should not be requested when the application warrants it.

2.4.6 The "Honeycomb" Connection

In some complex electronic equipments, two or more series heater strings have been tied together by connecting points of equal potential between the strings in a "honeycomb" arrangement. This is done to compensate for the variability of heater current in different tubes. It is self-evident that, when more than one tube is connected in parallel in each section of a series string, the total current will be the sum of the individual currents, and the voltage drop across the parallel arrangement will be given by the over-all average voltage drop of the various tubes.

In military electronic equipments, the honeycomb arrangement is frequently a source of unreliable performance. If one tube happens to burn out and the failure is not detected until several hours of operation have elapsed, the failed tube unbalances the heater strings in such a way as to apply an over-voltage to all other tubes in parallel with it. This over-voltage usually produces a much faster rate of deterioration in the tubes concerned -- deterioration which may not be detected at the time the burned-out tube is detected, and which will cause additional malfunctions in later periods of operation. The condition is particularly damaging to equipment reliability if maintenance personnel cannot understand the nature of the interactions taking place within the equipment.

2.5 <u>Warm-Up and Operation Time</u>

In some electron-tube applications, it is necessary to reduce to a minimum the time between the application of power and the operation of the equipment at full power. This time is usually relatively long -- about one-half minute -- because of the lag in the heating of the indirectly heated cathodes generally used in electron tubes.*

Before proceeding further, it is important to note that "warm-up" time and "operation" time are two entirely different tube characteristics.** "Warm-up" time is that time required for a heater which is originally at room temperature to reach 80 percent of its rated voltage after four times the rated voltage is applied to the heater in series with a fixed resistor. The value of the resistor is three times the value of the hot resistance of the heater.

"Operation" time is usually defined as that time required for a tube which is originally at room temperature to reach a specified proportion (X) of its stabilized plate current. The stabilized plate current is usually that value reached after three minutes of operation (two minutes according to another definition). The specified proportion (X) has not been standardized and may vary from 80 to 98 percent. Another definition requires that plate current be measured 20 seconds after power is applied. Tubes are accepted if their plate current exceeds 80 percent of the three-minute value.

It is instructive to examine in detail the mechanism of operation of the heater inside the cathode of an electron tube to understand more completely the limitations in the operating time of a circuit involving tubes.

The heater wire of most receiving tubes is made of pure tungsten, the only metal available in commercial quantities that can stand the high operating temperature attained (about $1500^{\circ}K$) without appreciable deterioration. Tungsten has a temperature coefficient of resistance of such value that at room ambient temperature, the actual resistance of the heater is approximately one-seventh the resistance at operating temperature.

The phenomenon described above helps to minimize warm-up time. In fact, if the source impedance of heater voltage is negligible, the power heating the cathode in the first instant of operation is seven times the steady-state power. In a typical receiving tube (in which heater current reaches a steady-state value after five seconds), more than three times as many calories are absorbed by the heater-cathode assembly in the first five seconds as in any subsequent five-second interval.

* Some rather interesting approaches to the design of heater-cathode systems, in which the reduction of operation time is the major objective, are discussed in the series of reports, <u>The Feasibility of Quick Warm-Up and Stabilization of Heater-Cathode Type Receiving Tubes</u>, (Raytheon Manufacturing Company, Newton, Mass., Contract NObsr-72718). These reports describe several structures capable of less than five seconds operating time.

** Thomas H. Briggs, <u>Electron Tube Operation as Influenced by Temperature</u>, WADC Technical Report 56-53, January 1956.

In tube type 5654, 4.5 calories are absorbed in the first five-second interval, in contrast to 0.25 calories per second in the steady state. The distribution of this energy is also interesting. Only two calories are needed to bring up to steady-state temperature the various parts of the assembly, such as the heater wire, the heater insulation, the nickel sleeve, and the emission coating of the cathode. Radiation absorbs very little of the energy in the first five seconds -- about one-half calorie -- the balance being dissipated through conduction by the mica and the cathode tab to the surrounding parts of the tube. Considering the fact that there is better thermal contact between cathode, mica, and tab than there is between heater and cathode, it is easy to understand the delay of 15 or 20 seconds between the application of voltage to the heater and the time when the cathode reaches operating temperature. In fact, all other factors being constant, heating time depends on the contact between heater and cathode. The shortest time is obtained with double helix heaters in which there is almost continuous contact between the cathode sleeve and the turns of the heater. The longest time occurs with short, folded heaters which fit very loosely in the cathode and therefore make only a few contacts with the cathode sleeve.

A few typical distributions of operation time are shown in Figure 31. It is important to notice that when operation time is not controlled, the distribution tends to be decidedly skewed, with a small percentage of tubes having very high values compared to the bulk of the population. In view of the harmful effect even these few tubes could have in equipments in which operation time seriously affects the operation of the equipment, it is important that specifications contain limits on this characteristic.

It is also important to analyze suggested methods for shortening operation time, and to consider the possible disadvantages -- for example, reduction in reliability -- which might accrue from adoption of various methods. The advantages and disadvantages of several techniques which have been used are discussed following a brief description of the tube characteristics which are affected by the full steady-state temperature of the cathode.

As indicated elsewhere, temperature is needed for obtaining full saturation emission from the cathode. Full saturation emission in turn, produces two effects -- it locates the space charge at its maximum distance from the cathode, and it produces the maximum grid equivalent voltage due to initial electron velocity. The characteristics affected by these two phenomena are the ones associated with grid-to-cathode spacing and grid bias -- that is, plate current, plate resistance, and transconductance. The amplification factor and the cut-off current are practically unaffected. Hence, a circuit based on these characteristics will come to full operation in a much shorter time than a circuit depending upon transconductance for its operation. Also, a circuit which can operate on, say, half the rated plate current will come to full operation in less time than a circuit which needs full rated plated current for proper operation. In the latter type of circuit, the space charge reaches nearly maximum position when the saturation current has a value approximately 10 times the value of the space current used, while in a tube in the usual circuit the saturation current reaches final values between 100 and 500 times the value

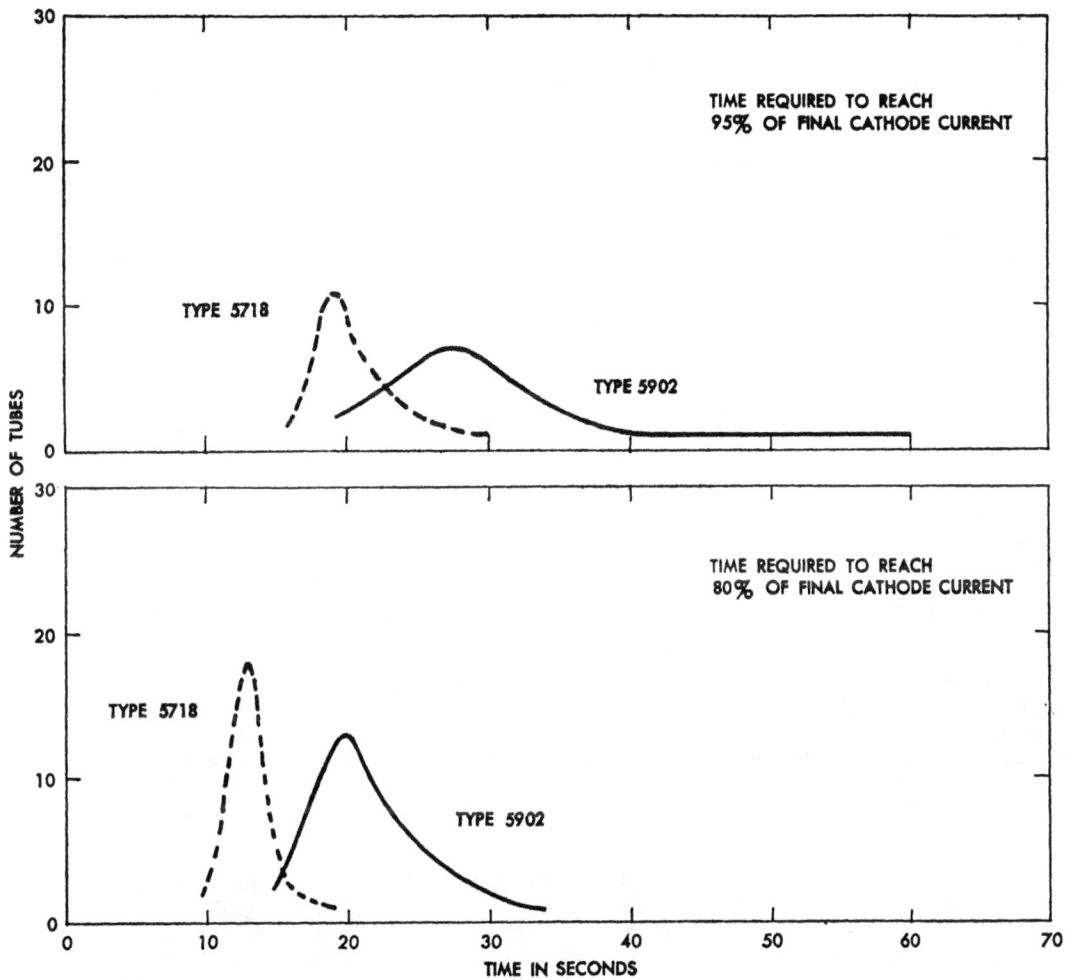

FIGURE 31
TYPICAL DISTRIBUTIONS OF OPERATING TIME

of space current. Conversely, the higher the saturation emission (that is, cathode activity), the shorter the operation time will be.

The first and most logical method of shortening operation time is to apply higher heater voltage for the first few seconds of operation. The higher power supplied to the assembly decreases operation time directly, because equilibrium temperature is reached as soon as the proper amount of energy has been accumulated in the various components. The desired effect can be achieved by properly programming the application of heater voltage, or by feeding the heaters through a resistor having the proper time-resistance characteristic.

The major disadvantage of this method derives from the fact that the heater already dissipates a much larger than rated power in the first few seconds after the start of operation (as was explained previously); hence, the chances of heater burn-out due to uneven heating, to a non-uniform cross-section, or simply to brittle spots in the crystal structure of the heater wire, are multiplied.

Because of these disadvantages, the programming of heater voltage may be superior to the method involving the series resistor, because the former method permits a limitation of the maximum current drawn initially and the retention of a high current for a longer period of time after the initial surge. The chances of burn-out produced by high current density are thus decreased.

Since the two methods described above require the application of voltages exceeding present maximum specifications, the tube manufacturers should be given the opportunity to assure the life of tubes thus treated. If necessary, the manufacturers should establish a test to determine the number of cycles of over-voltage which the tubes can stand and still provide acceptable quality.

Another method used to decrease operation time involves the application of direct-emission, filamentary-type tubes throughout the equipment. Such tubes are employed in proximity fuses and in military equipments whose reliability is very high in spite of a severe mechanical environment. For these reasons, direct-emission filamentary-type tubes cannot be dismissed lightly as being unsuitable for use in complex equipments.

The fact that filamentary tubes are thought to have lower performance than heater-cathode tubes may explain their lack of popularity. Actually, if transconductance per unit heater or filament power is used as the criterion of performance (rather than transconductance per tube), the filamentary-type tubes are far superior. They should be considered for use in applications where supply power is at a premium, as in guided missiles.

Another important reason for the unpopularity of filamentary-type tubes is the difficulty of isolating the cathode from ground, as can be easily accomplished in cathode-type tubes. However, by isolating transformers or other simple devices, especially in relatively simple equipments, the isolation of cathode from ground can be achieved without much effort.

An instructive comparison can be made by examining the characteristics of two tube types -- 5840 and 5678. Direct comparison of transconductance (5,000 versus 1,100 micromhos) would certainly favor tube type 5840, especially for applications requiring transconductance. However, the ratio of transconductance to heater power (5,300 versus 17,600 micromhos per watt) and the ratio of transconductance to cathode power (4,150 versus 7,150 micromhos per watt) are superior in tube type 5678, and would make this type more suitable for applications in which power supply requirements are important. The operation time of tube type 5678 is less than 1 second, while that of tube type 5840 is about 20 seconds. In any equipment requiring less than 15-second operation time, it would be unrealistic to expect such performance from present-day cathode-type tubes unless they were pre-heated by some method such as the application of heater voltage immediately before use.

2.6 Heater-Cathode Leakage

The flow of current between heater and cathode in electron tubes is a well determined phenomenon upon which most specifications place a maximum limit between 5 and 20 microamperes, depending on the tube type. It is a problem which must be dealt with in circuit design, especially in high-impedance circuitry in which the input circuit to the grid is connected with the heater-cathode circuit.

The insulation between heater and cathode consists of a thin coating of pure alumina, which has extremely high resistance even at the operating temperature of the heater. This coating is usually made extremely porous in order to obtain a good bond between the tungsten of the heater wire and the aluminum oxide used in the coating -- a bond that will hold even when the heater is bent, or when it is heated or cooled in relatively frequent cycles. The difference in the expansion coefficients of the two materials produces cracking of the insulation and peeling of the coating when the latter is too compact. Because of the porosity of the coating material, there is a possibility of contamination of the insulating body by foreign materials. This contamination may reduce the resistivity of the insulation or produce ionic or electronic emission between heater and cathode. Contamination and consequent leakage may not be present upon initial inspection of the tube, but may develop at any time during subsequent operation.

Metallic vapors and ions can easily be present at the operating temperatures used. The electric field and the emission of electrons can produce further ionization, and migration of ionized metals can cause metallic deposits to occur in the body of the insulator. After a period of time, heater-cathode leakage currents will be produced as a result of (a) ion conduction through contaminated alumina, (b) electron emission, and (c) positive or negative ion emission.

The heater-cathode current produced by the above three phenomena bears a decidedly non-linear relationship with the voltage applied. The current usually saturates at a very low voltage, as shown in Figure 32. All the currents shown in the figure are produced by impurity phenomena and are strongly temperature-dependent -- that is, they vary with heater voltage in the manner shown in Figure 33.

From Figure 33 it can be seen that if the center value of heater temperature varies from tube type to tube type, the characteristics of heater-cathode leakage are displaced to the right or to the left.

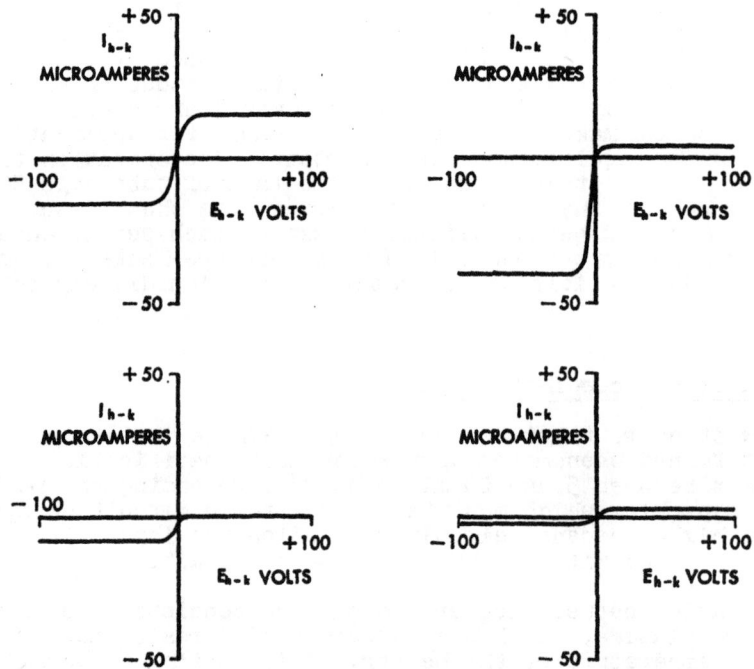

FIGURE 32
TYPICAL EXAMPLES OF HEATER-CATHODE
LEAKAGE CHARACTERISTICS

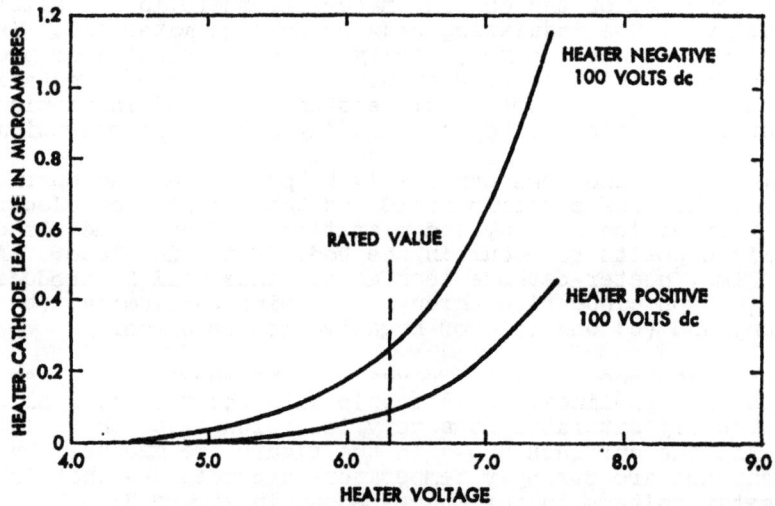

FIGURE 33
HEATER-CATHODE LEAKAGE AS A FUNCTION OF
HEATER VOLTAGE AND DC BIAS

46

Hence, the effect of a given difference in temperature can be greater
or less than the effect shown in the figure. In an ARINC test, it was
found that a change of 50°C in the temperature of the heater could
cause a change in heater-cathode leakage current which varied from a
ratio of 3-to-1 to a ratio of 50-to-1 in different tube types. This
range of variation indicates the degree of care which must be taken
by the circuit designer whenever there is interaction between input
and heater-cathode circuits.

If interaction is inevitable, several precautions can be taken:

(1) Since any decrease in the operating temperature of the
     heater will be beneficial, heater voltage should always
     be set at the minimum value compatible with other cir-
     cuit requirements.

(2) Whenever a common impedance is connected between cathode
     and grid, adequate bypasses for the frequency of the
     heater supply should be used.

(3) The heater supply voltage should be grounded as close as
     possible to its electrical center value, either by means
     of a center tap in the heater winding or by means of a
     center-tapped resistor-potentiometer connected across
     the heater terminals.

(4) In extreme cases in which there is no possibility of
     bypassing -- for example, when negative feedback, un-
     bypassed cathode resistors are used, or when the signal
     level is very low -- dc heater supply should be used.

(5) A method of decreasing leakage current can be inferred
     through inspection of Figure 32. If a d-c bias is
     applied between heater and cathode so as to bring the
     current characteristics beyond the saturation point
     to the flat region of operation, variations in voltage
     at supply frequency will produce currents of much
     lower amplitude than would occur if no bias were applied.

The d-c heater-cathode bias must be well within the rating of the
heater-cathode voltage and must be positive (heater positive with
respect to cathode) in order to avoid the possibility of embrittling
the heater wire. Embrittlement of the heater wire has not yet been
adequately explained, but appears to occur in tube types using negative bias. The effect seems to be produced by migration of metallic
ions from the cathode to the heater. These metallic ions alloy with
the heater material to produce a very brittle heater which will open
under mechanical or electrical shock.

In conditions very similar to those described above, heater-
cathode leakage may produce actual arc discharge between heater and
cathode and complete destruction of the tubes. This phenomenon has
been observed in almost all rectifier types that have a heater-cathode
structure. This group of rectifier tubes usually has a heater-cathode
voltage rating equal to the maximum d-c voltage rating obtained from
the supply. Such a rating is not in line with regular receiving-tube
ratings and does not indicate the actual capability of the tubes. In
fact, in these rectifier tubes, the field between heater and cathode

is more than twice as high as the field obtained in regular receiving
tubes at maximum ratings. The conditions in the heater-cathode space
are worse for rectifier tubes than for receiving tubes because of the
fact that the voltage applied to rectifiers has very low source impedance.

The heater-cathode voltage rating has been dictated by the requirements of circuits in which the rectifier heater is fed by the
same transformer winding that feeds all the other heaters in the equipment. In such circuits, the heater winding has to be grounded while
the cathode is at high positive d-c potential. This arrangement saves
a few pennies in the construction of the transformer, but the saving
is accompanied by too many disadvantages to be worthwhile in the production of military equipments. In military equipments, a properly
insulated transformer winding should be provided for the rectifier
heater supply alone, so that the stresses can be transferred from the
heater-cathode space to the insulation between the windings. Clearly,
it is much easier to design a wide margin of safety into a transformer
winding than into the small space between heater and cathode in a vacuum tube. The high heater-cathode voltage rating of rectifier tubes
should not be used if high reliability is of prime importance.

2.7 Parallel Operation of Electron Tubes

Extensive field experience with electron tubes has made apparent
the necessity of exerting due caution in the design of power-supply
rectifiers, series-pass circuits, and power amplifiers requiring the
parallel operation of electron tubes to obtain the necessary plate
dissipation.

Theoretically, if one tube will handle 5 watts, two tubes of the
same type connected in parallel should each be able to handle 5 watts.
Unfortunately, since few tubes are "bogie" tubes, it is rare that a
10-watt load would be divided equally. It has been found in practice
that one tube will carry more of the load than the other and will fail
before the other. The remaining tube will, of course, fail shortly
after it has been subjected to full load.

In addition to the load-dividing problem mentioned above, there
have been numerous cases of parasitic oscillations which cause
parallel-connected tubes to exceed dissipation ratings. To avoid both
problems, the following precautions should be taken:

(1) Employ tubes well below their ratings, particularly
    when fixed bias is used. If necessary, use extra
    tubes for conservative design.

(2) Where possible, use individual degenerative resistors
    for each tube. These resistors may be of the order
    of 50 to 100 ohms, and should be placed in the plate
    and screen circuits. In many applications, individual
    cathode-bias resistors may suffice.

(3) The small resistors mentioned above are also effective
     in suppressing UHF parasitic oscillations if care is
     taken to mount the resistors close to the tube sockets
     with very short leads. Occasionally, small RF chokes
     designed to be effective for the frequency range of the
     parasitic oscillations, will be effective suppressors
     if inserted in the individual plate or grid circuits.
     The chokes should be mounted close to associated cir-
     cuit parts with very short leads.

3. APPLICATION DATA ON 20 TUBE TYPES COVERED IN MIL-STD-200D

3.1 General Information

This part of the Appendix gives specification and application data for 20 receiving tubes which appear on the MIL-STD-200D preferred tube list. Twelve of these tube types did not appear on the MIL-STD-200C list and therefore were not covered in the 1958 edition of the handbook, Techniques for Application of Electron Tubes in Military Equipment.* The other eight tube types appeared on the list and were covered in the handbook. They are also treated here because more up-to-date information on their behavior has since become available.

MIL-STD-200 lists preferred types of electron tubes that have been selected jointly by the Army, the Navy, and the Air Force to fill the majority of equipment applications. The purpose of the standard is two-fold:

(a) To guide military equipment designers and manufacturers in the choice of tube types which represent the highest quality available for military use;

(b) To provide for a minimum stock of tubes for use by maintenance personnel, by making extensive use of a minimum number of tube types.

The current list of tubes included in MIL-STD-200D is given in Table 1. However, reference should always be made to the most recent issue of the standard, since it is subject to periodic revision.

The controls given in Military Specification MIL-E-1 are intended to provide assurance that the equipment designer using electron tubes can expect comparatively uniform initial characteristics, relatively stable characteristics throughout life, and a high level of attribute quality. It follows that there is no assurance of satisfactory operation when tubes are used under conditions which are incompatible with test conditions and ratings set forth in this specification. Both the quiescent operating point and the dynamic operating requirements must be considered in relation to these ratings.

Specification data which are applicable to the receiving-type tubes listed in MIL-STD-200D are presented in Tables 1 and 2. Tab¹ includes a summary of specification controls and a list of proper which are subject to variables testing. Table 2 lists a number o. specification changes affecting ratings and acceptance tests of 5 the 56 tube types treated in the original handbook.

---

* Military Handbook No. 211, U.S. Government Printing Office, Washington, D.C., December 31, 1958 (Price $3.25).

TABLE 1

TABLE 1-CHARACTERISTICS OF MIL-STD-200D RECEIVING TUBES

MIXERS AND CONVERTERS

Designation	Size	Technical Characteristics								Max. Altitude
		Ef	If	Gm	Ik	Pp Max.	Pg2 Max.	Ib	Ic2	
5636	Submin.	6.3	150	3200	16.0	0.55	0.45	5.3	4.1	60,000
5725/6AS6W	Min.	6.3	175	3200	20	1.65	0.55	5.5	3.5	60,000
5750/6BE6W	Min.	6.3	300	500	15.5	1.1	1.1	2.5	7.6	60,000
5784WA	Submin.	6.3	200	3200	20	0.79	0.6	5.2	3.2	60,000

POWER OUTPUT TRIODES

Designation	Size	Type	Technical Characteristics								Altitude
			Ef	If	Gm	Mu	Pp Max.	Ehk Max.	Ib	Ik	
5687WA	Min.	Twin	6.3/ 12.6	880	11500	18.5	3.75	100	36	65	60,000
6080WA	Octal	Twin	6.3	2.5A	7000	1.7	13	300	125	Not listed	60,000

POWER OUTPUT PENTODES

Designation	Size	Type	Ef	If	Gm	Ik	Pp Max.	Pg2 Max.	Ib	Ic2	Max. Altitude
2E30	Min.		3/6	600	4250	Not listed	10	2.5	60	5.5	10,000
3B4	Min.		1.4/ 2.8	165	1850	Not listed	3	1.1	25	6.2	10,000
3V4	Min.		1.4/ 2.8	100	2150	13	Not listed	Not listed	9.5	2.2	10,000
6AG7Y	Octal	Video Amp	6.3	650	11700	95	9.0	1.5	30	6.5	10,000
6AN5WA	Min.	Video Amp	6.3	450	8500	55	4.6	1.55	33	11	60,000
6BG6G	Octal	Deflection Amp	6.3	900	6000	110	25	3.5	110	4	10,000
6L6WGB	Octal		6.3	900	6000	Not listed	26	3.5	65	2.5	10,000
5639	Submin.		6.3	450	9000	40.5	3.5	3.5	21	40	60,000
5672	Submin.		1.25	50	650	5.5	Not listed	Not listed	3.1	.95	10,000
5686	Min.		6.3	350	3300	Not listed	8.25	3.3	28	3.5	Not listed
5902	Submin.		6.3	450	4200	50	3.7	0.4	30	2.0	60,000
6005/6AQ5W	Min.		6.3	450	4100	65	11	2.2	45	4.5	60,000
6094	Min.		6.3	150	4100	75	12.5	2.0	45	3.5	60,000
6088	Submin.		1.25	20	560	1.5	Not listed	Not listed	675 ua	150 ua	10,000
6384	Octal	Beam	6.3	1200	5400	125	30	3.5	77	3.5	60,000

RECTIFIERS

Designation	Size	Full or Half Wave	Ef	If	Max. Peak Inverse V	Max. Avg. I	Max. Altitude
1B3GT	Octal	Half	1.25	200	33,000	2.2	10,000
1Z2	Min.	Half	1.25	200	15,000	1.5	10,000
5R4WGA	Octal	Full	5.0	2.0A	2,900	190	30,000- 60,000
5Y3WGTA	Octal	Full	5.0	1.8A	1,550	140	65,000
5641	Submin.	Half	6.3	450	775	50	60,000
6203	Min.	Full	6.3	900	1,250	500	60,000

TABLE 1- CHARACTERISTICS OF MIL-STD-200D RECEIVING TUBES (Cont'd)

DIODES

Designation	Size	Type	Technical Characteristics					Max.
			Ef	If(ma)	Io/p	ib/p	Ehk Max.	Altitude
1A3	Min.	Single	1.4	150	.55	5.5	100	10,000
5647	Submin.	Single	6.3	150	10	15	360	60,000
5726/6AL5W	Min.	Twin	6.3	300	10	60	360	60,000
5829WA	Submin.	Twin	6.3	150	5.5	33	360	60,000
5896	Submin.	Twin	6.3	300	10	60	360	60,000
6110	Submin.	Twin	6.3	150	4.4	26.5	360	60,000

TRIODES

Designation	Size	Type	Technical Characteristics							
			Ef	If	Gm	Mu	Ehk Max.	Pp Max.	Ib	Altitude
6C4WA	Min.	Single	6.3	150	2200	17	Not listed	3.8	10.5	10,000
5703WA	Submin.	Single	6.3	200	5000	25	100	3.3	9.6	10,000
5703WB	Submin.	Single	6.3	200	5000	25.5	100	1.35	9.4	Not listed
5718	Submin.	Single	6.3	150	5800	27	200	0.9	22.0	60,000
5719	Submin.	Single	6.3	150	1700	70	200	0.10	3.3	60,000
5744WA	Submin.	Single	6.3	200	4000	70	200	1.1	4.2	60,000
5744WB	Submin.	Single	6.3	200	4000	70	200	1.3	4.2	60,000
6222	Submin.	Single	6.3	175	1700	70	200	0.3	0.7	80,000
6533	Submin.	Single	6.3	200	1750	54	200	0.5	2.5	60,000
3A5	Min.	Twin	1.4/2.8	220 / 1.4V	2800	15	Not listed	1.0	12.5	10,000
12AT7WA	Min.	Twin	6.3/12.6	150	5500	60	Not listed	2.8	10.0	10,000
5670	Min.	Twin	6.3	350	5500	35	100	1.35	8.2	60,000
5751	Min.	Twin	6.3/12.6	175	1200	70	100	0.8	1.0	60,000
5755	Min.	Twin	6.3/12.6	360	1550	70	75	1.0	2.1	10,000
5814A	Min.	Twin	6.3/12.6	175	2200	17.0	100	3.0	10.5	60,000
6021	Submin.	Twin	6.3	300	5400	35	200	0.7	22	60,000
6111	Submin.	Twin	6.3	300	5000	20	200	0.95	22	60,000
6112	Submin.	Twin	6.3	300	1800	70	200	0.10	3.3	60,000

PENTODES

Designation	Size	Cutoff	Technical Characteristics								Max.
			Ef	If	Gm	Ehk Max.	Pp Max.	Pg2 Max.	Ib	Ic2	Altitude
5749/6BA6W	Min.	Remote	6.3	300	4400	100	3.3	0.7	11.0	4.2	10,000
5899	Submin.	Remote	6.3	150	4500	200	0.75	0.35	7.2	2.0	60,000
6206	Submin.	Semi-Remote	6.3	150	4500	200	1.1	0.55	7.2	2.2	80,000
1AD4	Submin.	Sharp	1.25	100	1850	Not listed	Not listed	Not listed	3.0	0.9	10,000
1AH4	Submin.	Sharp	1.25	40	750	Not listed	0.3	0.1	.7	0.20	10,000
6AH6	Min.	Sharp	6.3	450	8500	90	3.3	0.45	9.5	2.25	10,000
6AU6WA	Min.	Sharp	6.3	300	5200	100	3.3	0.7	10.6	4.3	10,000
5654/6AK5W	Min.	Sharp	6.3	175	5000	130	165	0.55	7.5	2.5	60,000
5702WA	Submin.	Sharp	6.3	200	5000	200	0.9	0.3	7.5	2.6	60,000
5702WB	Submin.	Sharp	6.3	200	5000	200	1.1	0.4	7.5	Not listed	60,000
5840	Submin.	Sharp	6.3	150	5000	200	0.80	0.35	7.5	2.5	60,000
6205	Submin.	Sharp	6.3	150	5000	200	0.8	0.35	7.5	2.4	60,000

TABLE 2

SPECIFICATION CHANGES AFFECTING RATINGS OR ACCEPTANCE TESTS FOR TUBE TYPES COVERED IN 1958 EDITION OF MIL HDBK 211, "TECHNIQUES FOR APPLICATION OF ELECTRON TUBES IN MILITARY EQUIPMENT"

Tube Type	Inspection Item	Old	New
5726/6AL5W	Specification	7A - 3 May 54	7B - 30 Nov. 56
	Bulb Temperature	140°C	165°C
	Altitude	10,000 ft.	60,000 ft.
	Operation Current (Initial; end of life limits removed)	----	20 mAdc Max
	Emission		
	500-hr. life test	----	35 mAdc
	1000-hr. life test	----	30 mAdc
	Heater-Cathode Leakage		
	Initial	10 µAdc Max	5 µAdc Max
	Life Test	20 µAdc Max	10 µAdc Max
5750/6BE6W	Specification	9A - 26 Dec. 56	No change
	Conversion Transconductance		
	500 Hours	----	250µ mhos Min; 700µ mhos Max
	1000 Hours	----	230µ mhos Min; 700µ mhos Max
	Grid Current Ic3		
	Initial	----	-1µ Adc Max
	500 Hours	----	-1µ Adc Max
	1000 Hours	----	-1µ Adc Max
5751	Specification	10 - 13 June 53	10A - 30 Nov. 56
	Heater Voltage	6.3 V ± 10% 12.6 V ± 10%	6.3 ± .6V 12.6 ± 1.3 V
	Control-Grid Voltage	-50 Vdc Min	-55 Vdc Min
	Control-Grid Series Resistance per Section	----	0.5 Meg
	Altitude	10,000 ft.	60,000 ft.
	Heater-Cathode Leakage		
	Initial	10 µAdc	7 µAdc
	500 Hours	10 µAdc	7 µAdc .
	1000 Hours	10 µAdc	7 µAdc
	Change in Amplification Between Sections	----	± 15%
5899	Specification	97C - 23 Jan. 55	97D - 22 Oct. 57
	Plate Dissipation	0.85 W	0.75 W
	Screen Dissipation	0.35 W	0.25 W
	Capacitance C-in	3.8µµf Min; 4.8µµf Max	3.5µµf Min; 4.5µµf Max
6005/6AQ5W	Specification	13A - 20 May 53	13B - 30 Nov. 56
	Heater Voltage	6.3 V ±10%	6.3 ±0.6 V
	Control-Grid Voltage	----	0 Min; -55 Vdc Max
	Plate Dissipation	13.2 W	11.0 W
	Control-Grid Series Resistance	----	0.1 Meg (may be 0.5 Meg if RT is used)
	Cathode Current	----	65 mAdc Max
	Control-Grid Current	----	3 mAdc Max
	Heater Current		
	500 Hours	490 mAdc	500 mAdc
	1000 Hours	490 mAdc	510 mAdc
	Power Output (1)		
	Initial	3.4 W Min	3.6 W Min
	500 Hours	2.3 W Min	Change in individuals, 15% Max
	1000 Hours	2.1 W Min	Change in individuals, 20% Max
	Power Output (Change with Ef)		
	Initial	3.2 Min (Ef = 5.5 V)	15% Change Max for Ef = 5.7 V
	500 Hours	----	15% Change Max for Ef = 5.7 V
	Capacitance		
	Cglp	0.7µµf Max	0.8µµf Max
	C-in	6.6µµf Min; 10µµf Max	6.6µµf Min; 9.6µµf Max
	C-out	6.0µµf Min; 9µµf Max	6.0µµf Min; 11.0µµf Max
	Grid Current		
	Initial	-2.0 µAdc Max	-1.0µ Adc Max
	500 Hours	-4.0 µAdc Max	-10 µ Adc Max
	Grid Emission, Initial	-2.0 µ Adc Max	-4.0µ Adc Max
	Heater-Cathode Leakage		
	Initial	30 µ Adc Max	20µ Adc Max
	500 Hours	30 µ Adc Max	20µ Adc Max
	Insulation of Electrodes		
	500 Hours	----	50 Meg

3.2 Types of Data Presented in Remainder of Supplement

Five types of application and specification information are presented on specific tube types: (1) essential information abstracted from the current specification, including outline drawing, pertinent ratings, and electrical characteristics; (2) permissible operating areas outlined on the plate-characteristic curves; (3) product variability in electrical characteristics permitted by the specification, outlined on plate-characteristic curves and, where possible, on transfer characteristics; (4) life-test behavior in the form of histograms of several important electrical characteristics, for either 500 or 1000 hours, as required by the individual specification; and (5) reliability functions derived from life-test data, showing survival characteristics (with 95-percent confidence limits) for tubes failing either as a result of shorts or opens or as a result of deterioration as determined by specification end-points.

3.3 Life-Test Requirements

Current military specifications for reliable electron tubes require that tubes be subjected to 1000-hour intermittent life tests, unless the manufacturer can qualify for reduced-hours (500-hour) life tests. The following three subsections describe regular life-test requirements, requirements for reduced-hours testing, sampling plans in current use, and the quality level guaranteed by specifications.

3.3.1 Regular Life-Test Requirements for Reliable Tubes

Regular life tests are those which are conducted for 1000 hours, with measurements being made at specified end-points of 500 hours and 1000 hours. Ordinarily, 20 tubes are randomly selected from each lot for regular life testing. At 500 hours, approximately 7 electrical characteristics are checked against specification limits, and no more than a total of 4 defective tubes are permitted, the maximum number in any category being 1 or 2 defectives, depending on the category. At 1000 hours, approximately 5 electrical characteristics are tested, and no more than 1 or 2 defectives per category are permitted, the total number of defectives not to exceed 5.

Should the lot fail, a double sampling plan is instituted -- that is, the manufacturer is allowed to place 40 additional randomly selected tubes on life test. At 500 hours the total number of defectives from both test groups cannot exceed 8 tubes, and the defectives in any one category cannot exceed 3 or 5 tubes, depending on the category. At 1000 hours, the total number of defectives from both test groups cannot exceed 10 tubes, and no more than 3 or 5 defectives are permitted in any one category, depending on the category.

3.3.2 Reduced-Hours Requirements

To permit manufacturers with good production history to ship prior to completion of the 1000-hour intermittent life tests (i.e., at the end of 500 hours), the specifications for reliable tubes usually require that lots be tested on a 1000-hour basis until 3 consecutive lots have been accepted, at which time eligibility for 500-hour

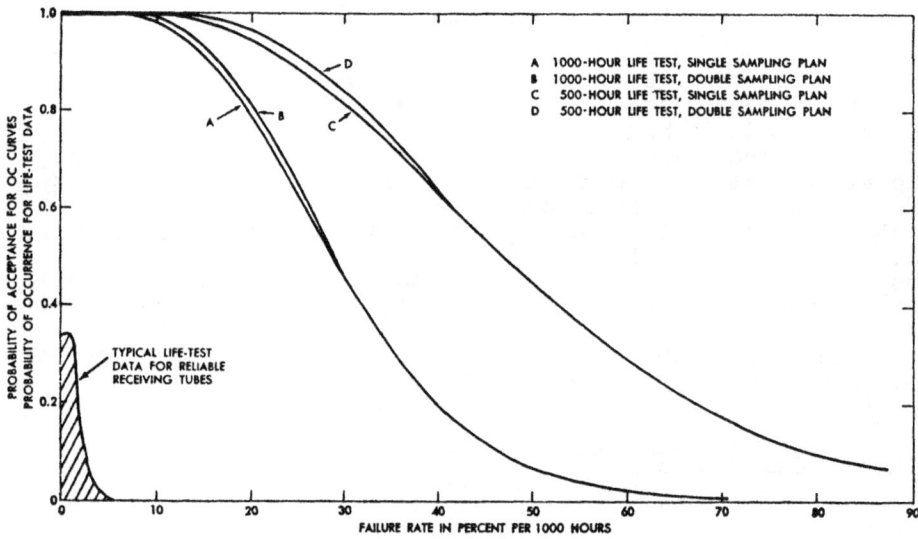

FIGURE 34

OPERATING CHARACTERISTICS FOR TYPICAL RELIABLE RECEIVING-TUBE
LIFE-TEST SAMPLING PLANS; ALSO COMPARABLE LIFE-TEST DATA

TABLE 3			
MAXIMUM FAILURE RATES PERMITTED BY RELIABLE-TUBE SPECIFICATIONS			
Curve	Description of Test	Acceptable Failure Rate, in % per 1000 hours	Lot Tolerance Failure Rate, in % per 1000 hours
A	1000-hr., single sampling	13	46
B	1000-hr., double sampling	14	46
C	500-hr., single sampling	20	80
D	500-hr., double sampling	22	80

TABLE 4		
MINIMUM SAMPLE SIZES REQUIRED FOR SPECIFIED ACCEPTABLE FAILURE RATE (AFR) AND LOT TOLERANCE FAILURE RATE (LTFR) DURING 1000-HOUR LIFE TESTS (Abstracted from MIL-S-19500B)		
Minimum Sample Size (1 failure permitted)	Acceptable Failure Rate, in % per 1000 hours	Lot Tolerance Failure Rate in % per 1000 hours
9	4.2	50.0
40	0.9	10.0
80	0.45	5.0
390	0.09	1.0
779	0.045	0.5
3,891	0.009	0.1

testing is conferred. The 1000-hour life test is reinstituted whenever 2 or more of the last 3 lots tested fail.

Even when reduced-hours testing is in effect, the first lot accepted each month must be life-tested for 1000 hours.

3.3.3 Quality Guaranteed by Current Specifications

In computing the maximum failure rates permitted by specifications, it has been assumed that tubes fail at a constant rate over the period of the life test, even though it is known that most tube types are likely to fail at an increasing rate as their age increases.* There is little error in the exponential assumption for periods of 500 or 1000 hours if tubes have mean lives greatly in excess of 1000 hours.

Figure 34 presents the maximum failure rates permitted by specifications, and includes typical data taken from recent life tests to give a measure of comparison. The four types of Operating Characteristic (OC) curves are related to the four types of life tests listed in Table 3. The Acceptable Failure Rate (AFR) is the quality level which an average of 19 out of 20 inspection lots must meet in order to be accepted. The Lot Tolerance Failure Rate (LTFR) is the quality level for which the probability of lot acceptance is 10 percent. The LTFR, therefore, gives a measure of the amount of consumer protection which is afforded by the specification.

The typical life-test failure rates shown in Figure 34 indicate that the quality of current electron tubes is considerably better than that required by the specification. On the other hand, the sample sizes shown in Table 4 indicate that improvement of specifications to the point where a high level of quality were assured with a high level of confidence would prove extremely costly.

3.4 Abbreviations and Symbols

Since the symbols and notations used in electron-tube specifications are somewhat different from those used in other standards employed in the eletronics industry, a list taken from the original handbook is given here. For the most part, the symbols used in the remainder of the supplement are those given in the MIL-E-1 specification.

---

* See Publication No. 110, A Selection of Electron Tube Reliability Functions, ARINC Research Corporation, January 8, 1958.

MIL-E-1D SYMBOLS

Abbreviations and symbols. For the purpose of simplification, the following abbreviations and symbols are used herein and on the tube specification sheets wherever practicable.

A	Angstrom unit
A	Amperes (may be either ac rms or dc)
a	Amperes (peak value) or anode
Aac	ac amperes (rms)
α (alpha)	Attenuation constant
ac	Alternating current
Adc	dc amperes
ALD	Acceptance limit for sample dispersion
AQL	Acceptable quality level
β (beta)	Phase constant
B/Yo	Tuning susceptance
c	Velocity of light
C	Capacitance
°C	Degrees centigrade
cb	Centibels
Cgk, Cgp, Cpk, etc. ...	Tube capacitance between the electrodes indicated
Cin	Input capacitance
Ck	Capacitor between cathode and ground
CL	Load capacitance
cm	Centimeter
Cout	Output capacitance
cps	Cycles per second
CRO	Cathode ray oscilloscope
ct	Center tap
CW	Continuous wave
Δ (delta)	A change in the value of the indicated variable. When expressed in percent the difference in readings is divided by the initial reading and multiplied by 100
db	Decibels
D1,2,3,4	Deflection plates
dc	Direct current
DF	Deflection factor in volts per inch
$\frac{dik}{dt}$	Rate of rise of cathode current pulse
Du	The product of time of pulse and pulse repetition rate (duty cycle)
dy	Dynode
EB	Ballistic deflection
Eb,Eb1,2,3 ..	dc voltage on respective anodes or plates. In the case of multiplex tubes containing more than one operating unit, the number of the unit concerned is inserted between the voltage symbol and the element symbol. For example, E2b, E1p, E1c2, etc. The number of the unit is the number of the plate in that unit
eb	Peak dc anode or plate voltage
Ebb	dc anode or plate supply voltage
Eb/Ib	Adjust plate voltage to produce the specified plate current
Ec,Ec1,2,3	dc voltage on respective grids
Ecc,Ecc1,2,3	ac supply voltage to respective grids
Ec/Ib	Adjust grid voltage for the specified plate current
Eco	dc cut-off grid voltage
ed	Voltage peak between anode No. 2 and any deflection plate in cathode ray tubes
Edy	dc voltage of anode producing secondary emission
Ee	End-of-plateau voltage
Ef	Filament or heater voltage
Ef/Po	Adjust filament potential (with other potentials held constant) to reduce the power output obtained on oscillation by the amount specified
Eg1,2,3	rms value of ac component of input voltage for respective grids
egk	Peak voltage drop between grid and cathode
egy,egy1,2,3	Peak forward grid voltage.
egx	Peak inverse grid voltage
Ehk	Heater-cathode voltage (sign to indicate polarity of heater with respect to cathode)
Eid	Ignitor voltage drop
Eo	dc component of output voltage of rectifiers
EO	Overvoltage for radiation counter tubes
eo	Pulse amplitude
Ep	rms value of the ac component of plate voltage with respect to cathode
Epp	ac anode or plate supply voltage
epx	Peak plate inverse voltage
epy	Peak forward anode or plate forward voltage.

Er	Reflector voltage	ic	Peak grid current
ER	Reservoir voltage	Idy	Current of anode producing secondary emission
Ers	Resonator voltage	If	Filament or heater current
Es	dc emission voltage	if	Intermediate frequency
Es	Starting voltages for radiation counter tubes	Ig	rms value of ac component of grid current
Esd	External shield voltage	Ihk	Heater-cathode leakage current
Esh	Shell voltage	Ii	Ignitor current
Esig	Applied signal voltage	Ik	dc cathode current
Eta	Target voltage	ik	Peak cathode current
Etd	Average voltage drop between anode and cathode	iL	Peak load current
		int.con.	Internal connection
etd	Peak voltage drop between anode and cathode	Io	dc component of output current of rectifiers per tube
Ez	Ionization, breakdown, or striking voltage	Ip	rms value of ac component of plate current
f	Filament	Ir	Reflector current
F	Frequency (in cps)	IR	Reservoir current
FA	Maximum frequency above which receiving tube performance deteriorates seriously and sharply	Irs	Resonator current
		Is	dc emission current
		is	Peak emission current
F1	Maximum frequency at which maximum ratings apply	Isg	dc component of primary emission from grid indicated
F2	Frequency at which maximum plate voltages and plate input are limited to 50 percent of the ratings for F1. For frequencies between F1 and F2 the maximum plate voltage and plate input will be reduced in the correct proportion so that at the frequency F2 these factors will not exceed 50 percent of their maximum ratings	Ita	dc target current
		Iz	Ionization current
		°K	Degrees Kelvin
		k	Cathode
		kc	Kilocycles
		kMc	Kilo-megacycles
		KTB	Theoretical resistance noise power
		kv	Peak kilovolts
		kVA	Kilovolt-amperes
fct	Filament center tap	kva	Peak kilovolt-amperes
fk	Filament-cathode return	kVac	ac kilovolts (rms)
Fsg	Frequency of signal generator	kVdc	dc kilovolts
ft. L.	Foot lamberts	kW	Kilowatts
G	Acceleration of gravity	kw	Peak kilowatts
G/Yo	Equivalent conductance	L	Lamberts
γ (gamma)	Propagation constant	LAL	Lower acceptance limit for sample average or sample median
g, g1,2,3	Grid (number to identify grids, starting from cathode)	λ (lambda)	Wavelength
		λo	Resonant wavelength
g2+4	Grids having common pin connection	Lc	Conversion loss or gain (ratio of available signal power to the available intermediate frequency power)
GA	Gas amplification		
Gr	Gas ratio		
H	Field strength in gauss	LIb	Leakage current
hct	Heater center tap	Li	Insertion loss
ht	Heater tap	lm	Lumens
Ia	Anode current	LRLM	Lower reject limit median for a sample of tubes
Ib, Ib1,2,3	dc current of respective anodes or plates	LS1	Standardized light source supplied by a coiled tungsten lamp with a lead or lime glass envelope operated at a color temperature of 2,870°K
ib	Peak value of dc anode or plate current. When used in reference to pulses, the maximum peak current excluding spike		
Ic, Ie1,2,3	dc current of respective grid	LSLA	Lower specification limit for average of acceptable lots

M . .	Figure of merit, or one million		QL	Loaded Q
m . .	Meter, or one-thousandth		Qo	Intrinsic Q or quality of a circuit without external loading
mA	ac (rms) or dc milliamperes		r	Reflector
ma	Peak milliamperes		r	Roentgen
mAac	ac milliamperes (rms)		R	Resistance
mAdc .	dc milliamperes		Rb	dc resistance of external plate circuit (by-passed)
Mc	Megacycles			
Meg	Megohms		Rc	dc resistance of external grid circuit (by-passed)
mftL	Millifoot lamberts			
mH	Millihenry		Rc	Reference resistor for noise ratio measurements (for crystal rectifiers)
mL	Millilamberts			
mr	Milliroentgen			
MRSD	Maximum rated standard deviation		rf	Radio frequency
			Rf	Resistance in series with filament or heater
ms	Milliseconds			
Mu or u	Amplification factor		Rg	Resistance in series with grid
mVac	ac millivolts (rms)		rg	Dynamic internal grid resistance
mVdc	dc millivolts		Rk	Resistance in series with cathode
mv	Peak millivolts		Rka1, Rka2,	
MW	Megawatts		Rkrs, Rfrs,	Tube resistance between the electrodes indicated
Mw	Peak megawatts		etc.	
mW	Milliwatts		RL	Load resistance (Unity power factor. Negligible dc resistance.)
mw	Peak milliwatts			
N	Counts for radiation counter tubes		rms	Root mean square
nc	No connection		Rp	Resistance in series with plate or anode
NF	Noise figure			
Npm	Counts per minute		rp	Dynamic internal plate resistance of tube
Nps	Counts per second			
Nr	Output noise ratio (ratio of noise power output to resistance noise power)		rs	Resonator
			Rv	Video impedance
			S	Static sensitivity (phototubes)
p	Plate		s	Dynamic sensitivity (phototubes)
/p	Per plate		se	Starter electrode
Pb	Plate breakdown factor (epx x prr x lb)		Sc	Conversion transconductance
			Sd	Spectral distribution
Pd	Average drive power		sd	Shield
pd	Peak drive power		Sg1, g2, etc.	Transconductance between the elements indicated
Pg1,2,3	Power dissipation of respective grids			
			sh	Shell
Pi	Power input (plate)		σ (sigma)	"Input" standing-wave ratio in voltage
pi	Peak power input			
Pj	Reactive power in watts		σ' (sigma prime)	"Output" standing-wave ratio in voltage
Pl	Plateau length			
Pn	Noise output		Sm	Transconductance (control grid-plate)
P'o	Intrinsic P			
Po	Average power output		ΔSm, etc. ..... Ef	Change in Sm, etc. of an individual tube, caused by the specified change in Ef
Po	Peak leakage power			
$\overline{Du}$				
ΔPo, etc ...... Ef	Change in Po, etc. of an individual tube, caused by the specified change in Ef		ΔSm, etc. ..... t	Change in Sm, etc. caused by a test (life, shock, fatigue, etc.)
			Sr	Sensitivity ratio (max. Ib to min. Ib)
ΔPo, etc ...... t	Change in Po, etc. caused by a test (life, shock, fatigue, etc.)			
po	Peak power output		T	Temperature (degrees centigrade)
Pp	Plate or anode power dissipation		t	Test duration (seconds, unless otherwise specified)
prr	Pulse recurrence rate in pulses per second			
			TA	Ambient temperature
Ps	Relative plateau slope		ta	Target
Q	Quality of a circuit			

tad .... Anode delay time. A time interval between the point on the rising portion of the grid pulse which is 26 percent of the maximum unloaded pulse amplitude and the point where anode conduction takes place

Δtad .... Anode delay time drift

TE .... Envelope temperature

tf .... Time of fall. The time duration of pulse to fall from 70.7 percent of the maximum pulse amplitude to 26 percent of the maximum pulse amplitude, excluding spike, in microseconds

THg .... Temperature of condensed mercury in °C

tj .. .. Variation in firing time

tk .... Cathode conditioning time (in seconds) necessary before the application of high voltage. In TR tubes, time delay between application of ignitor voltage and rf power

tp .... Pulse duration (excluding magnetrons). The time interval between the points on the trace envelope at which the instantaneous amplitudes are equal to 70.7 percent of the maximum amplitude excluding spike. For magnetrons, see 4.16.3.3

tr .... Time constant of rise (excluding magnetrons). The time duration of a pulse to rise from 26 percent of the maximum pulse amplitude to 70.7 percent of the maximum pulse amplitude, excluding spike, in microseconds

trc .... Time of rise of current pulse in microseconds (for magnetrons, see 4.16.3.3)

trv .... Time of rise of voltage pulse in microseconds (for magnetrons, see 4.16.3.3)

u .... Amplification factor

ua .... Microamperes, peak value

uAac .... ac microamperes (rms)

uAdc .... dc microamperes

UAL . ... Upper acceptance limit for sample average or sample median

umhos .... Micromhos

uf .... Microfarads

uh .... Microhenries

URLM . ..... Upper reject limit median of a sample of tubes

us .... Microseconds

USLA .... Upper specification limit for averages of acceptable lots

uuf .... Micromicrofarads

uVac .... ac microvolts (rms)

uVdc .... dc microvolts

uW .... Microwatts

V .... Volts (may be either ac rms or dc)

v .... Volts, peak value

VA .... Volt-amperes

va .... Peak volt-amperes

Vac .... ac volts (rms)

Vdc .... dc volts

v/in .... Volts, peak value, per inch of deflection

Vj .... Amplitude jitter

VSWR .... Voltage standing wave ratio

VU .... Volume units

Vx .... Extinguishing voltage

W .... Watts

w .... Peak watts

Ws .... Spike leakage energy

X1 .. .. The orientation of a tube rigidly mounted for mechanical tests with the main axis of the tube and the major cross-section of the tube elements normal to the direction of the accelerating force

X2 .... The orientation of a tube rigidly mounted for mechanical tests with the main axis of the tube normal and the major cross-section parallel to the accelerating force

x .... Denoting peak inverse value

Y1 .... The orientation of a tube rigidly mounted for mechanical tests with the main axis of the tube parallel to the direction of the accelerating force. (When Y1 is referred to for shock tests, the principal base of the tube is toward the hammer)

Y2 ... . The orientation of a tube (for shock test only) which is the same as Y1 except that the principal base of the tube is away from the hammer

y .... Denoting peak forward value

Z . ... Impedance

Zd .... Impedance to anode of deflection plate circuit at power-supply impedance

62

Zg .... Impedance of the grid circuit

Zgg .. .. Impedance between grids of push-pull circuit

Zgk ... . Impedance between grid and cathode

Zi .... Input impedance

ZL .... Load reactance (with negligible dc resistance)

Zm .... Modulator frequency load impedance

Zo .... Output impedance and characteristic impedance

Zp .... Impedance in plate circuit

Zpp .... Impedance between plates in push-pull circuit

1D2 .... Deflection produced by the deflection plates near the screen (for cathode-ray tubes)

3D4 .... Deflection produced by the deflection plates near the base (for cathode-ray tubes)

** .... Qualification test

* .... Standard design test

.... Special design test

† .... Test to be performed at the conclusion of the holding period (See 4.5)

← .... Indicates change on tube specification sheet

←o .... Indicates deletion from the tube specification sheet

TUBE TYPE JAN-6AN5WA

DESCRIPTION:

The JAN-6AN5WA[1] is a 7 pin miniature, RF beam power pentode having a transconductance in the range, 7000 to 10000 micromhos.

ELECTRICAL: The electrical characteristics are as follows:
Heater Voltage..6.3 V
Heater Current..420-480 mA
Cathode...Coated Unipotential

MOUNTING: Any type mounting is adequate.

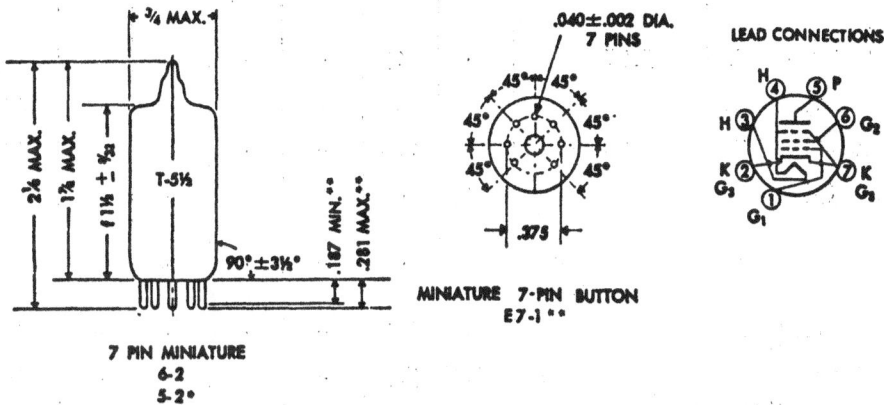

7 PIN MINIATURE
6-2
5-2*

MINIATURE 7-PIN BUTTON
E7-1**

LEAD CONNECTIONS

*REFERS TO JETEC PUBLICATION JO-G2-2, MARCH 1955 SUPERSEDED BY JO-G2-2, MARCH 1958

**REFERS TO JETEC PUBLICATION JO-G3-1, MARCH 1955 SUPERSEDED BY JO-G2-2, MARCH 1958

f MEASURE FROM BASE SEAT TO BULB TOP-LINE AS DETERMINED BY RING GAGE OF 7/16 I.D.

ALL DIMENSIONS IN INCHES

RATINGS:	Ef	Eb	Ec1*	Ec2	Ehk	Rk	Rg1	Ik **	Pp*	Pg2*	T Envelope	Alt
Design	V	Vdc	Vdc	Vdc	v	ohms	Meg	mAdc	W	W	°C	ft
Maximum NOTE	6.9	135	---	135	200	---	0.1	55	4.6	1.55	200	60,000
Minimum	5.7	---	---	---	---	---	---	---	---	---	---	---
Test Cond:	6.3	120	0	120	---	125	---	---	---	---	---	---

[1] The values and specification comments presented in this section are related to MIL-E-1/839A dated 30 Nov 1956.

* No test at this rating exists in the specification.

** No specification assurance of life exists under conditions of cathode current approaching the maximum.

NOTE: The voltage at the plate or screen may be as high as 330 Vdc provided the following condition is met: when the average voltage at the electrode (taken over any (1) second interval) exceeds the design maximum rating of dc voltage for that electrode, the average dissipation for that electrode shall not exceed the design maximum rating of dissipation divided by the ratio of the average voltage to the design maximum rating of dc voltage.

ACCEPTANCE TEST LIMITS SUMMARY

PROPERTY	MEASUREMENT CONDITIONS	INITIAL		500 HR LIFE TEST		1000 HR LIFE TEST		UNITS
		MIN	MAX	MIN	MAX	MIN	MAX	
Heater Current If		420	480	410	490	410	490	mA
Transconductance Sm		7000	10000	-	-	-	-	umhos
Change in individual ΔSmt		-	-	-	20	-	20	%
Change in average Avg ΔSmt		-	-	-	15	-	-	%
Transconductance Change with Ef ΔSmEf	Ef=5.7V	-	15	-	-	-	-	%
Plate Current (1) Ib		25	43	25.0	-	25.0	-	mAdc
Plate Current (2) Ib	Ec1=-20Vdc	-	1.0	-	-	-	-	mAdc
Plate Current (3) Ib	Eb=Ec2=60Vdc; Ec1=0; Rk=0	25	-	-	-	-	-	mAdc
Emission Is	Eb=Ec2=Ec1=15 Vdc;Esig=500 mVac	100	-	-	-	-	-	mAdc
Screen Grid Current Ic2 (1)		6.0	16.0	-	-	-	-	mAdc
Screen Grid Current Ic2 (2)	Eb=Ec2=60Vdc; Ec1=0; Rk=0	6.5	15.5	-	-	-	-	mAdc
Power Output Po	Esig=4.25Vac; RL=2500 ohms	1.0	-	0.75	-	0.5	-	W
Capacitance	.75 in. dia. shield, Ef=0							
C g1p		-	0.075	-	-	-	-	uuf
C in		6.0	12.0	-	-	-	-	uuf
C out		4.0	7.0	-	-	-	-	uuf
Control Grid Current Ic1		0	-2.0	0	-2.0	0	-4.0	uAdc
Control Grid Emission Ic1	Ef=7.5V;Rg1= 0.01Meg min; Ec1=-45Vdc; NOTE 1	0	-4.0	-	-	-	-	uAdc
Heater Cathode Leakage Ihk	Ehk=+100Vdc	-	20	-	75	-	75	uAdc
Ihk	Ehk=-100Vdc	-	20	-	75	-	75	uAdc
Insulation of R(g1-all)	Eg1-all=-100V;	100	-	50	-	40	-	Meg
Electrodes R(p-all)	Ep-all=-300V;	100	-	50	-	40	-	Meg
Interface ri	Ef=5.7V;Ec1/Ib =4mAdc;NOTE 2	-	-	-	25	-	-	ohms

Measurement conditions are the same as stated under Test Conditions, unless otherwise indicated.

NOTE 1: The tube shall be preheated a minimum of five minutes at test conditions for this test (except Ec1=0) prior to this test.

NOTE 2: Preheat 3 minutes with Ef=5.7V and all other tube elements disconnected. No other test shall be made from the start of the cathode interface life test until the specified minimum number of hours has been completed.

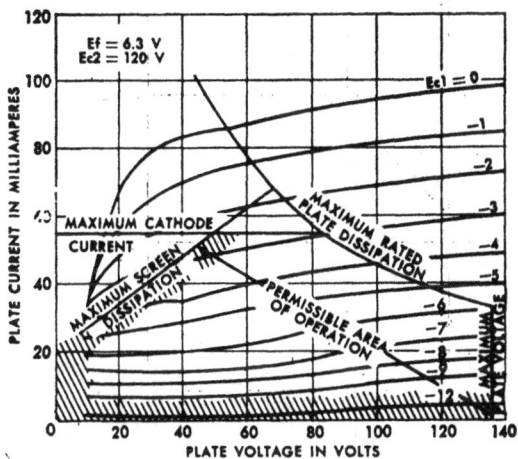

TYPICAL STATIC-PLATE CHARACTERISTICS;
PERMISSIBLE AREA OF OPERATION

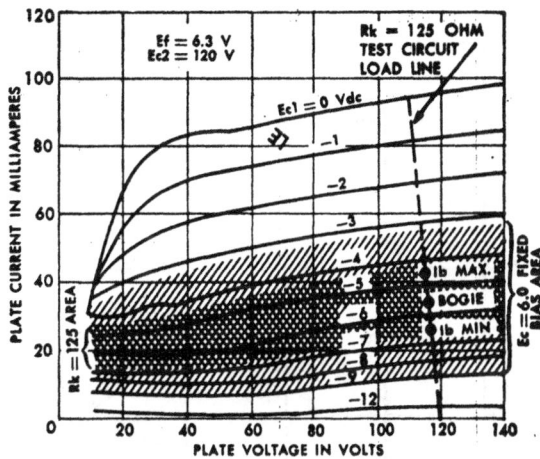

LIMIT BEHAVIOR STATIC-PLATE DATA;
VARIABILITY OF Ib

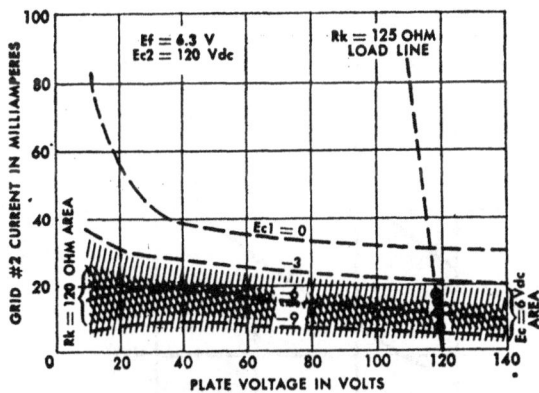

LIMIT BEHAVIOR STATIC-PLATE DATA;
VARIABILITY OF Ic2

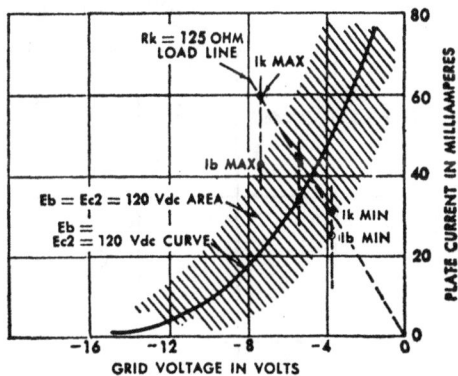

LIMIT BEHAVIOR TRANSFER DATA;
VARIABILITY OF Ib

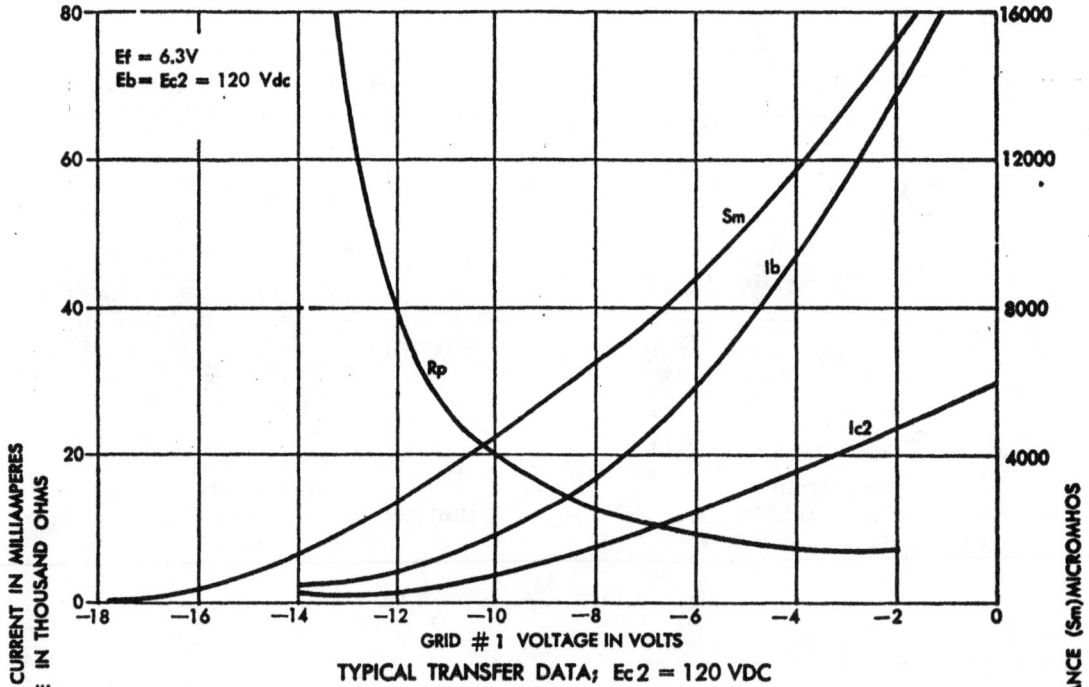

TYPICAL TRANSFER DATA; Ec2 = 120 VDC

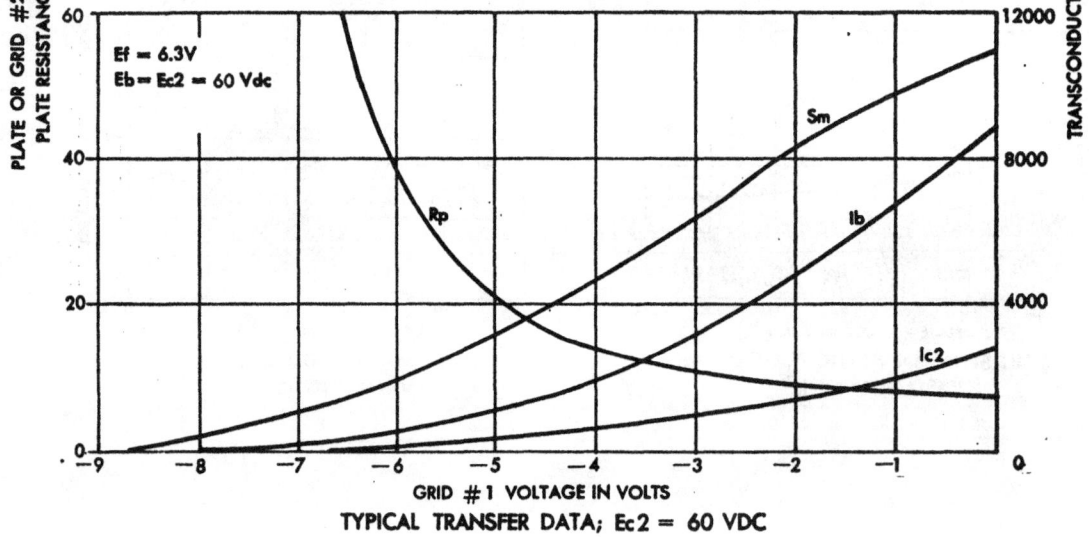

TYPICAL TRANSFER DATA; Ec2 = 60 VDC

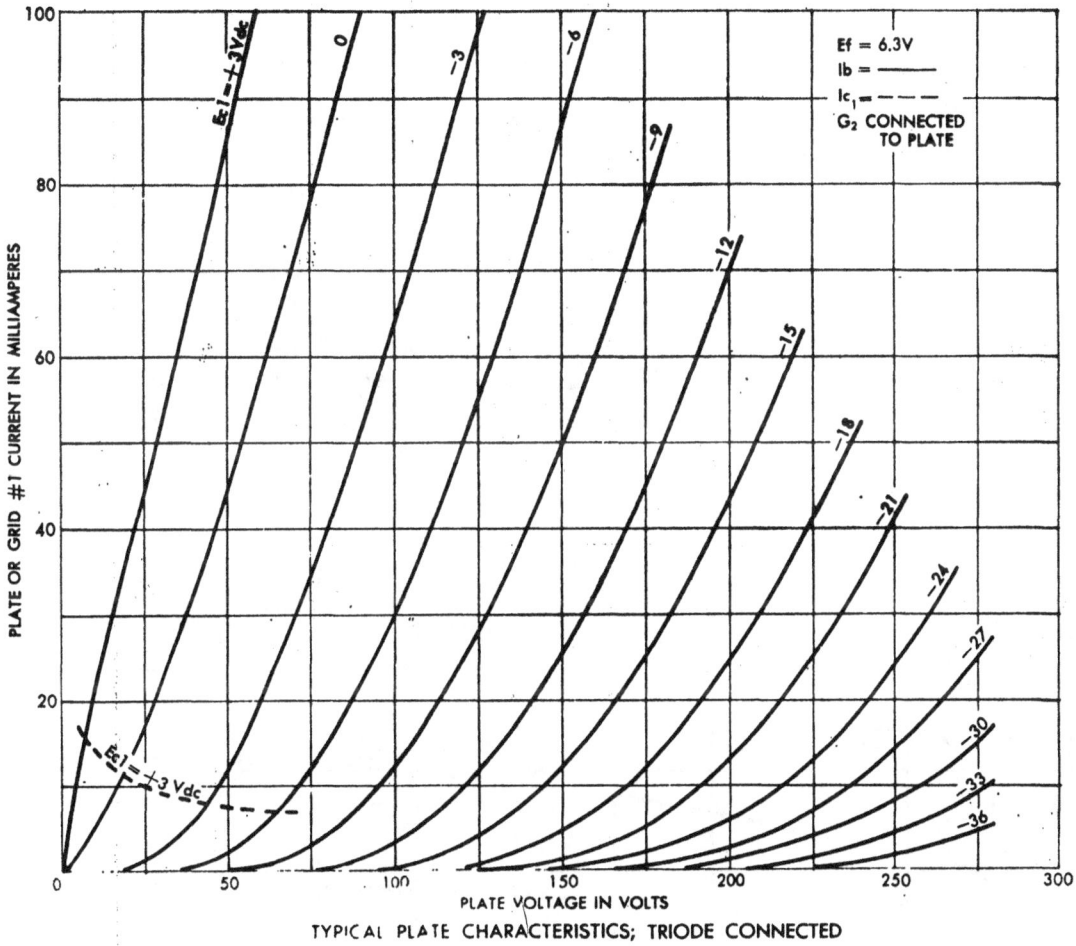

TYPICAL PLATE CHARACTERISTICS; TRIODE CONNECTED

DESIGN CENTER
OBTAINED FROM DATA PUBLISHED BY ORIGINAL
RETMA REGISTRANT

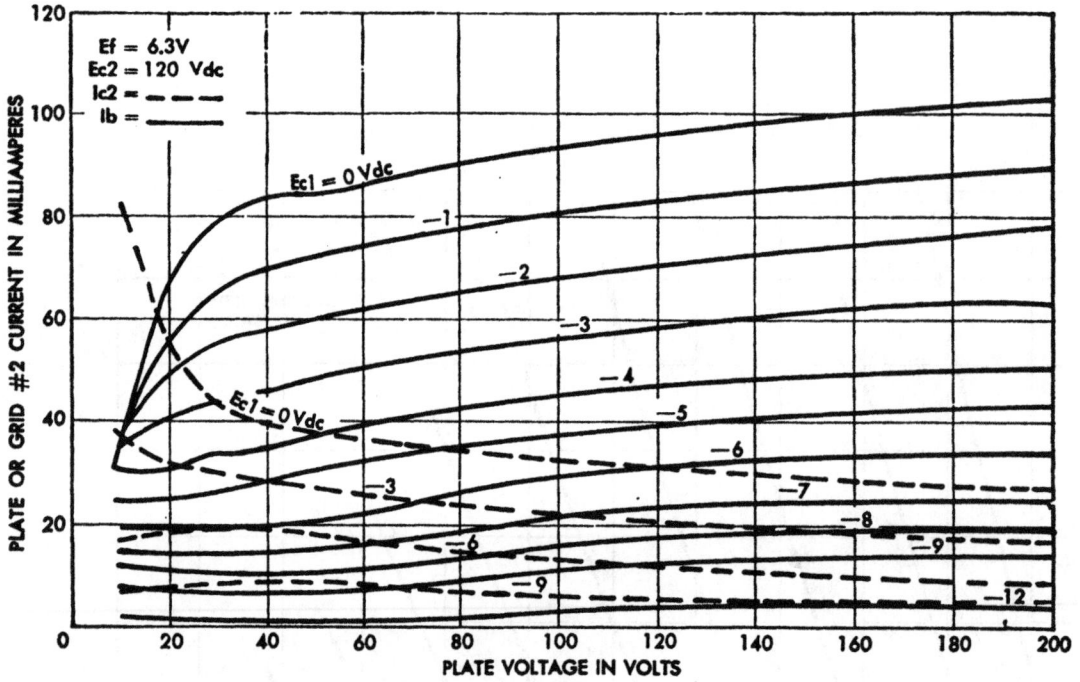

Ef = 6.3V
Ec2 = 120 Vdc
Ic2 =
Ib =

Ec1 = 0 Vdc
—1
—2
—3
—4
—5
—6
—7
—8
—9
—12

Ec1 = 0 Vdc
—3
—6
—9

PLATE OR GRID #2 CURRENT IN MILLIAMPERES

PLATE VOLTAGE IN VOLTS

TYPICAL STATIC–PLATE CHARACTERISTICS; Ec2 = 120

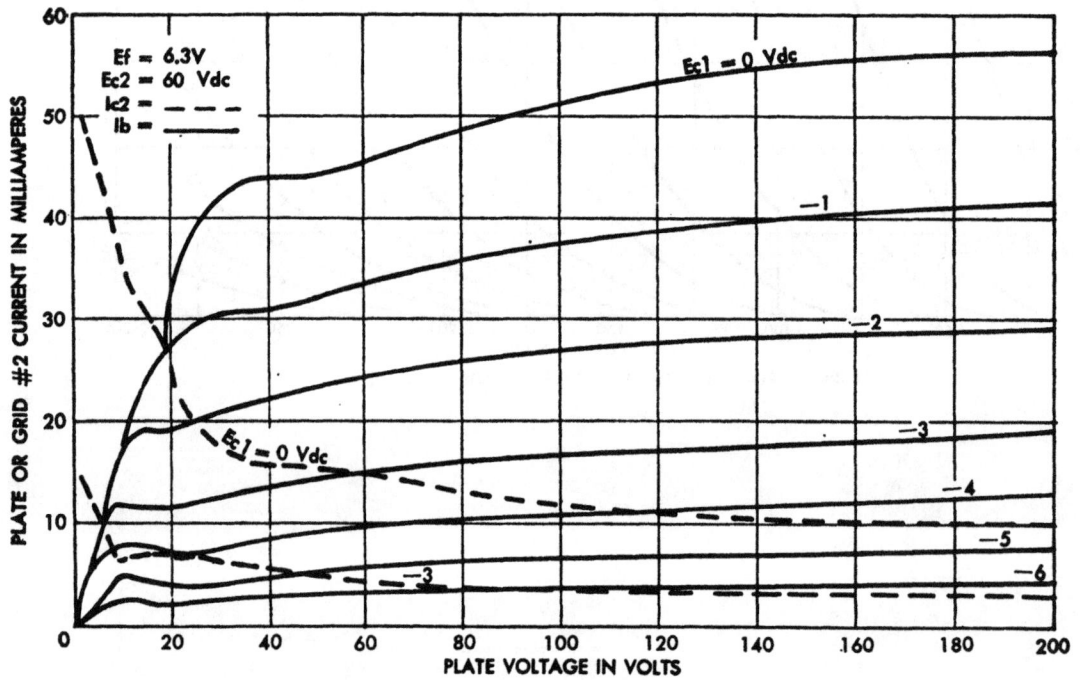

Ef = 6.3V
Ec2 = 60 Vdc
Ic2 =
Ib =

Ec1 = 0 Vdc
—1
—2
—3
—4
—5
—6

Ec1 = 0 Vdc
—3

PLATE OR GRID #2 CURRENT IN MILLIAMPERES

PLATE VOLTAGE IN VOLTS

TYPICAL STATIC–PLATE CHARACTERISTICS; Ec2 = 60

70

LIFE TEST PROPERTY BEHAVIOR
MIL-E-1/839A   30 NOV. '56
PRODUCED IN 1958 BY ONE MANUFACTURER

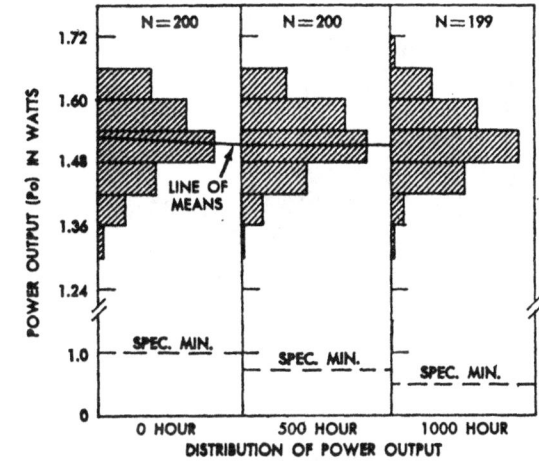

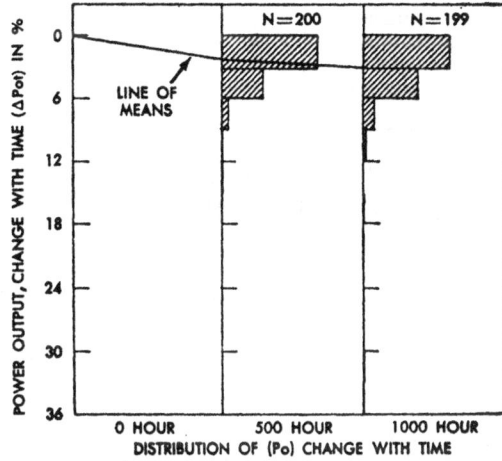

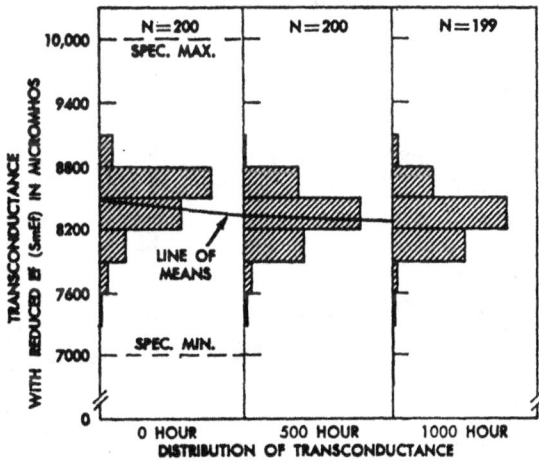

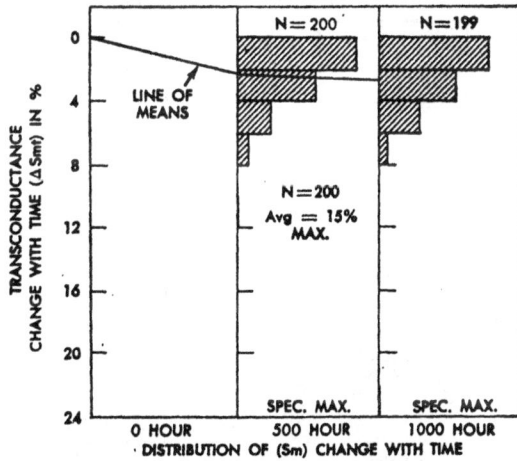

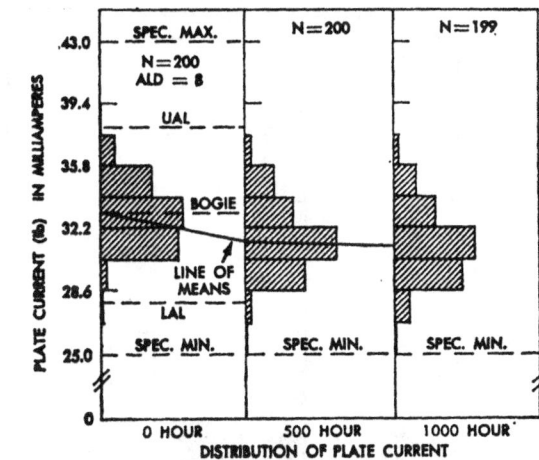

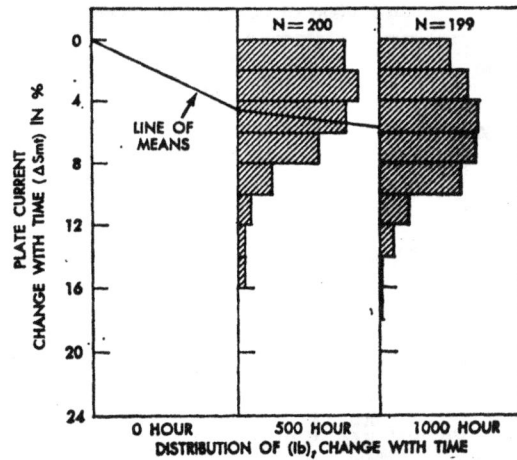

71

LIFE TEST PROPERTY BEHAVIOR
MIL-E-1/839A   30 NOV. '56
PRODUCED IN 1958 BY ONE MANUFACTURER

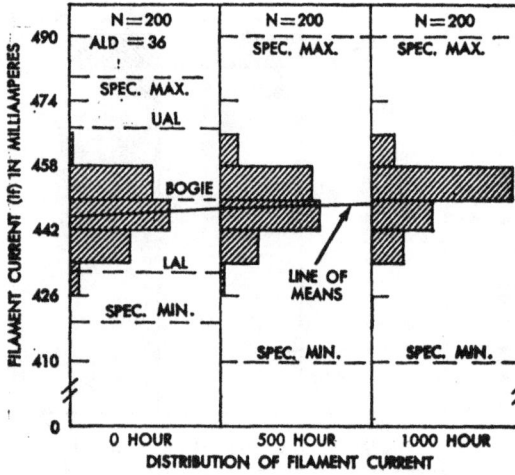

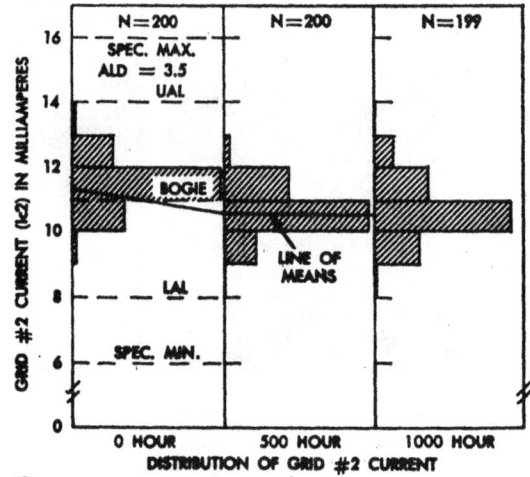

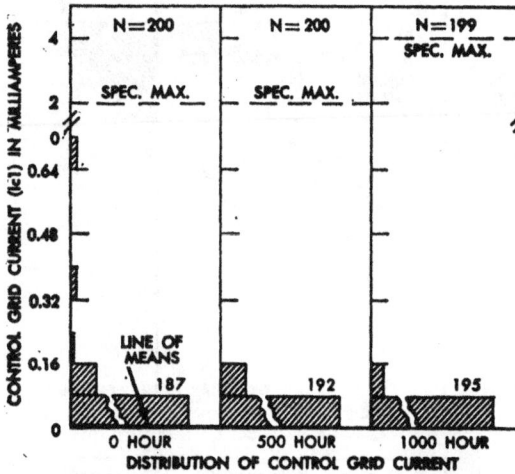

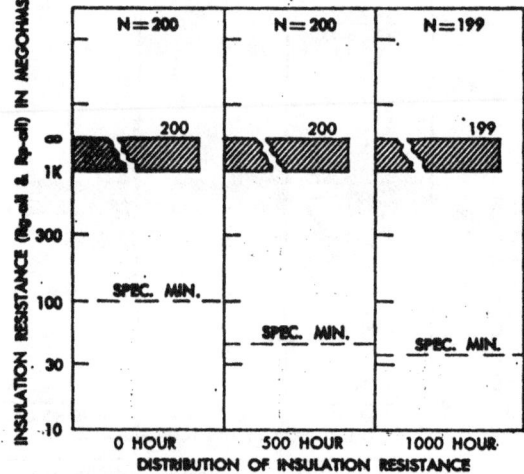

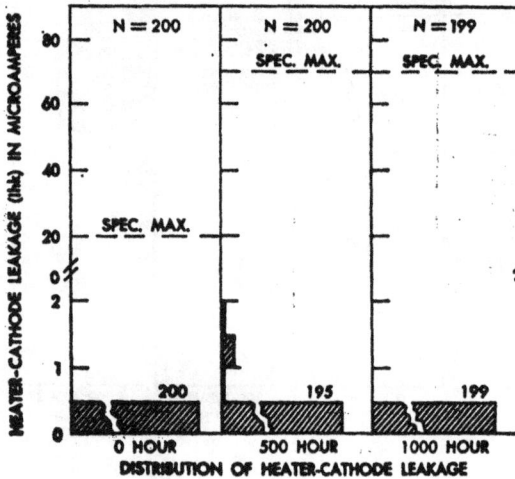

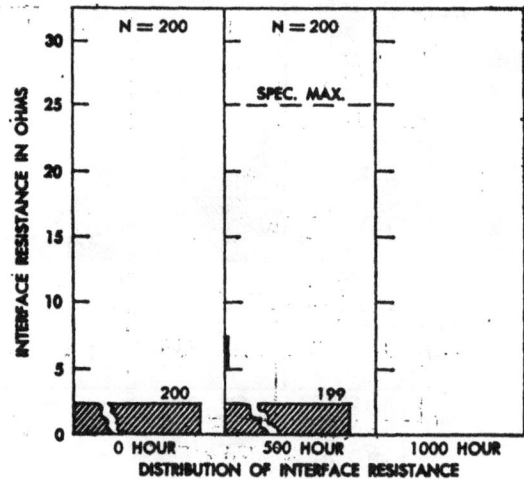

DISTRIBUTION OF FILAMENT CURRENT

DISTRIBUTION OF GRID #2 CURRENT

DISTRIBUTION OF CONTROL GRID CURRENT

DISTRIBUTION OF INSULATION RESISTANCE

DISTRIBUTION OF HEATER-CATHODE LEAKAGE

DISTRIBUTION OF INTERFACE RESISTANCE

LIFE TEST PROPERTY BEHAVIOR
PROBABILITY OF SURVIVAL
MIL-E-1/839A   30 NOV. '56
PRODUCED IN 1958 BY ONE MANUFACTURER

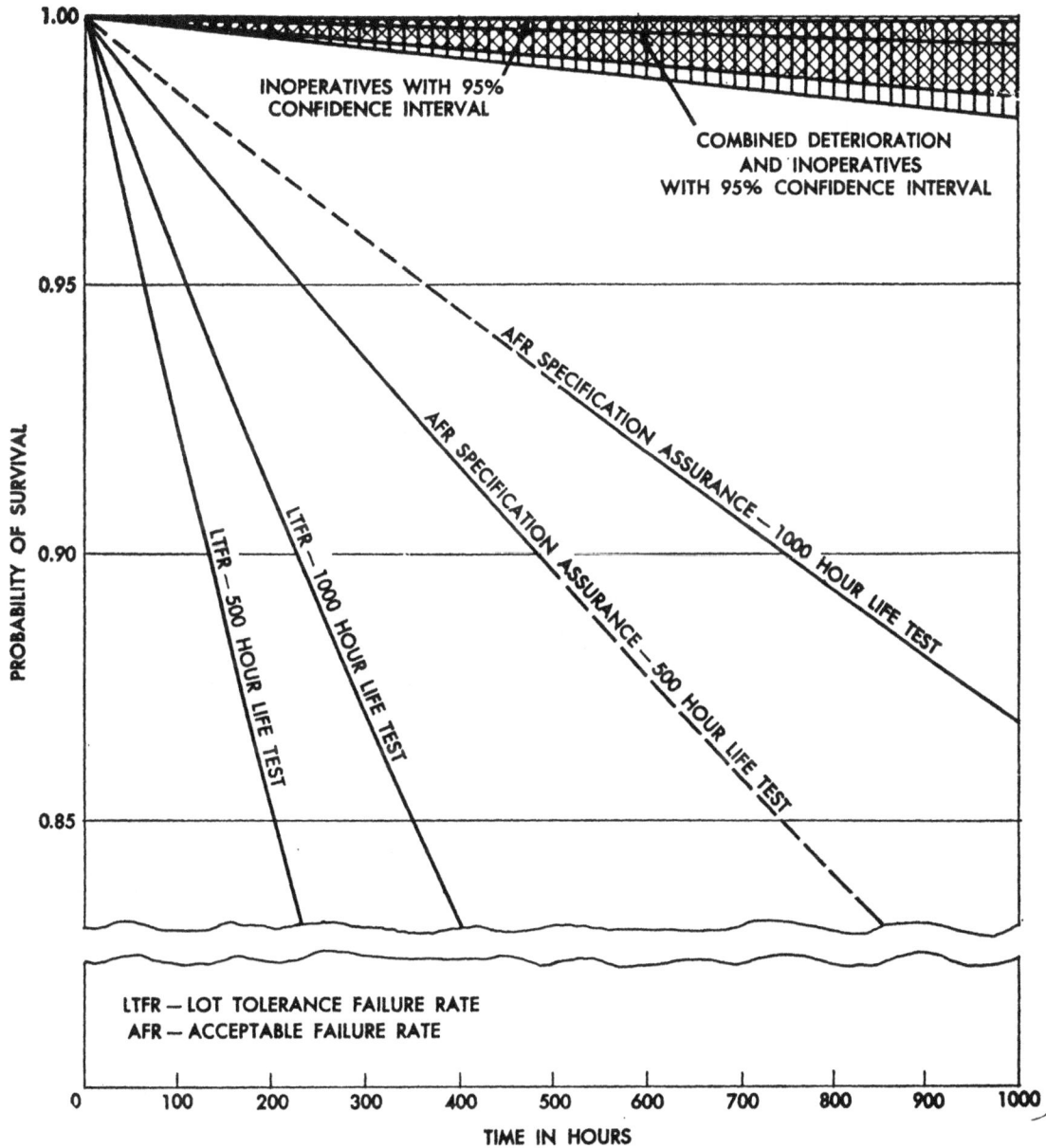

INOPERATIVES WITH 95%
CONFIDENCE INTERVAL

COMBINED DETERIORATION
AND INOPERATIVES
WITH 95% CONFIDENCE INTERVAL

AFR SPECIFICATION ASSURANCE – 1000 HOUR LIFE TEST

AFR SPECIFICATION ASSURANCE – 500 HOUR LIFE TEST

LTFR – 500 HOUR LIFE TEST

LTFR – 1000 HOUR LIFE TEST

1.00

0.95

0.90

0.85

PROBABILITY OF SURVIVAL

LTFR – LOT TOLERANCE FAILURE RATE
AFR – ACCEPTABLE FAILURE RATE

0   100   200   300   400   500   600   700   800   900   1000

TIME IN HOURS

73

TUBE TYPE JAN-6C4WA

DESCRIPTION:

The JAN-6C4WA[1] is a 7 pin miniature triode with a Mu in the range 15.5 to 18.5 with a transconductance ranging from 1750 to 2650 micromhos depending upon the choice of operating point.

ELECTRICAL: The electrical characteristics are as follows:

Heater Voltage..6.3 V
Heater Current..138-162 mA
Cathode...Coated Unipotential

MOUNTING: Any type mounting is adequate.

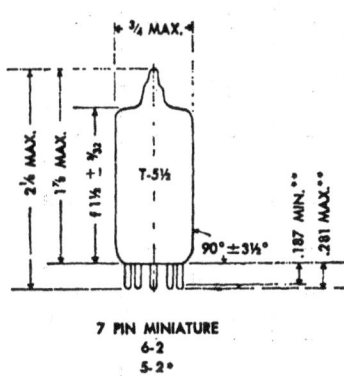

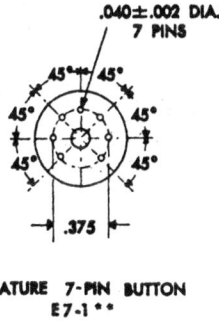

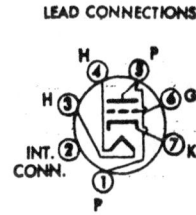

7 PIN MINIATURE
6-2
5-2*

.040±.002 DIA.
7 PINS

MINIATURE 7-PIN BUTTON
E7-1**

LEAD CONNECTIONS

*REFERS TO JETEC PUBLICATION JO-G2-2, MARCH 1955 SUPERSEDED BY JO-G2-2, MARCH 1958
**REFERS TO JETEC PUBLICATION JO-G3-1, MARCH 1955 SUPERSEDED BY JO-G2-2, MARCH 1958
f MEASURE FROM BASE SEAT TO BULB TOP-LINE AS DETERMINED BY RING GAGE OF 7/16 I.D.
ALL DIMENSIONS IN INCHES

RATINGS:	Ef	Eb	Ec	Ehk	Rg	Ik**	Ic*	Pp*	F1*	T Envelope	Alt
Absolute	V	Vdc	Vdc	v	Meg	mAdc	mAdc	W	Mc	°C	ft
Maximum	6.9	330	0	135	0.5	20	5.5	3.8	150	+165	60,000 Note
Minimum	5.7	---	-55	---	---	---	---	---	---	---	---
Test Cond:	6.3	250	-8.5	0	---	---	---	---	---	---	---

1/ The values and specification comments presented in this section are related to MIL-E-1/857A dated 5 Dec 1955.

*   No test at this rating exists in the specification.

** Difficulty may be encountered if this tube is operated for long periods of time with very small values of cathode current. No specification assurance of life exists under conditions of cathode current approaching the maximum.

Note: If altitude rating is exceeded, reduction of instantaneous voltages (Ef excluded) may be required.

ACCEPTANCE TEST LIMITS SUMMARY

PROPERTY		MEASUREMENT CONDITIONS	INITIAL		500 HR LIFE TEST		1000 HR LIFE TEST		UNITS
			MIN	MAX	MIN	MAX	MIN	MAX	
Heater Current	If		138	162	138	162	138	162	mA
Transconductance (1)	Sm		1750	2650	-	-	-	-	umhos
Change in individual	Δ Smt		-	-	-	20	-	25	%
Change in average	AvgΔ Smt		-	-	-	15	-	-	%
Transconductance change with Ef	Δ SmEf	Ef=5.7V	-	15	-	15	-	-	%
Transconductance (3)	Sm	Eb = 100 Vdc; Ec = 0 Vdc	2500	4000	-	-	-	-	umhos
Amplification Factor	Mu		15.5	18.5	-	-	-	-	-
Plate Current (1)	Ib		6.5	14.5	-	-	-	-	mAdc
Plate Current (2)	Ib	Ec=-25Vdc	-	20.	-	-	-	-	uAdc
Plate Current (3)	Ib	Ec=-18Vdc	5	-	-	-	-	-	uAdc
Power Oscillation	Po	F=150Mc;Eb=300 Vdc;Rg=8500; Note 2	1.8	-	-	-	-	-	W
Capacitance	C gp	No shield,Ef=0	1.2	2.0	-	-	-	-	uuf
	C in		1.35	2.25	-	-	-	-	uuf
	C out		0.98	1.62	-	-	-	-	uuf
Control Grid Current	Ic	Rg=0.5Meg Max.	0	-0.5	0	-0.5	0	-0.5	uAdc
Control Grid Emission	Ig	Ef=7.5V;Ec=-25 Vdc;Rg=0.5Meg; Note 1	0	-1.0	-	-	-	-	uAdc
Heater Cathode Leakage	Ihk	Ehk=+100Vdc	-	10	-	10	-	10	uAdc
	Ihk	Ehk=-100Vdc	-	10	-	10	-	10	uAdc
Insulation of Electrodes	R(g1-all)	Eg1-all=-100V	100	-	50	-	-	-	Meg
	R(p-all)	Ep-all=-300V	100	-	50	-	-	-	Meg

Measurement conditions are the same as stated under Test Conditions, unless otherwise indicated.

Note 1: The tube shall be preheated a minimum of five minutes at test conditions for this test (except Ec1=0; Rk=470 ohms) prior to this test.

Note 2: A plate current of 25 mAdc on a bogie tube shall be obtained by adjusting the coupling between the load and the tank circuit while the load is simultaneously tuned to yield maximum power output.

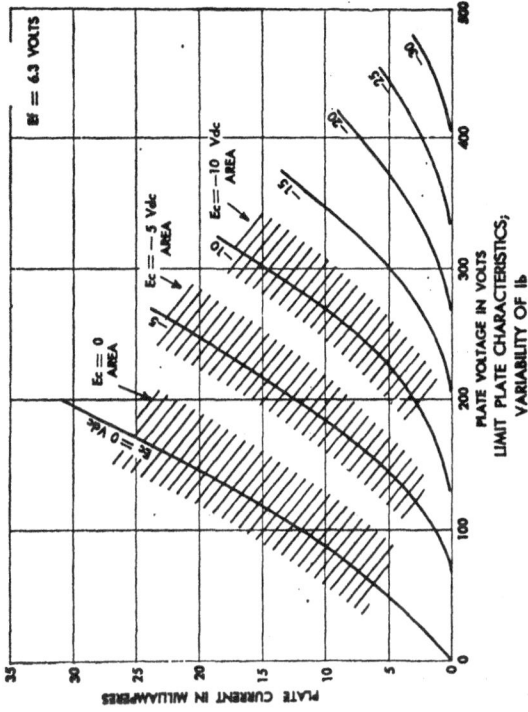

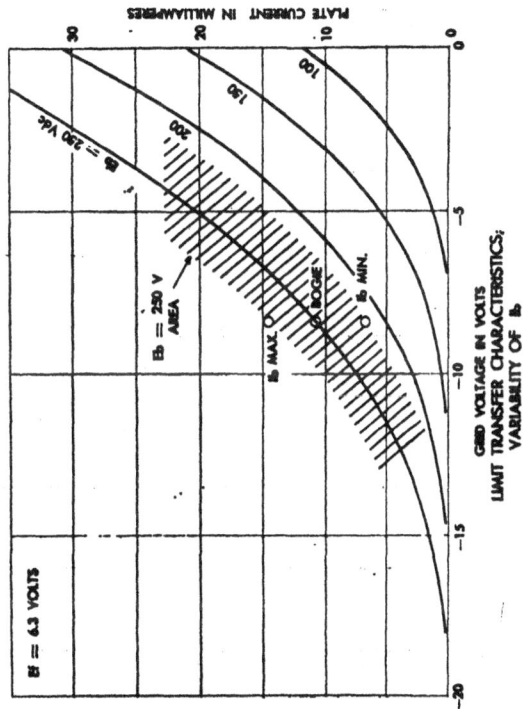

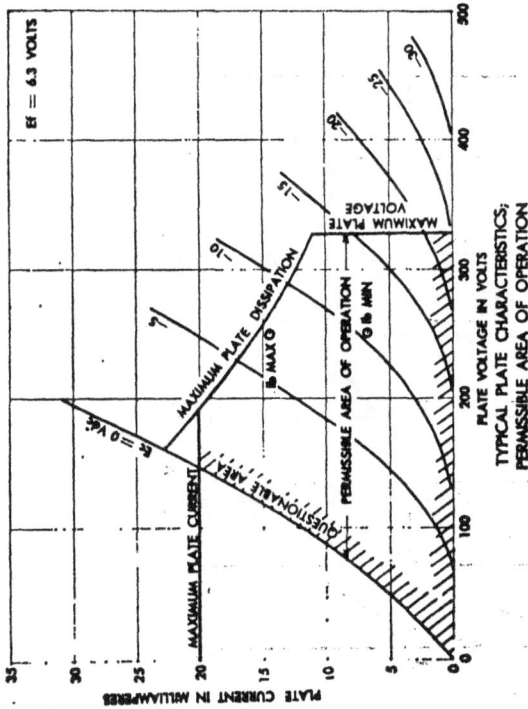

DESIGN CENTER CHARACTERISTICS
OBTAINED FROM DATA PUBLISHED BY ORIGINAL
RETMA REGISTRANT

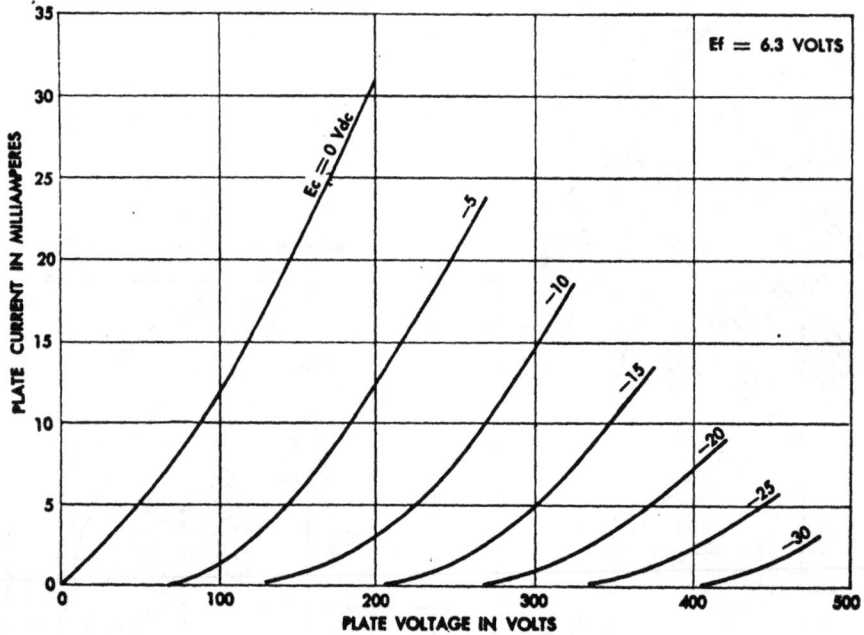

Ef = 6.3 VOLTS

PLATE CURRENT IN MILLIAMPERES

Ec = 0 Vdc

-5

-10

-15

-20

-25

-30

PLATE VOLTAGE IN VOLTS

TYPICAL PLATE CHARACTERISTICS

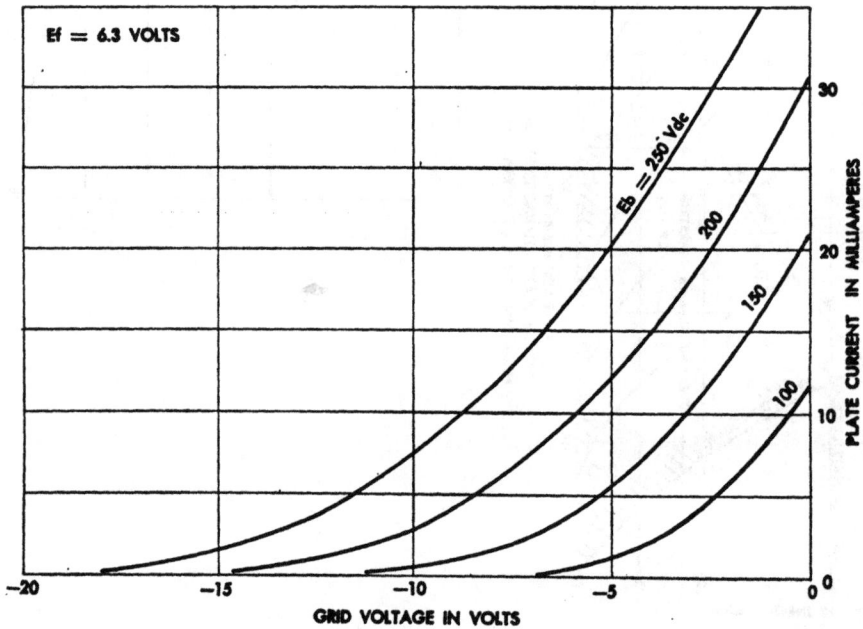

Ef = 6.3 VOLTS

Eb = 250 Vdc

200

150

100

PLATE CURRENT IN MILLIAMPERES

GRID VOLTAGE IN VOLTS

TYPICAL TRANSFER CHARACTERISTICS

78

TUBE TYPE JAN-5687WA

DESCRIPTION:

The JAN-5687WA[1/] is a 9 pin, miniature, general purpose twin triode having a Mu in the range of 16.0 to 21.0 and a transconductance in the range of 8500 to 14500 micromhos. Each triode is electrically independent, although the two heaters have a common connection.

ELECTRICAL: The electrical characteristics are as follows:

	Series	Parallel
Heater Voltage..	12.6 V	6.3 V
Heater Current......................................	.41-.47 A	.82-.94 A
Cathode...	Coated Unipotential	

MOUNTING: Any type mounting is adequate.

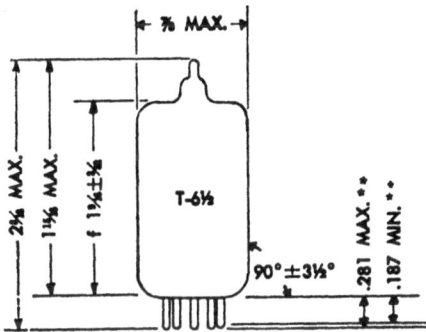

LEAD CONNECTIONS

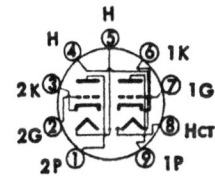

9-PIN MINIATURE
6-7
6-2*

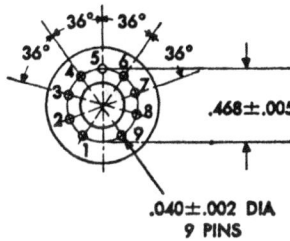

MINIATURE 9-PIN BUTTON
E9-1**

*REFERS TO JETEC PUBLICATION JO-G2-2, MARCH 1955 SUPERSEDED BY JO-G2-2, MARCH 1958
**REFERS TO JETEC PUBLICATION JO-G3-1, MARCH 1955 SUPERSEDED BY JO-G2-2, MARCH 1958
f MEASURE FROM BASE SEAT TO BULB TOP-LINE AS DETERMINED BY RING GAGE OF 7/16 I.D.
ALL DIMENSIONS IN INCHES

RATINGS:	Ef	Eb	Ec	Ehk	Rg/g	Ik/k*	Ic/g	Pp/p	T Envelope	Alt
Design	V	Vdc	Vdc	v	Meg	mAdc	mAdc	W	°C	ft
Maximum	6.6 13.2	330	0	±100	0.1	65	6	3.75 Note	+225	60,000
Minimum	6.0 12.0	---	-200	---	---	---	---	---	---	---
Test Cond:	12.6	120	-2	0	---	---	---	---	---	---

1/ The values and specification comments presented in this section are related to MIL-E-1/779B dated 13 Mar 1958.

Note: Pp/p on one section may be as great as 4.2 watts provided that maximum dissipation for both sections does not exceed 7.5 watts.

* Difficulty may be encountered if this tube is operated for long periods of time with very small values of cathode current. No specification assurance of life exists under conditions of cathode current approaching the maximum.

ACCEPTANCE TEST LIMITS SUMMARY

PROPERTY		MEASUREMENT CONDITIONS	INITIAL		500 HR LIFE TEST		1000 HR LIFE TEST		UNITS
			MIN	MAX	MIN	MAX	MIN	MAX	
Heater Current	If		0.82	0.94	0.80	0.96	0.80	0.96	A
Transconductance	Sm		8500	14500	-	-	-	-	umhos
Change in individual	Δ Smt		-	-	-	20	-	25	%
Transconductance change with Ef	Δ SmEf	Ef=11.4V	-	15	-	25	-	30	%
Amplification Factor	Mu		16	21	-	-	-	-	-
Plate Current (1)	Ib		27	45	-	-	-	-	mAdc
Plate Current (2)	Ib	Eb=300Vdc; Ec=-20Vdc;	-	6.0	-	-	-	-	mAdc
Plate Current (3)	Ib	Eb=300Vdc; Ec=-25Vdc;	-	1.0	-	-	-	-	mAdc
Plate Emission	Ib	Eb=195Vac;Rk/ Ib=10.5mAdc; Ec=0	-	25	-	-	-	-	uAdc
Emission	Is	Eb=Ec=15Vdc	125	-	-	-	-	-	mAdc
Capacitance		No shield, Ef=0							
	C	gp	2.8	5.2	-	-	-	-	uuf
	C	in	2.8	5.2	-	-	-	-	uuf
	C	out (1)	0.42	0.78	-	-	-	-	uuf
	C	out (2)	0.34	0.66	-	-	-	-	uuf
	C	hk	-	9.7	-	-	-	-	uuf
Control Grid Current	Ic		0	-1.5	0	-2.0	0	-2.5	uAdc
Control Grid Emission	Ic	Ef=14.0V;Rg/g =1.0Meg; NOTE	0	-5.0	-	-	-	-	uAdc
Heater Cathode Leakage	Ihk	Ehk=+100Vdc	-	30	-	50	-	50	uAdc
	Ihk	Ehk=-100Vdc	-	30	-	50	-	50	uAdc
Insulation of Electrodes	R(g1-all)	Eg1-all=-300V	100	-	50	-	25	-	Meg
	R(p-all)	Ep-all=-500V	100	-	50	-	25	-	Meg

Measurement conditions are the same as stated under Test Conditions, unless otherwise indicated.

NOTE: The tube shall be preheated a minimum of five minutes at test conditions for this test, prior to this test.

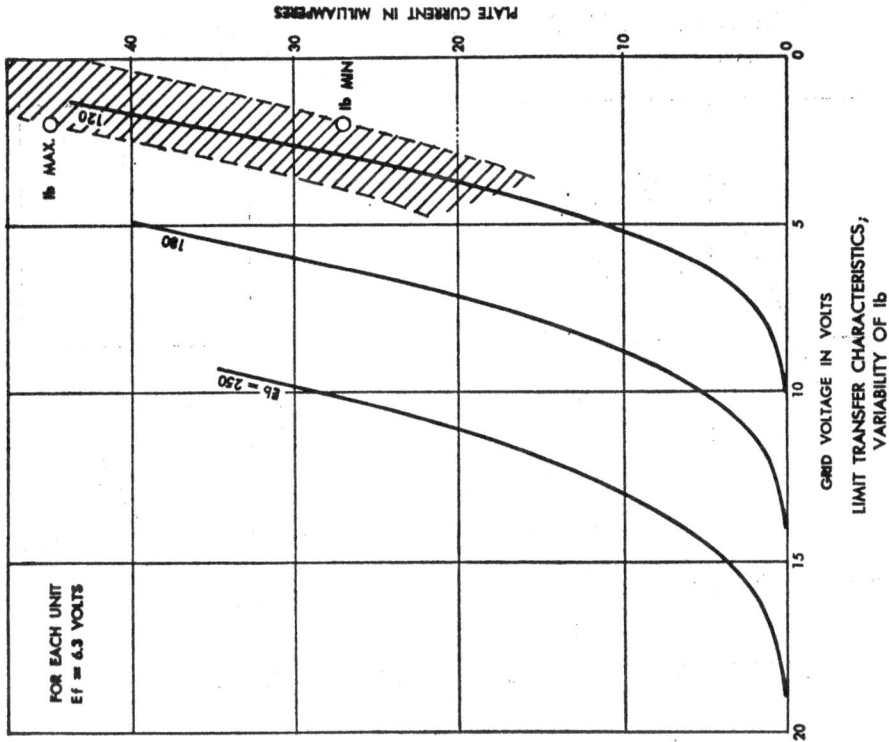

LIMIT TRANSFER CHARACTERISTICS;
VARIABILITY OF Ib

FOR EACH UNIT
Ef = 6.3 VOLTS

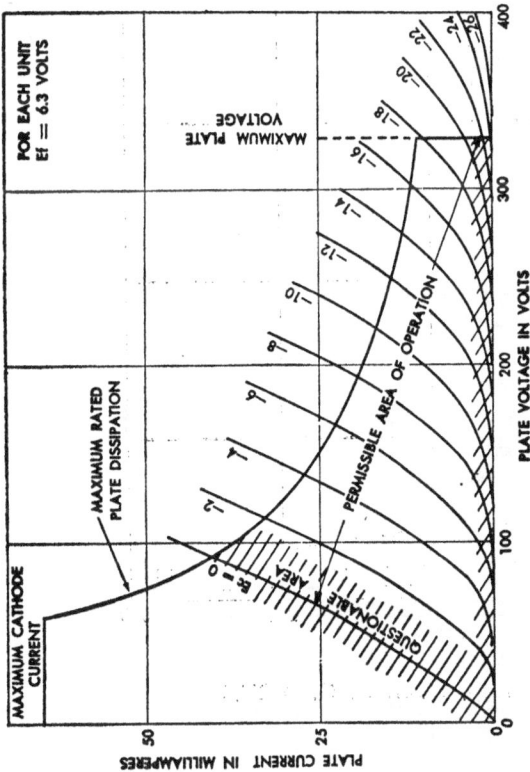

TYPICAL PLATE CHARACTERISTICS;
PERMISSIBLE AREA OF OPERATION

FOR EACH UNIT
Ef = 6.3 VOLTS

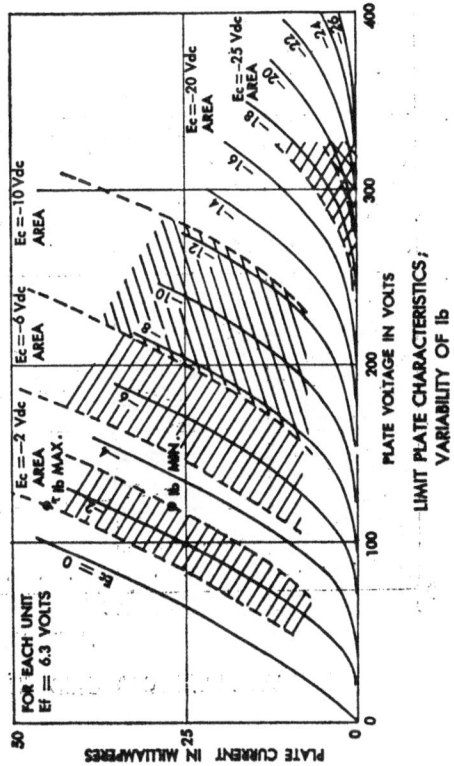

LIMIT PLATE CHARACTERISTICS;
VARIABILITY OF Ib

FOR EACH UNIT
Ef = 6.3 VOLTS

DESIGN CENTER CHARACTERISTICS OBTAINED FROM
DATA PUBLISHED BY ORIGINAL RETMA REGISTRANT

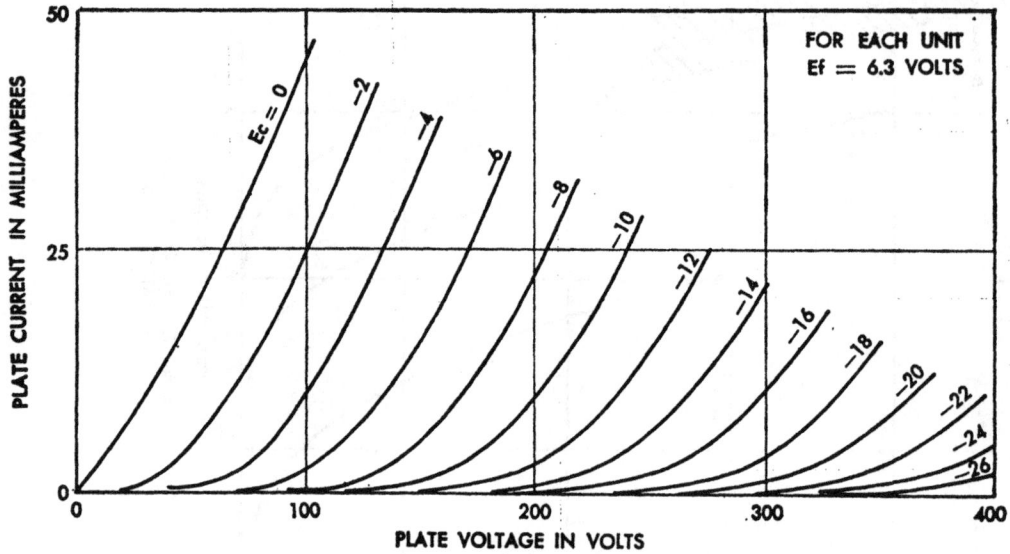

TYPICAL PLATE CHARACTERISTICS; NEGATIVE GRID BIAS

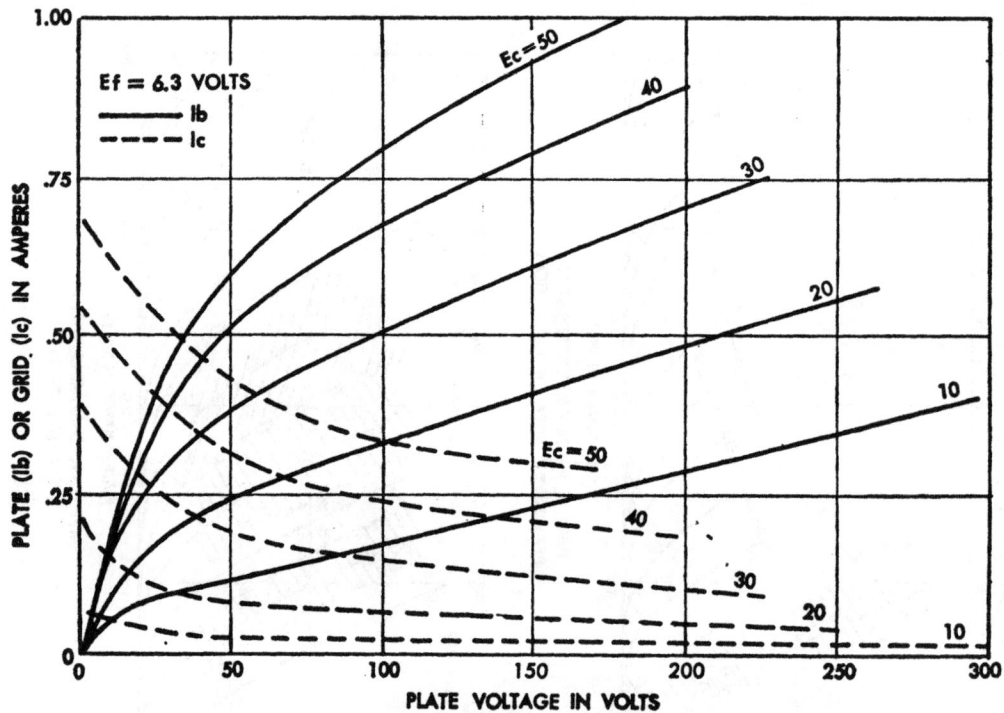

TYPICAL PLATE AND GRID CHARACTERISTICS; POSITIVE GRID BIAS

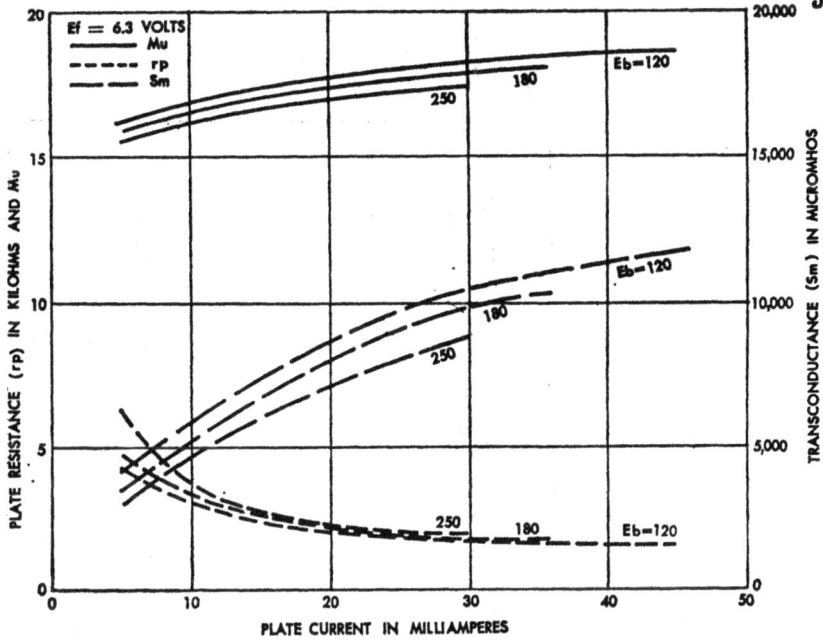

TYPICAL Sm, Mu, AND rp CHARACTERISTICS

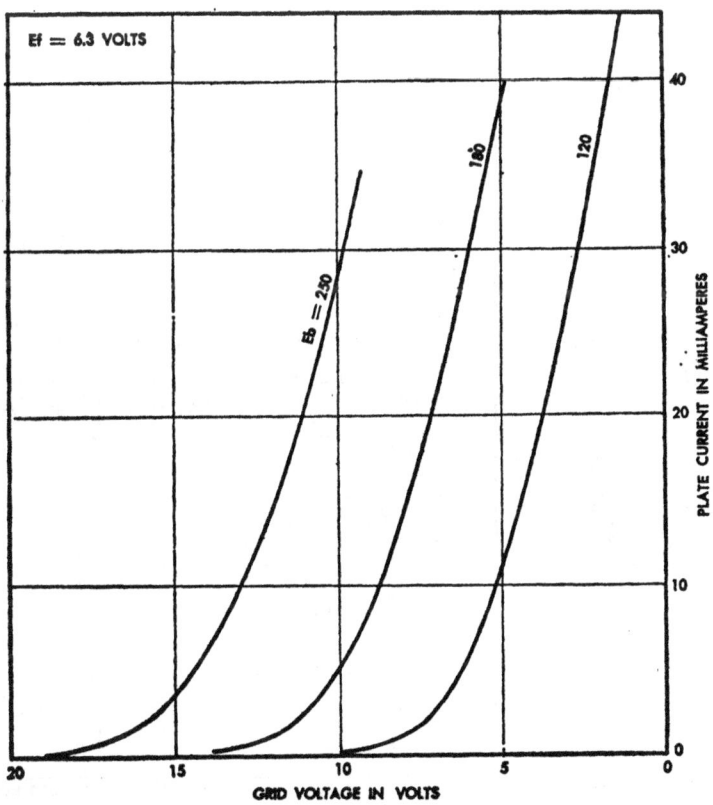

TYPICAL TRANSFER CHARACTERISTICS

83

LIFE TEST PROPERTY BEHAVIOR
MIL-E-1/779B   13 MAR. '58
PRODUCED IN 1957-'58 BY ONE MANUFACTURER

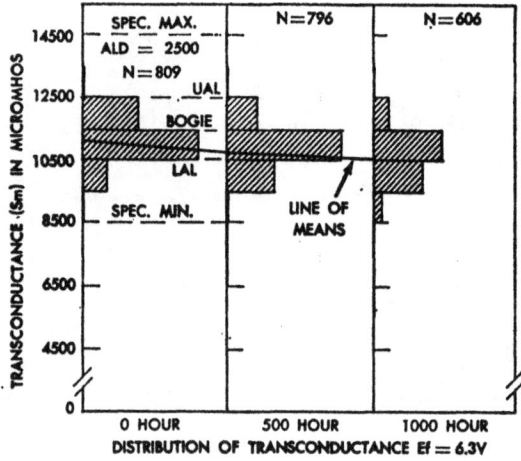

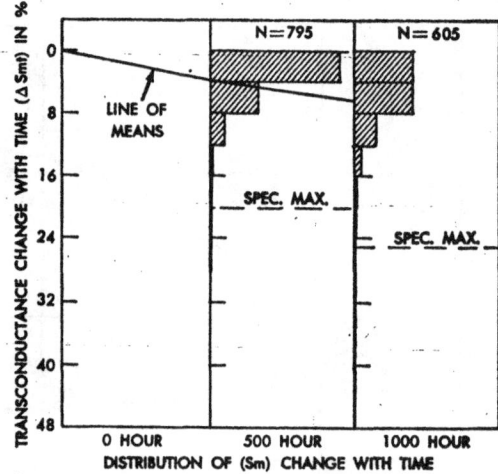

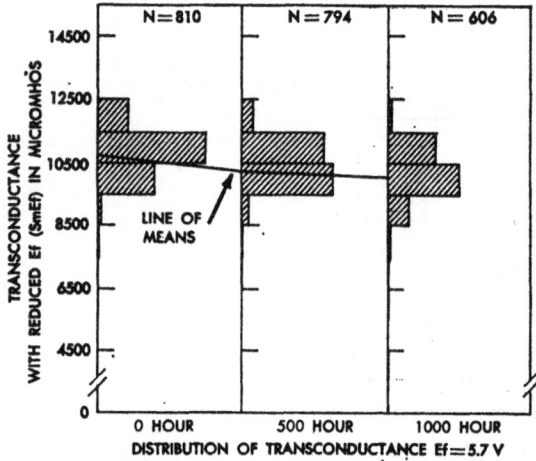

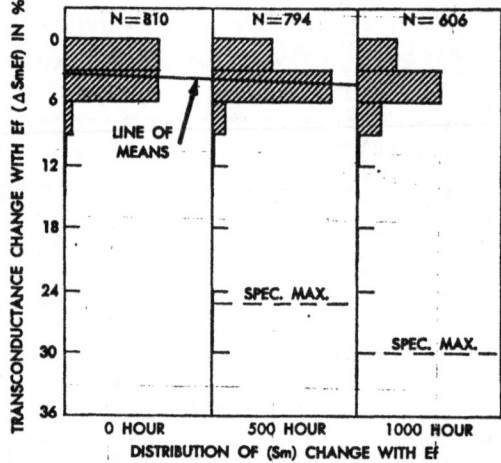

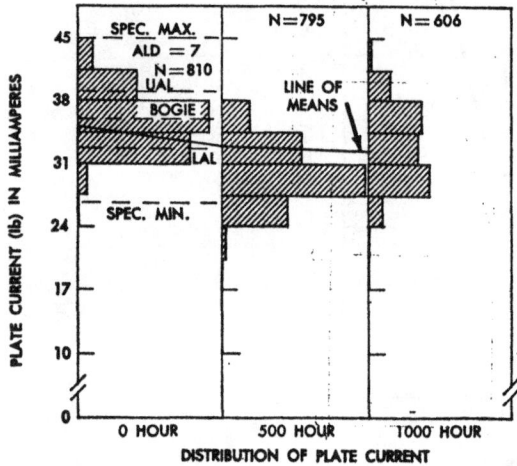

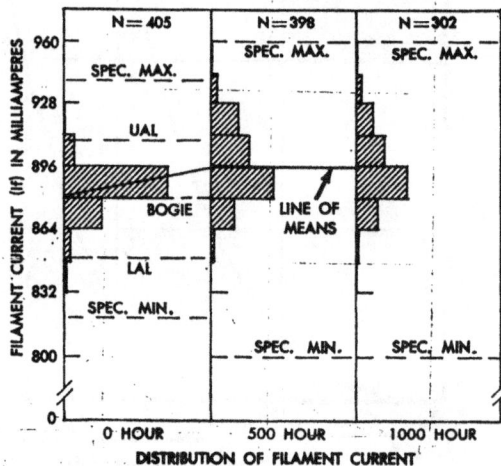

DISTRIBUTION OF TRANSCONDUCTANCE Ef = 6.3V

DISTRIBUTION OF (Sm) CHANGE WITH TIME

DISTRIBUTION OF TRANSCONDUCTANCE Ef = 5.7 V

DISTRIBUTION OF (Sm) CHANGE WITH Ef

DISTRIBUTION OF PLATE CURRENT

DISTRIBUTION OF FILAMENT CURRENT

LIFE TEST PROPERTY BEHAVIOR
MIL-E-1/779B   13 MAR. '58
PRODUCED IN 1957-'58 BY ONE MANUFACTURER

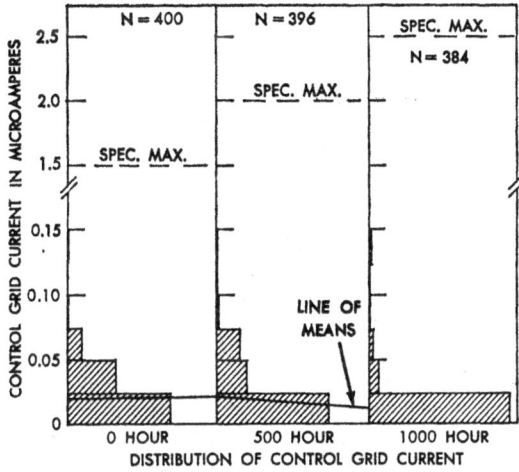

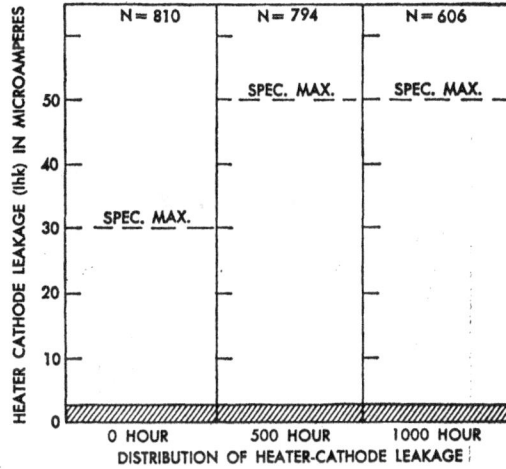

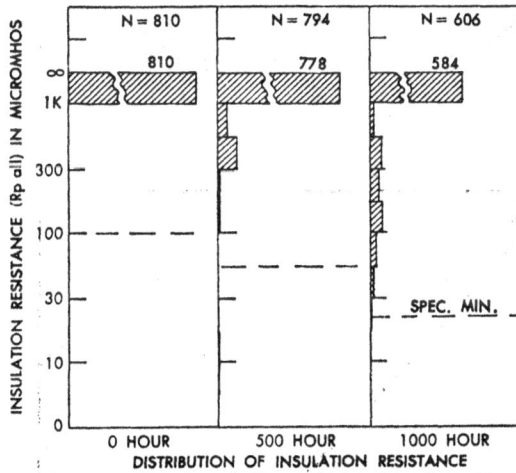

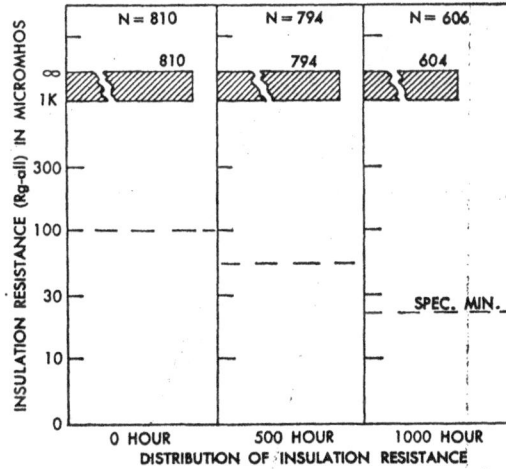

MIL-HDBK-211
APPENDIX-A
12 January 1960
JAN-5687WA

LIFE TEST PROPERTY BEHAVIOR
PROBABILITY OF SURVIVAL
MIL-E-1/779B   13 MAR. '58
PRODUCED IN 1957-'58 BY TWO MANUFACTURERS

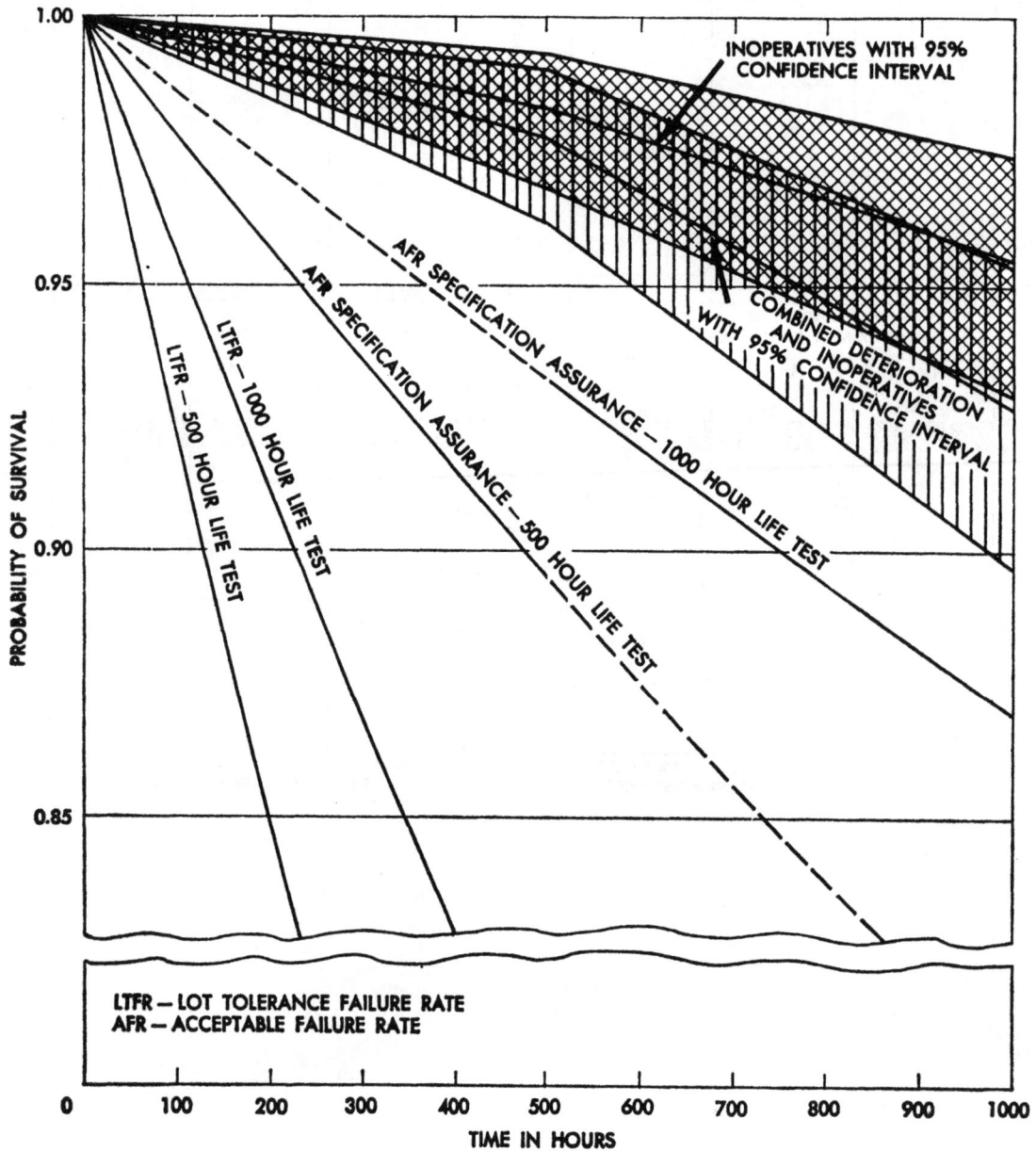

86

TUBE TYPE JAN-5702WA

DESCRIPTION:

The JAN-5702WA[1] is a 7 lead, pinch press, subminiature, sharp cutoff pentode having a design center transconductance of 5000 micromhos. The JAN-5702WA is similar in plate characteristics to JAN-5840 and the miniature type JAN-5654/6AK5W.

ELECTRICAL: The electrical characteristics are as follows:
    Heater Voltage .... 6.3 V
    Heater Current ....183-217 mA
    Cathode...Coated Unipotential

MOUNTING: Any type of mounting is adequate.

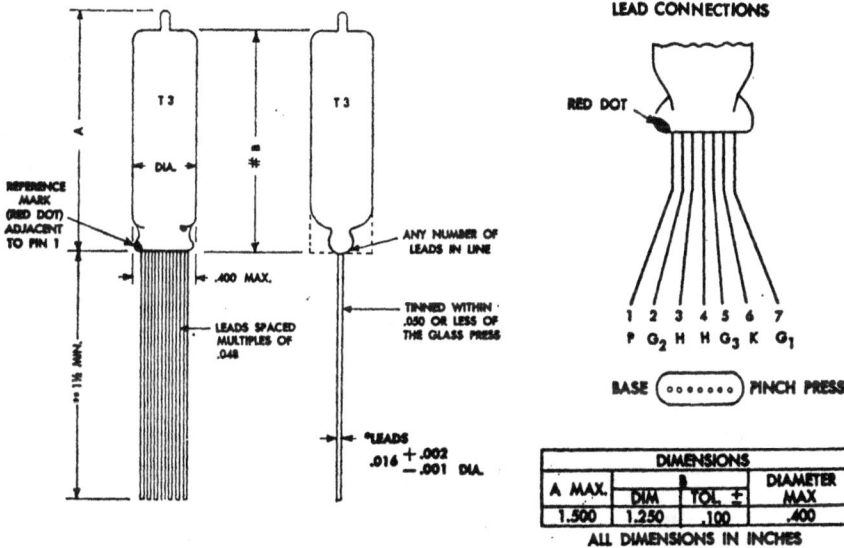

LEAD CONNECTIONS

RED DOT

```
1  2  3  4  5  6  7
P  G₂ H  H  G₃ K  G₁
```

BASE (∘∘∘∘∘∘) PINCH PRESS

DIMENSIONS			
A MAX.	DIM	TOL ±	DIAMETER MAX
1.500	1.250	.100	.400

ALL DIMENSIONS IN INCHES

\# MEASURE FROM BASE SEAT TO BULB TOP-LINE AS DETERMINED BY RING GAGE OF .210 ± .001.

* LEAD DIAMETER TOLERANCE SHALL GOVERN BETWEEN .050 FROM THE GLASS TO .250 FROM THE GLASS.

** ALTERNATIVE LEAD LENGTH SHALL BE .200 ± .015 WHEN CUT LEADS ARE REQUIRED BY PROCUREMENT CONTRACT OR TSS. CUT LEADS SHALL BE ESSENTIALLY SQUARE CUT AND THE MAXIMUM BURR SHALL BE .003 INCREASE OVER THE ACTUAL LEAD DIAMETER.

RATINGS:	Ef	Eb	Ecl*	Ec2	Ec3	Ehk	Rk	Rgl	Ik**	Pp*	Pg2*	T Envelope	Alt
Absolute	V	Vdc	Vdc	Vdc	Vdc	v	ohms	Meg	mAdc	W	W	°C	ft
Maximum	6.9	165	---	155	0	200	---	1.2	16.5	---	---	265	60,000
Design Maximum	---	---	---	---	---	---	---	---	---	1.10	0.40	---	---
Minimum	5.7	---	-55	---	---	---	---	---	---	---	---	---	---
Test Cond:	6.3	120	0	120	0	0	200	---	---	---	---	---	---

1/ The values and specification comments presented in this section are related to MIL-E-1/82C dated 4 Dec 1957.

* No test at this rating exists in the specification.

** Difficulty may be encountered if this tube is operated for long periods of time with very small values of cathode current. No specification assurance of life exists under conditions of cathode current approaching the maximum.

ACCEPTANCE TEST LIMITS SUMMARY

PROPERTY		MEASUREMENT CONDITIONS	INITIAL		500 HR LIFE TEST		1000 HR LIFE TEST		UNITS
			MIN	MAX	MIN	MAX	MIN	MAX	
Heater Current	If		183	217	180	220	177	223	mA
Transconductance	Sm		4200	5800	–	–	–	–	umhos
Change in individual	ΔSmt		–	–	–	20	–	30	%
average	Avg ΔSmt		–	–	–	15	–	–	%
Transconductance Change with Ef	ΔSmEf	Ef=5.5V	–	10	–	15	–	–	%
Plate Resistance	rp		0.15	–	–	–	–	–	Meg
Plate Current(1)	Ib		5.5	9.5	–	–	–	–	mAdc
Plate Current(2)	Ib	Ecl=-9.0Vdc; RK=0	–	50	–	–	–	–	uAdc
Screen Grid Current	Ic2		1.7	3.5	–	–	–	–	mAdc
Capacitance		0.405 in. dia. shield, Ef=0							
	C glp		–	0.03	–	–	–	–	uuf
	C in		4.1	5.5	–	–	–	–	uuf
	C out		2.9	4.1	–	–	–	–	uuf
Control Grid Current	Icl	Rgl=1.0 Meg	0	-0.1	0	-0.5	0	-1.0	uAdc
Control Grid Emission	Icl	Ef=7.5;Rgl=1.0 Meg;Ecl=-10Vdc NOTE	0	-0.5	–	–	–	–	uAdc
Heater-Cathode Leakage	Ihk	Ehk=+100Vdc	–	5	–	10	–	15	uAdc
	Ihk	Ehk=-100Vdc	–	5	–	10	–	15	uAdc
Insulation of Electrodes R(gl-all)		Egl-all=-100V	100	–	50	–	–	–	Meg
R(p-all)		Ep-all=-300Vdc	100	–	50	–	–	–	Meg

Measurement conditions are the same as stated under Test Conditions, unless otherwise indicated.

NOTE: The tube shall be preheated a minimum of five minutes at test conditions for this test (except Ecl=0) prior to this test.

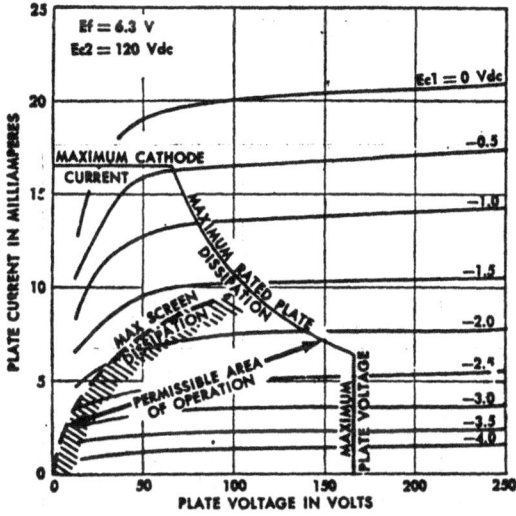

TYPICAL STATIC-PLATE CHARACTERISTICS;
PERMISSIBLE AREA OF OPERATION

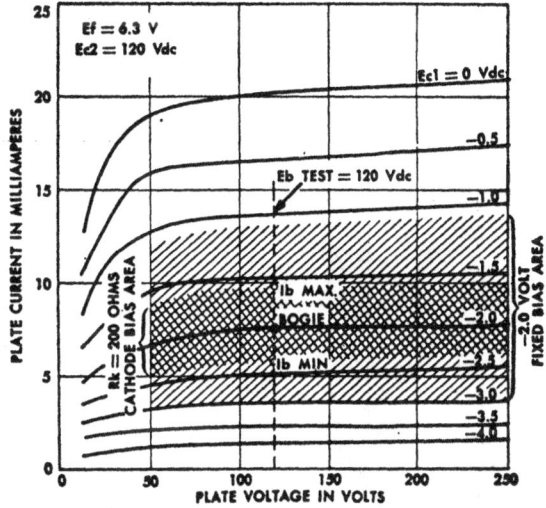

LIMIT BEHAVIOR STATIC—PLATE DATA;
VARIABILITY OF Ib

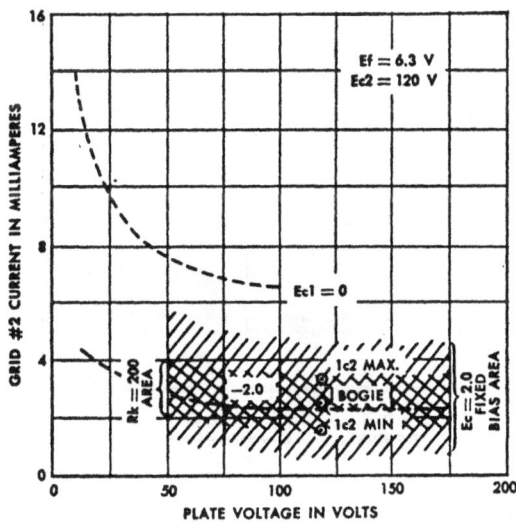

LIMIT BEHAVIOR STATIC—PLATE DATA;
VARIABILITY OF Ic2

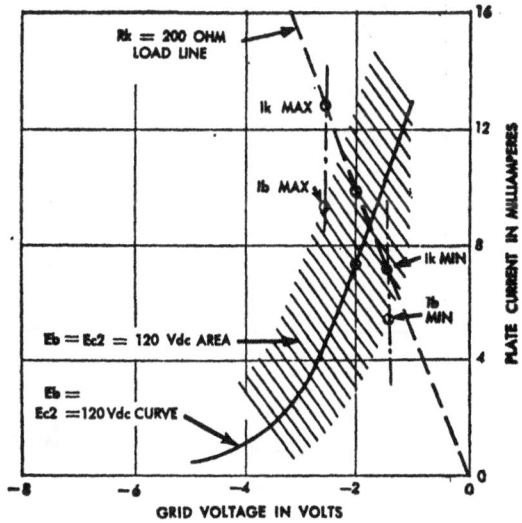

LIMIT BEHAVIOR TRANSFER DATA;
VARIABILITY OF Ib

DESIGN CENTER CHARACTERISTICS
OBTAINED FROM DATA PUBLISHED BY ORIGINAL
RETMA REGISTRANT

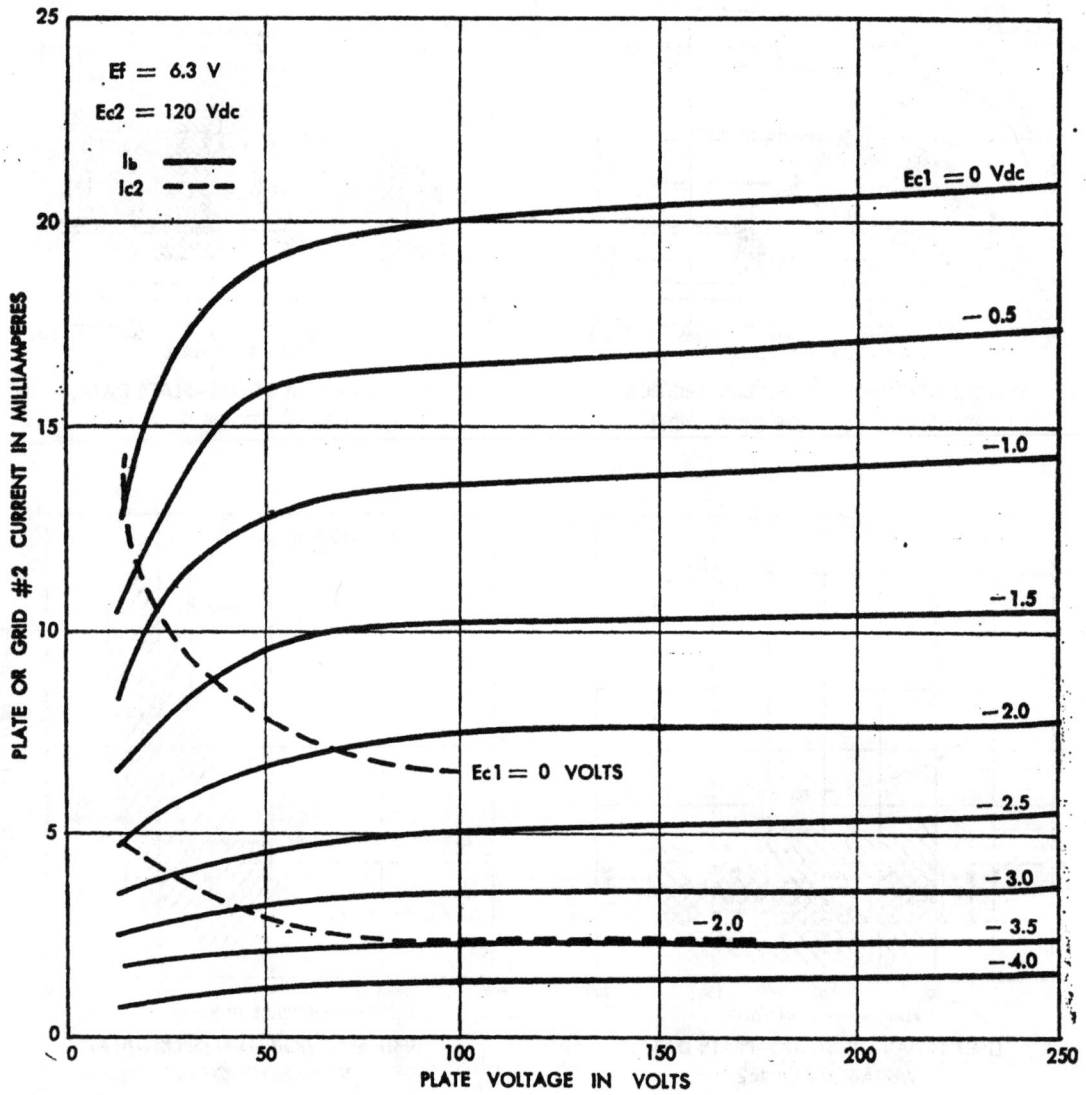

TYPICAL STATIC-PLATE CHARACTERISTICS

90

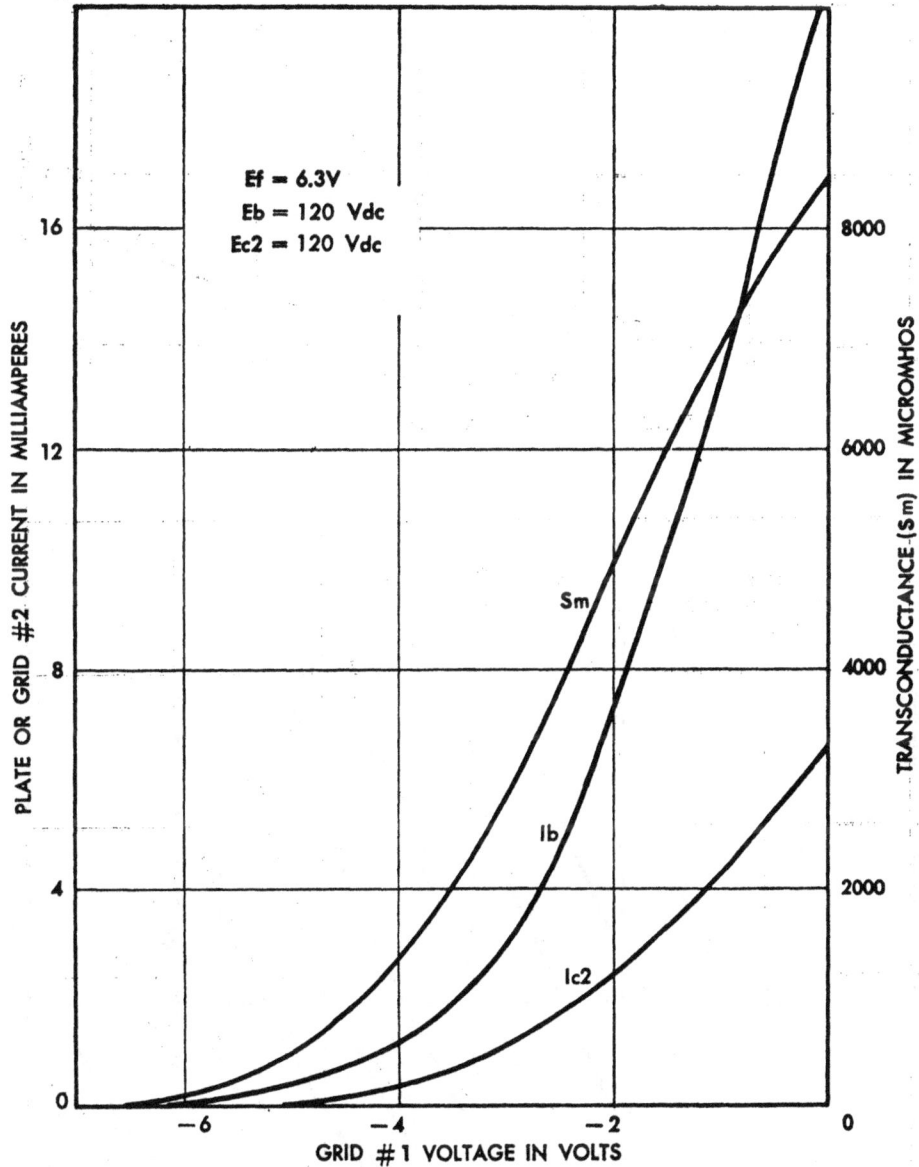

Ef = 6.3V
Eb = 120 Vdc
Ec2 = 120 Vdc

TYPICAL TRANSFER DATA Ec2 = 120 Vdc

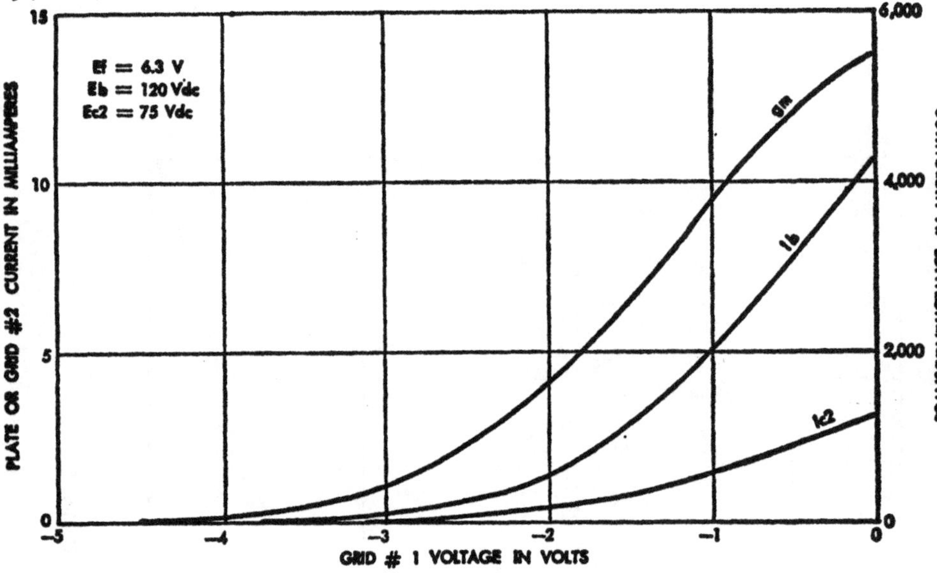
Ef = 6.3 V
Eb = 120 Vdc
Ec2 = 75 Vdc
gm
Ib
Ic2
PLATE OR GRID #2 CURRENT IN MILLIAMPERES
TRANSCONDUCTANCE IN MICROMHOS
GRID # 1 VOLTAGE IN VOLTS
15
10
5
0
6,000
4,000
2,000
0
−5
−4
−3
−2
−1
0

TYPICAL TRANSFER DATA, Ec2=75 Vdc

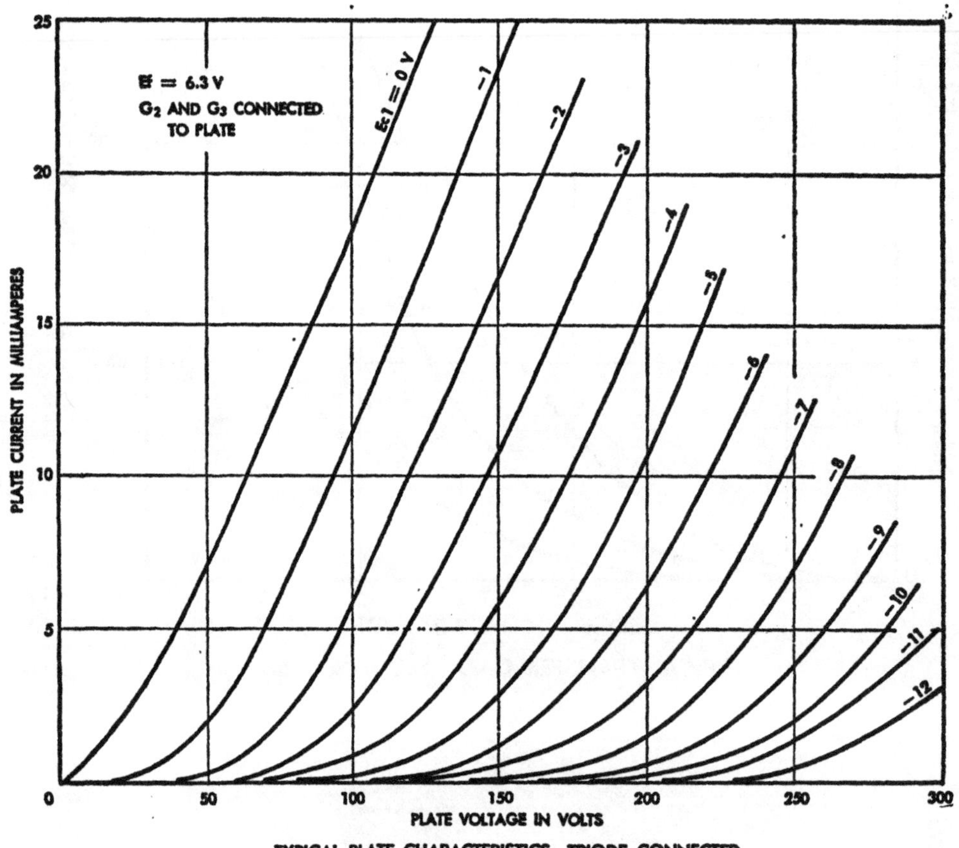
Ef = 6.3 V
G2 AND G3 CONNECTED TO PLATE
Ec1 = 0 V
−1
−2
−3
−4
−5
−6
−7
−8
−9
−10
−11
−12
PLATE CURRENT IN MILLIAMPERES
PLATE VOLTAGE IN VOLTS
25
20
15
10
5
0
50
100
150
200
250
300

TYPICAL PLATE CHARACTERISTICS, TRIODE CONNECTED

LIFE-TEST PROPERTY BEHAVIOR
MIL-E-1/82 C   4 DEC 57
PRODUCED IN 1958 BY ONE MANUFACTURER

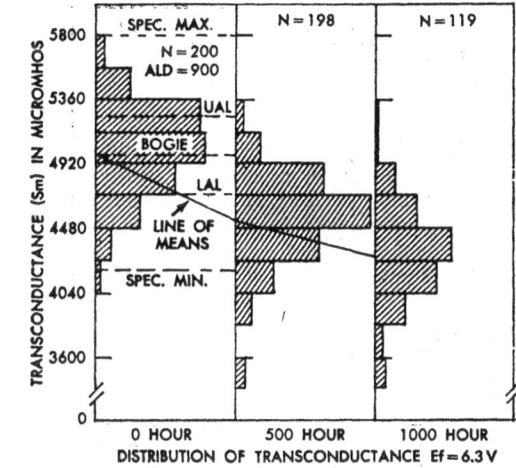

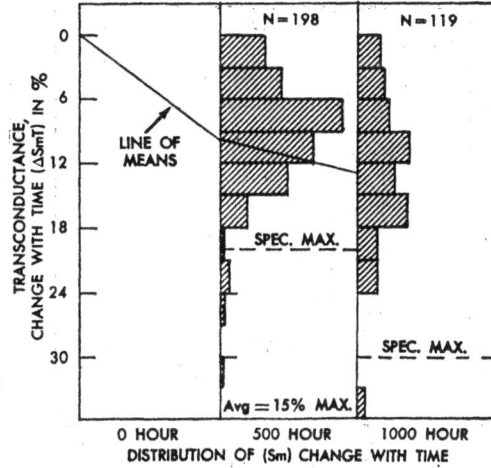

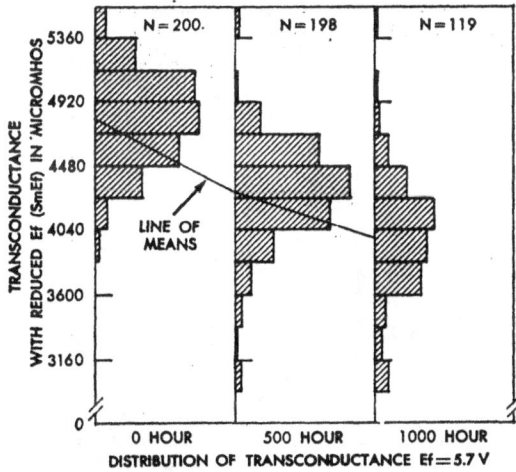

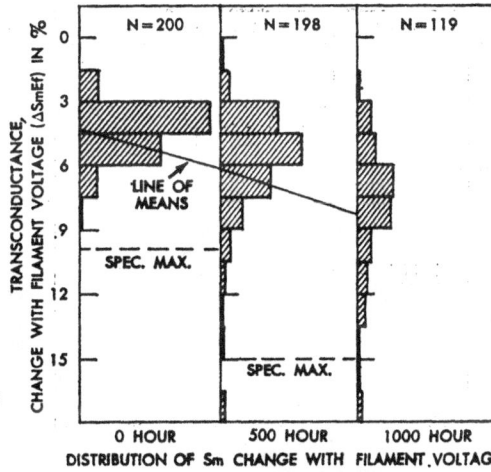

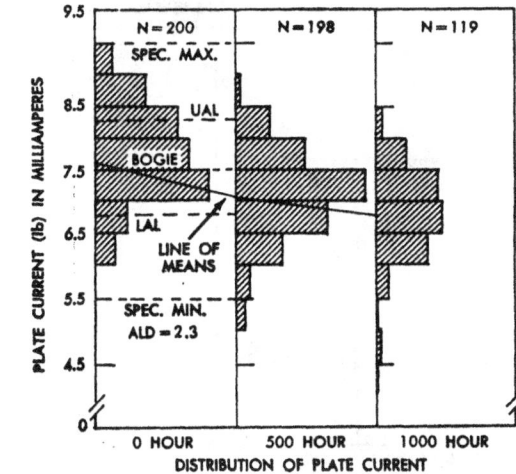

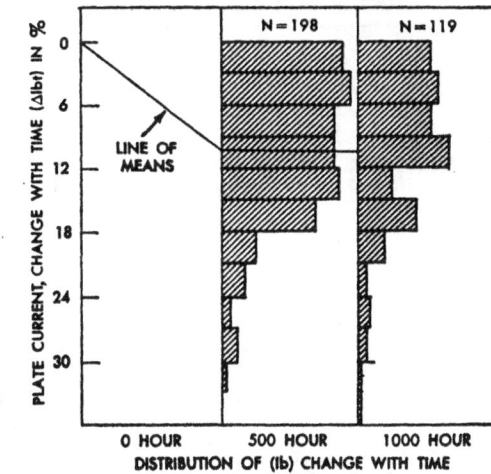

DISTRIBUTION OF TRANSCONDUCTANCE Ef = 6.3 V

DISTRIBUTION OF (Sm) CHANGE WITH TIME

DISTRIBUTION OF TRANSCONDUCTANCE Ef = 5.7 V

DISTRIBUTION OF Sm CHANGE WITH FILAMENT VOLTAGE

DISTRIBUTION OF PLATE CURRENT

DISTRIBUTION OF (Ib) CHANGE WITH TIME

93

LIFE TEST PROPERTY BEHAVIOR
MIL-E-1/82 C   4 DEC 57
PRODUCED IN 1958 BY ONE MANUFACTURER

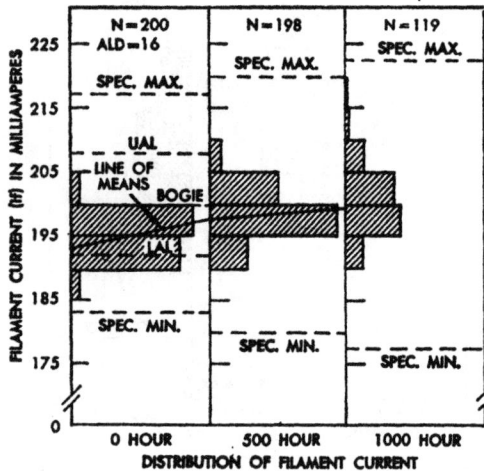

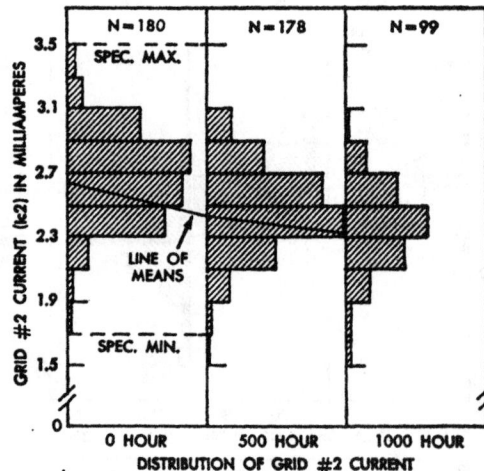

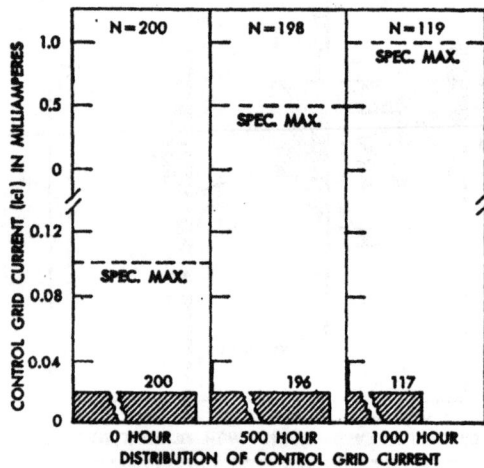

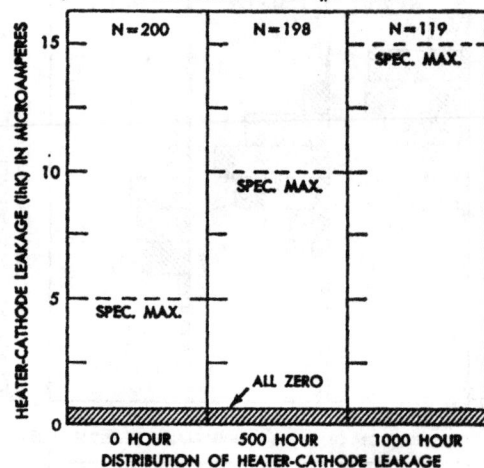

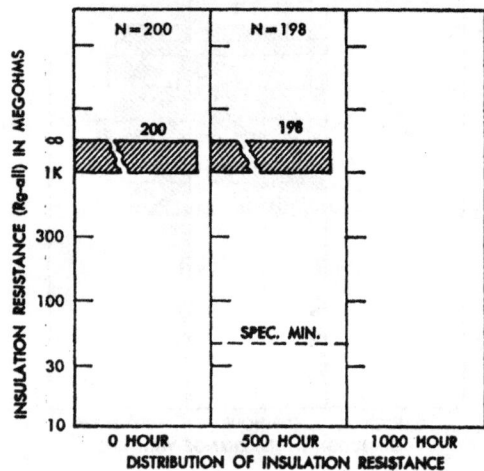

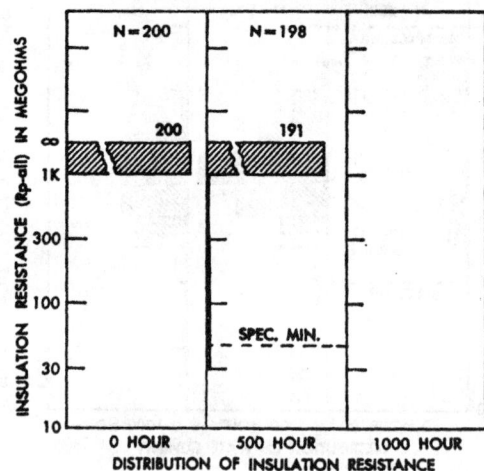

94

LIFE TEST PROPERTY BEHAVIOR
PROBABILITY OF SURVIVAL
MIL-E-1/82C   4 DEC '57
PRODUCED IN 1958 BY ONE MANUFACTURER

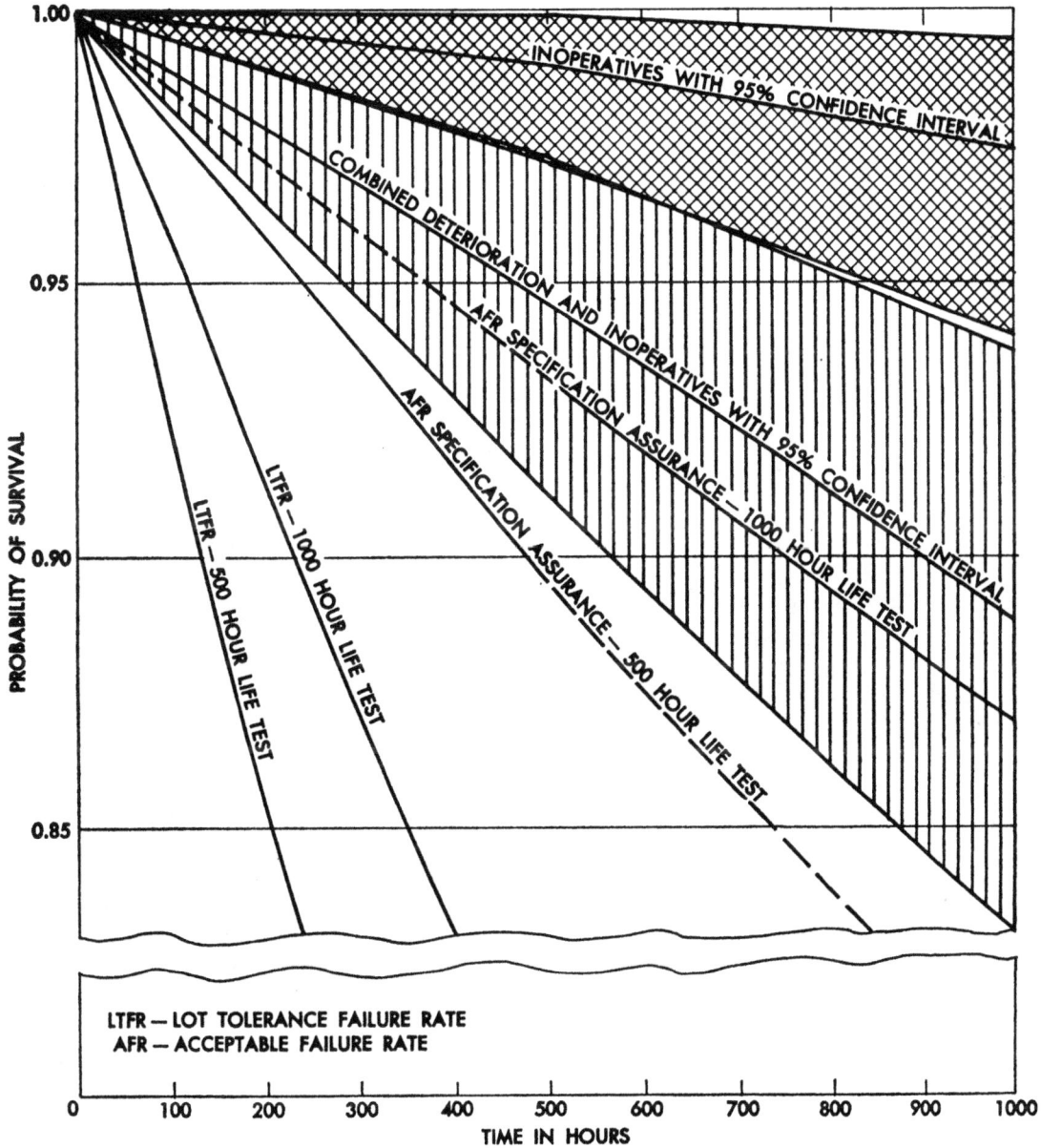

95

TUBE TYPE JAN-5702WB

DESCRIPTION:

The JAN-5702WB[1] is a 7 lead, pinch press, subminiature, sharp cutoff pentode having a design center transconductance of 5000 micromhos. The JAN-5702WB is similar in plate characteristics to JAN-5840 and the miniature type JAN-5654/6AK5W.

ELECTRICAL: The electrical characteristics are as follows:
    Heater Voltage .... 6.3 V
    Heater Current ....190-210 mA
    Cathode ....Coated Unipotential

MOUNTING: Any type of mounting is adequate.

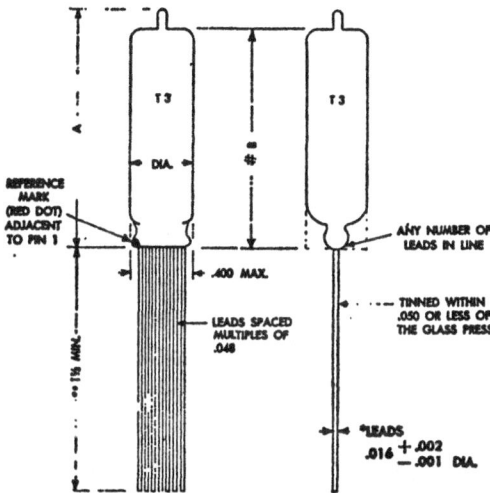

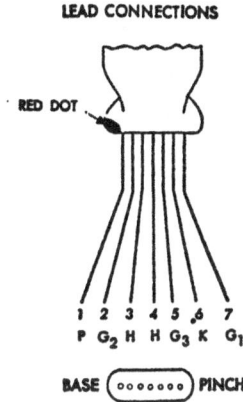

	DIMENSIONS			
A MAX.	DIM	TOL. ±	DIAMETER MAX	
1.500	1.250	.100	.400	

ALL DIMENSIONS IN INCHES

\# MEASURE FROM BASE SEAT TO BULB TOP-LINE AS DETERMINED BY RING GAGE OF .210 ± .001.

* LEAD DIAMETER TOLERANCE SHALL GOVERN BETWEEN .050 FROM THE GLASS TO .250 FROM THE GLASS.

** ALTERNATIVE LEAD LENGTH SHALL BE .200 ± .015 WHEN CUT LEADS ARE REQUIRED BY PROCUREMENT CONTRACT OR TSS. CUT LEADS SHALL BE ESSENTIALLY SQUARE CUT AND THE MAXIMUM BURR SHALL BE .003 INCREASE OVER THE ACTUAL LEAD DIAMETER.

RATINGS:	Ef	Eb	Ec1*	Ec2	Ec3	Ehk	Rk	Rg1	Ik**	Pp*	Pg2*	T Envelope*	Alt
Absolute	V	Vdc	Vdc	Vdc	Vdc	v	ohms	Meg	mAdc	W	W.	°C	ft
Maximum Design	6.9	1.5	---	155	0	200	---	1.2	16.5	---	---	220	60,000
Maximum	---	---	---	---	---	---	---	---	---	1.10	0.40	---	---
Minimum	5.7	-55	---	---	---	---	---	---	---	---	---	---	---
Test Cond:	6.3	120	0	120	0	0	200	---	---	---	---	---	---

1/ The values and specification comments presented in this section are related to MIL-E-1/1C99A dated 4 Dec 1957.

* No test at this rating exists in the specification.

** Difficulty may be encountered if this tube is operated for long periods of time with very small values of cathode current. No specification assurance of life exists under conditions of cathode current approaching the maximum.

ACCEPTANCE TEST LIMITS SUMMARY

PROPERTY		MEASUREMENT CONDITIONS	INITIAL		500 HR LIFE TEST		1000 HR LIFE TEST		UNITS
			MIN	MAX	MIN	MAX	MIN	MAX	
Heater Current	If		190	210	187	217	177	223	mA
Transconductance	Sm		4200	5800	-	-	-	-	umhos
Change in individual	Δ Smt		-	-	-	20	-	30	%
Change in average	AvgΔ Smt		-	-	-	15	-	-	%
Transconductance Change with Ef	Δ SmEf	Ef=5.7V	-	5	-	15	-	-	%
Plate Resistance	rp		0.15	-	-	-	-	-	Meg
Plate Current (1)	Ib		5.5	9.5	-	-	-	-	mAdc
Plate Current (2)	Ib	Ecl=-9.0Vdc; RK=0	-	50	-	-	-	-	uAdc
Screen Grid Current	Ic2		1.7	3.5	-	-	-	-	mAdc
Capacitance		0.405 in. dia. shield, Ef=0							
	C glp		-	0.03	-	-	-	-	uuf
	C in		4.1	5.5	-	-	-	-	uuf
	C out		2.9	4.1	-	-	-	-	uuf
Control Grid Current	Icl	Rgl= 1 Meg	0	-0.1	0	-0.5	0	-1.0	uAdc
Control Grid Emission	Icl	Ef=7.5V;Rgl= 1.0 Meg Ecl= -10V;NOTE	0	-0.5	-	-	-	-	uAdc
Heater-Cathode Leakage	Ihk	Ehk=+100Vdc	-	5	-	10	-	15	uAdc
	Ihk	Ehk=-100Vdc	-	5	-	10	-	15	uAdc
Insulation of Electrodes	R(gl-all)	Egl-all=-100V	250	-	50	-	-	-	Meg
	R(p-all)	Ep-all=-300Vdc	250	-	50	-	-	-	Meg

Measurement conditions are the same as stated under Test Conditions, unless otherwise indicated.

NOTE: The tube shall be preheated a minimum of five minutes at test conditions for this test (except Ecl=0) prior to this test.

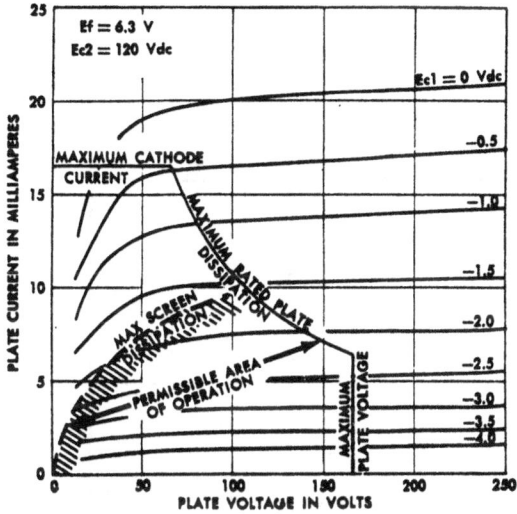

TYPICAL STATIC-PLATE CHARACTERISTICS;
PERMISSIBLE AREA OF OPERATION

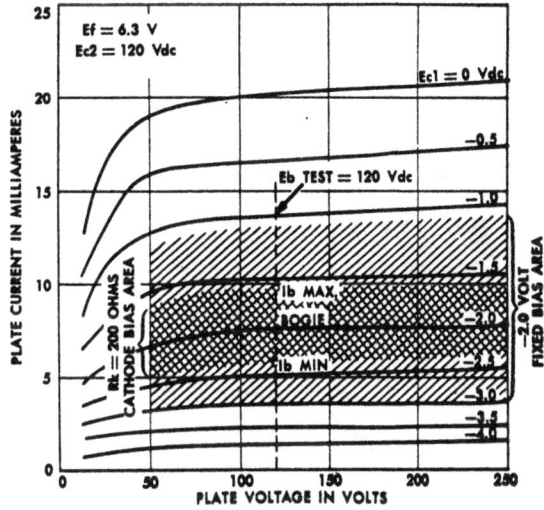

LIMIT BEHAVIOR STATIC-PLATE DATA;
VARIABILITY OF Ib

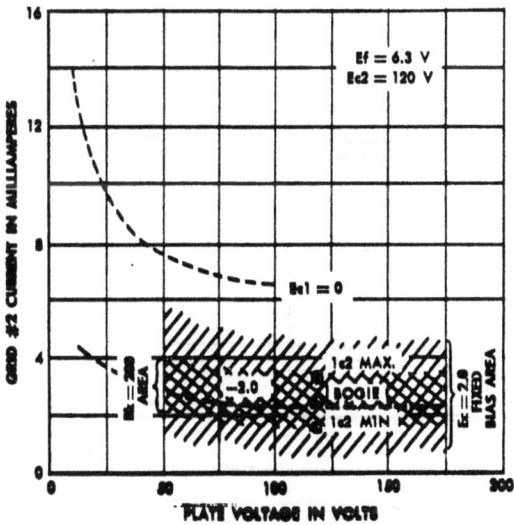

LIMIT BEHAVIOR STATIC-PLATE DATA;
VARIABILITY OF Ic2

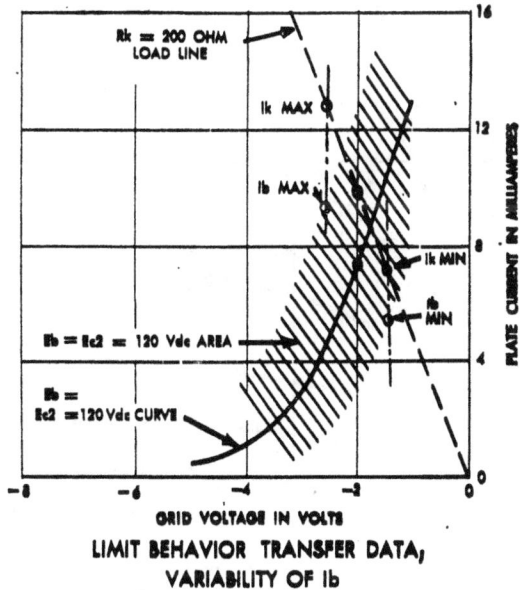

LIMIT BEHAVIOR TRANSFER DATA;
VARIABILITY OF Ib

DESIGN CENTER CHARACTERISTICS
OBTAINED FROM DATA PUBLISHED BY ORIGINAL
RETMA REGISTRANT

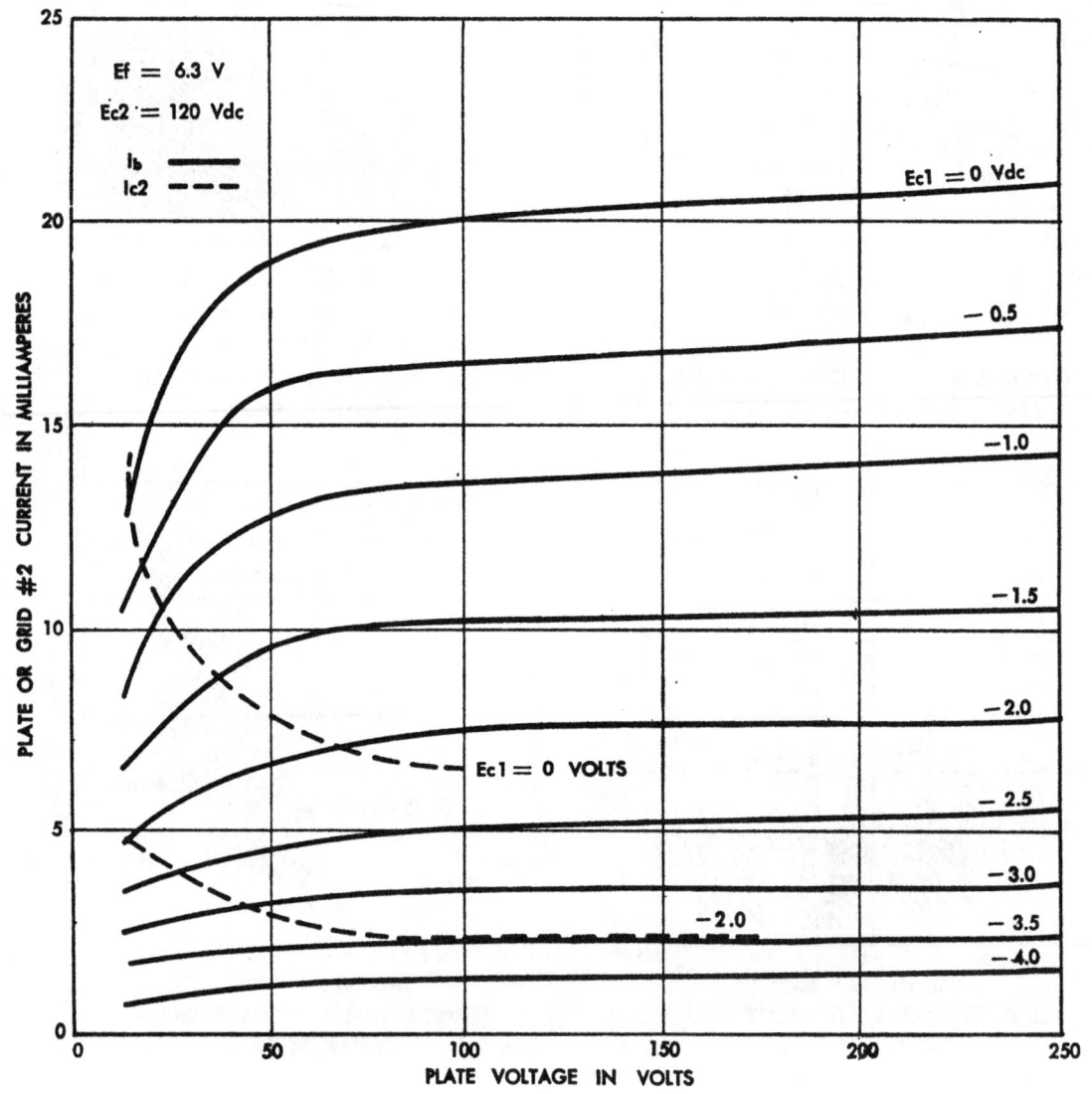

$Ef = 6.3$ V

$Ec2 = 120$ Vdc

I_b ———

I_{c2} - - -

$Ec1 = 0$ Vdc

— 0.5

—1.0

—1.5

—2.0

$Ec1 = 0$ VOLTS

— 2.5

—3.0

—2.0

— 3.5

—4.0

PLATE OR GRID #2 CURRENT IN MILLIAMPERES

PLATE VOLTAGE IN VOLTS

TYPICAL STATIC PLATE CHARACTERISTICS

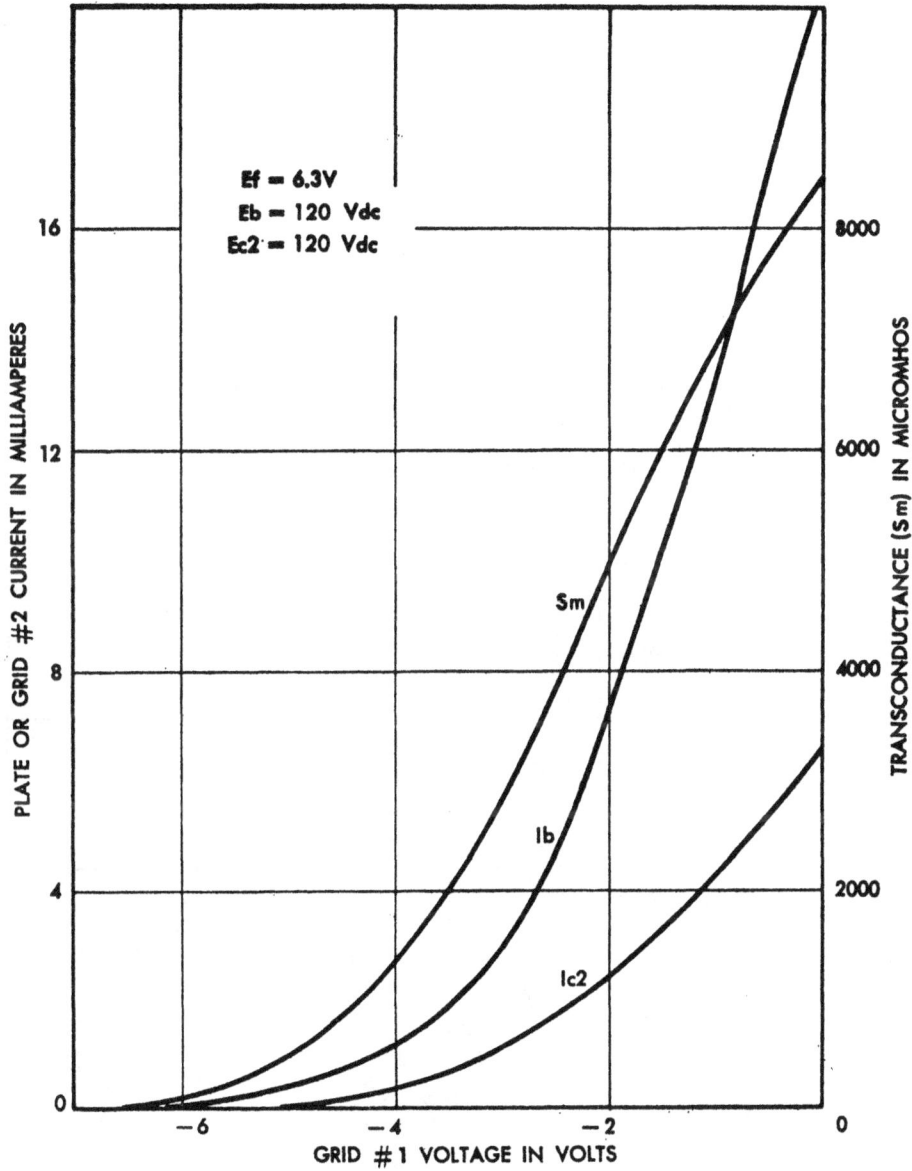

TYPICAL TRANSFER DATA Ec2 = 120 Vdc

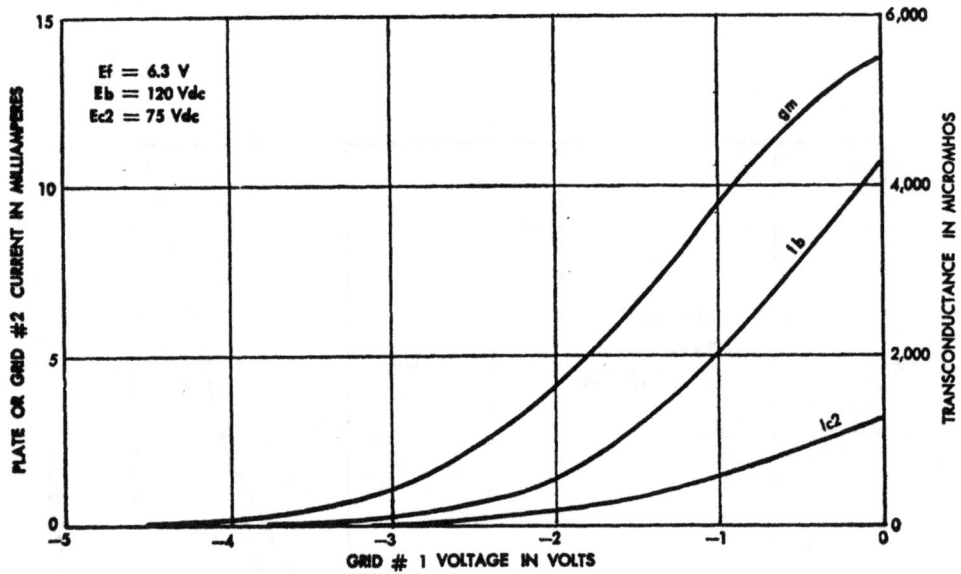

TYPICAL TRANSFER DATA, Ec2=75 Vdc

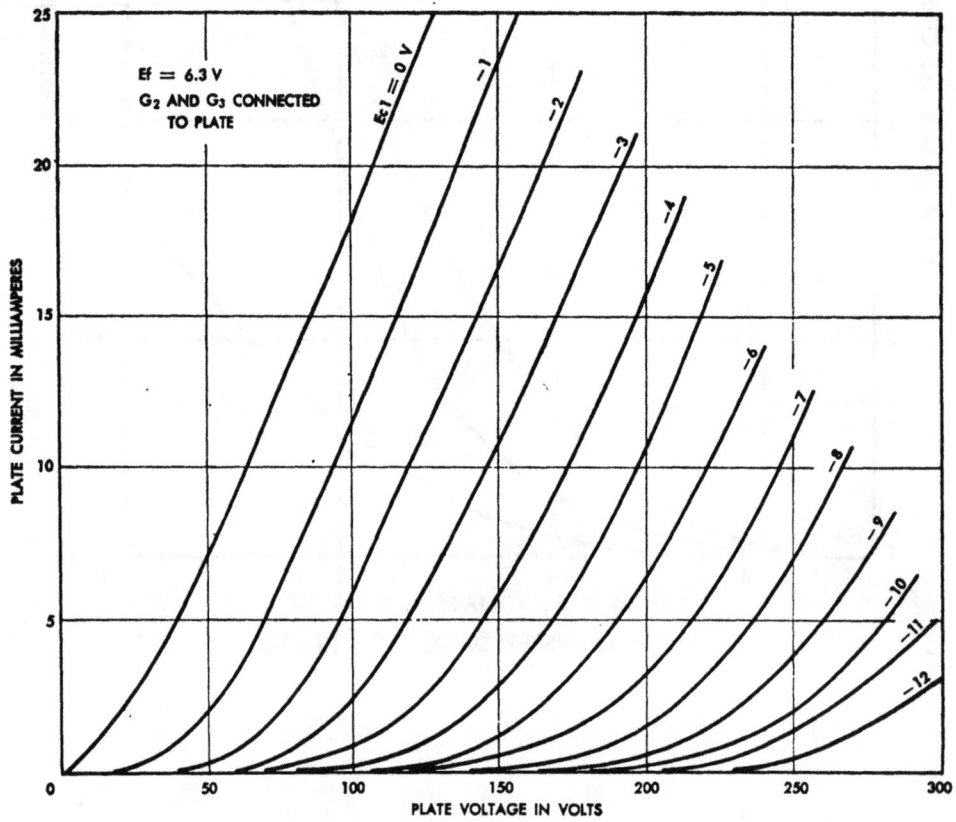

TYPICAL PLATE CHARACTERISTICS, TRIODE CONNECTED

LIFE TEST PROPERTY BEHAVIOR
MIL-E-1/1069 A   4 DEC 56
PRODUCED IN 1957-'58   BY ONE MANUFACTURER

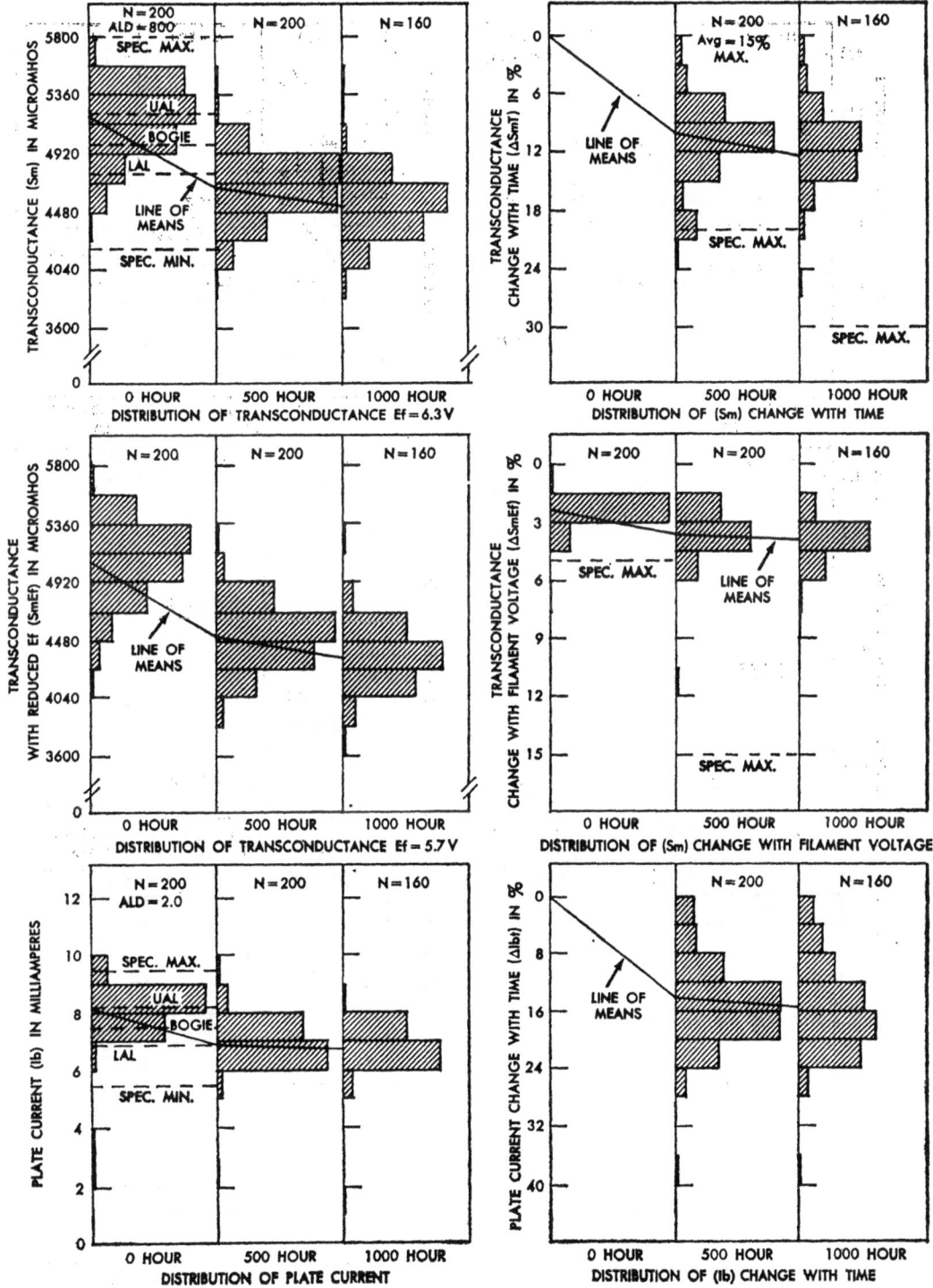

DISTRIBUTION OF TRANSCONDUCTANCE Ef=6.3V

DISTRIBUTION OF (Sm) CHANGE WITH TIME

DISTRIBUTION OF TRANSCONDUCTANCE Ef=5.7V

DISTRIBUTION OF (Sm) CHANGE WITH FILAMENT VOLTAGE

DISTRIBUTION OF PLATE CURRENT

DISTRIBUTION OF (Ib) CHANGE WITH TIME

103

LIFE TEST PROPERTY BEHAVIOR
MIL-E-1/1069 A   4 DEC 57
PRODUCED IN 1957-'58   BY ONE MANUFACTURER

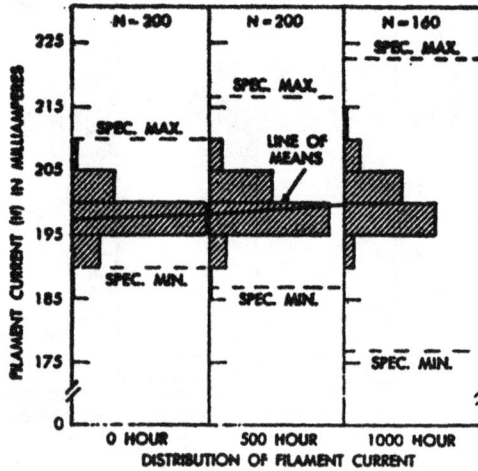

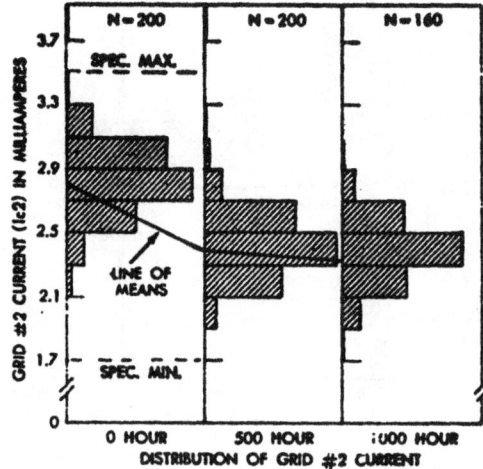

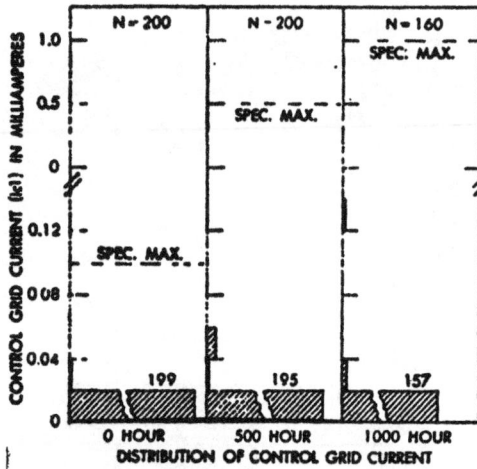

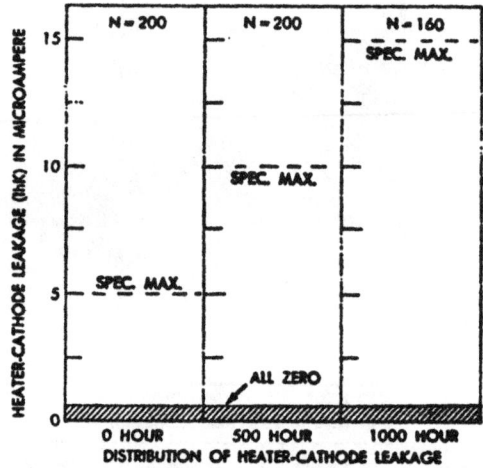

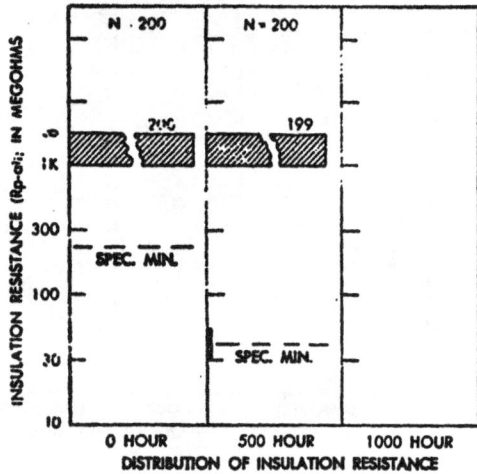

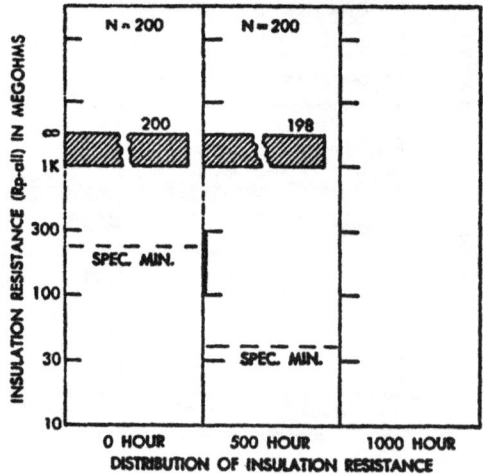

DISTRIBUTION OF FILAMENT CURRENT

DISTRIBUTION OF GRID #2 CURRENT

DISTRIBUTION OF CONTROL GRID CURRENT

DISTRIBUTION OF HEATER-CATHODE LEAKAGE

DISTRIBUTION OF INSULATION RESISTANCE

DISTRIBUTION OF INSULATION RESISTANCE

LIFE TEST PROPERTY BEHAVIOR
PROBABILITY OF SURVIVAL
MIL-E-1/1069A   4 DEC '57
PRODUCED IN 1958 BY ONE MANUFACTURER

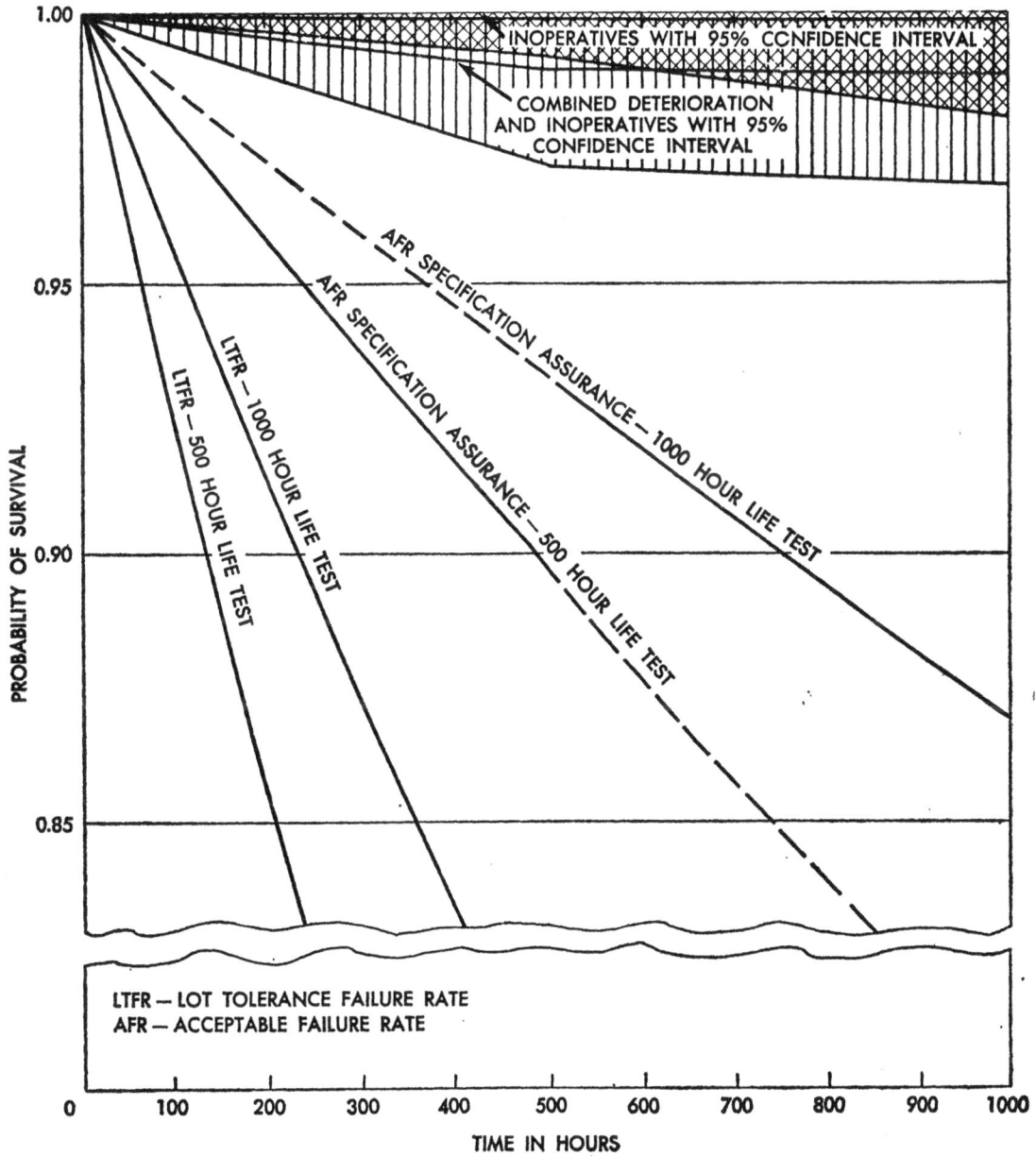

TUBE TYPE JAN-5703WA

DESCRIPTION:

The JAN-5703WA[1] is a 5 lead, pinch press, subminiature, triode having a Mu in the range of 21 to 30 and a transconductance in the range 4200 to 6000 micromhos. The JAN-5703WA is similar in plate characteristics to the JAN-5718 and the JAN-6111. This tube type has given satisfactory service in a variety of applications including oscillator circuits at 500 Mc.

ELECTRICAL: The electrical characteristics are as follows:
Heater Voltage...6.3 V
Heater Current...183-217 mA
Cathode..Coated Unipotential

MOUNTING: Any type mounting is adequate.

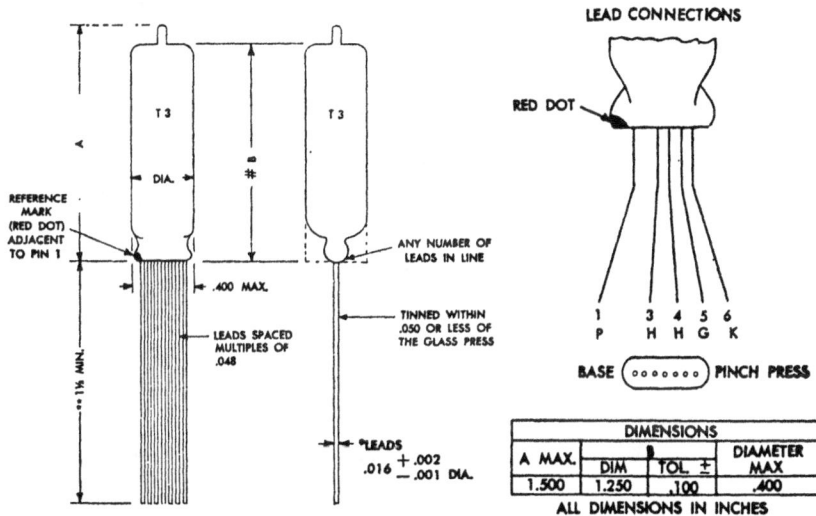

	DIMENSIONS		
A MAX.	DIM	TOL. ±	DIAMETER MAX
1.500	1.250	.100	.400

ALL DIMENSIONS IN INCHES

\# MEASURE FROM BASE SEAT TO BULB TOP-LINE AS DETERMINED BY RING GAGE OF .210 ± .001.

*. LEAD DIAMETER TOLERANCE SHALL GOVERN BETWEEN .050 FROM THE GLASS TO .250 FROM THE GLASS.

** ALTERNATIVE LEAD LENGTH SHALL BE .200 ± .015 WHEN CUT LEADS ARE REQUIRED BY PROCUREMENT CONTRACT OR TSS. CUT LEADS SHALL BE ESSENTIALLY SQUARE CUT AND THE MAXIMUM BURR SHALL BE .003 INCREASE OVER THE ACTUAL LEAD DIAMETER.

RATINGS:	Ef	Eb	Ec	Ehk.	Rk	Rg	Ib**	Ic*	Pp*	T Envelope	Alt
Design	V	Vdc	Vdc	v	ohms	Meg	mAdc	mAdc	W	°C	ft
Maximum	6.9	200	---	.200	---	1.2	15	5.5	1.35	220	60,000
Minimum	5.7	---	---	---	---	---	---	---	---	---	---
Test Cond:	6.3	120	0	0	220	---	---	---	---	---	---

1/ The values and specification comments presented in this section are related to MIL-E-1/293C dated 17 Sep 1956.

*   No test at this rating exists in the specification.

** Difficulty may be encountered if this tube is operated for long periods of time with very small values of cathode current. No specification assurance of life exists under conditions of cathode current approaching the maximum.

ACCEPTANCE TEST LIMITS SUMMARY

PROPERTY		MEASUREMENT CONDITIONS	INITIAL		500 HR LIFE TEST		1000 HR LIFE TEST		UNITS
			MIN	MAX	MIN	MAX	MIN	MAX	
Heater Current	If		183	21.7	1.80	220	1.77	223	mA
Transconductance	Sm		4200	6000	-	-	-	-	umhos
Change in individual	Δ Smt		-	-	-	20	-	30	%
Change in average	Avg Δ Smt		-	-	-	15	-	-	%
Transconductance Change with Ef	Δ SmEf	Ef=5.5V	-	10	-	15	-	-	%
Amplification Factor	Mu		21	30	-	-	-	-	-
Plate Current (1)	Ib		6.8	12.0	-	-	-	-	mAdc
Plate Current (2)	Ib	Ec1=-8.5Vdc	-	50	-	-	-	-	uAdc
Plate Current (3)	Ib	Ec=-5Vdc	20	-	-	-	-	-	uAdc
Power Oscillation	Po	F=500Mc;Eb=150 Vdc;Rg/Ib=20 mAdc; NOTE 2	600	-	-	-	-	-	mW
Capacitance		No shield,Ef=0							
	C g1p		0.9	1.6	-	-	-	-	uuf
	C in		2.0	3.2	-	-	-	-	uuf
	C out		0.5	0.9	-	-	-	-	uuf
Control Grid Current	Ic1		0	-0.3	0	-0.6	0	-1.0	uAdc
Control Grid Emission	Ic	Ef=7.5;Rg=1.0 Meg;Ec1=-10Vdc NOTE 1	0	-0.4	-	-	-	-	uAdc
Heater Cathode Leakage	Ihk	Ehk=+100Vdc	-	5	-	10	-	15	uAdc
	Ihk	Ehk=-100Vdc	-	5	-	10	-	15	uAdc
Insulation of Electrodes	R(g1-all)	Eg1-all=-100	100	-	50	-	-	-	Meg
	R(p-all)	Vdc;Ep-all= -300Vdc	100	-	50	-	-	-	Meg

Measurement conditions are the same as stated under Test Conditions, unless otherwise indicated.

NOTE 1: The tube shall be preheated a minimum of five minutes at test conditions for this test (except Ec1=0) prior to this test.

NOTE 2: Test per drawing 226-JAN.

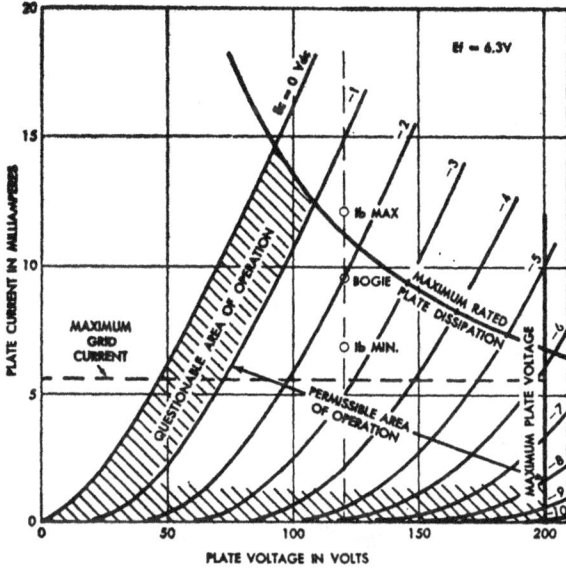

TYPICAL STATIC-PLATE CHARACTERISTICS; PERMISSIBLE AREA OF OPERATION

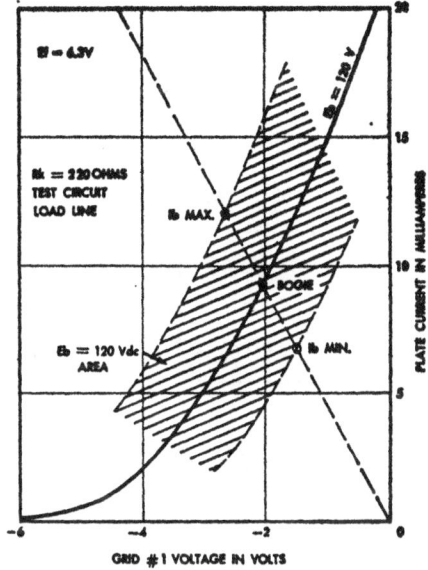

LIMIT BEHAVIOR TRANSFER DATA; VARIABILITY OF Ib

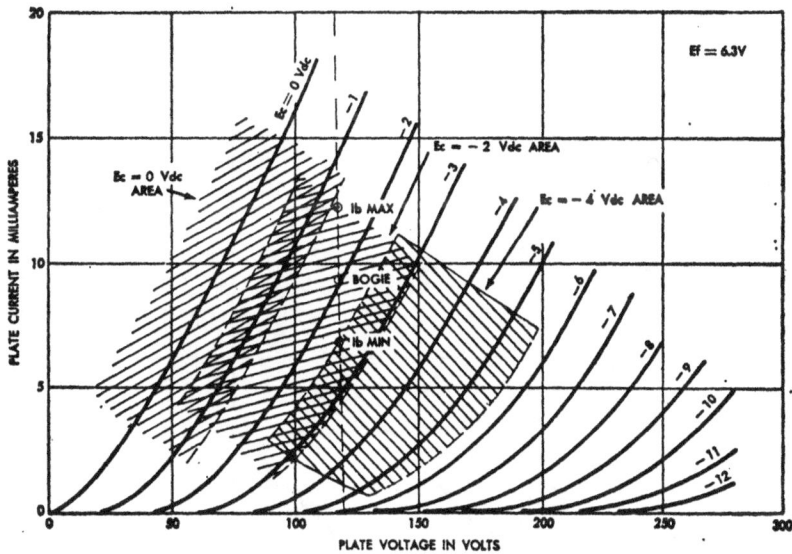

LIMIT BEHAVIOR STATIC-PLATE DATA; VARIABILITY OF Ib

DESIGN CENTER CHARACTERISTICS
OBTAINED FROM DATA PUBLISHED BY ORIGINAL
RETMA REGISTRANT FOLLOWS

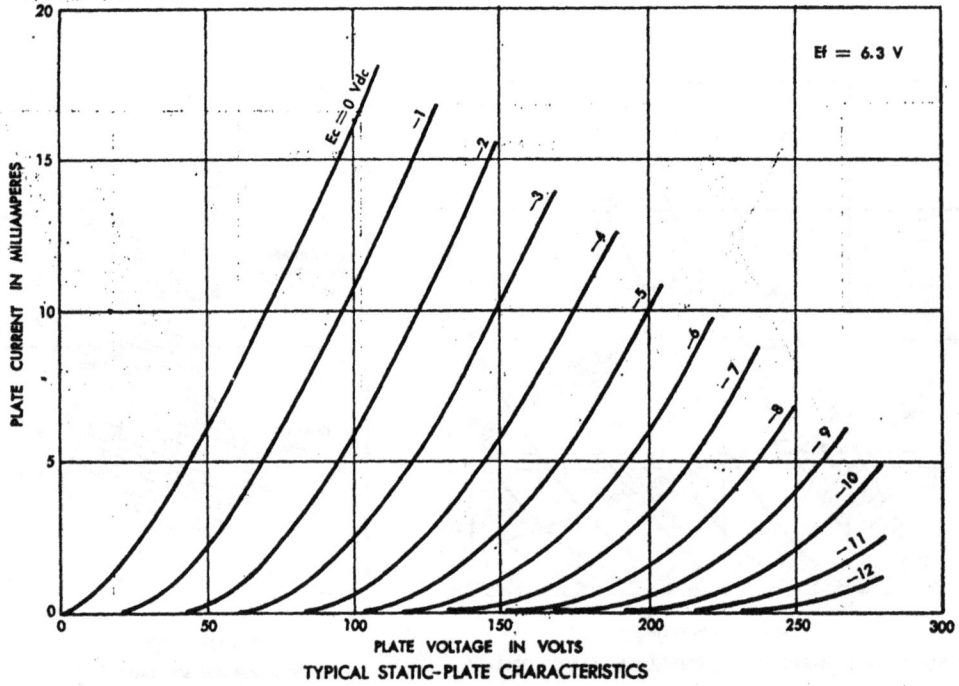

TYPICAL STATIC-PLATE CHARACTERISTICS

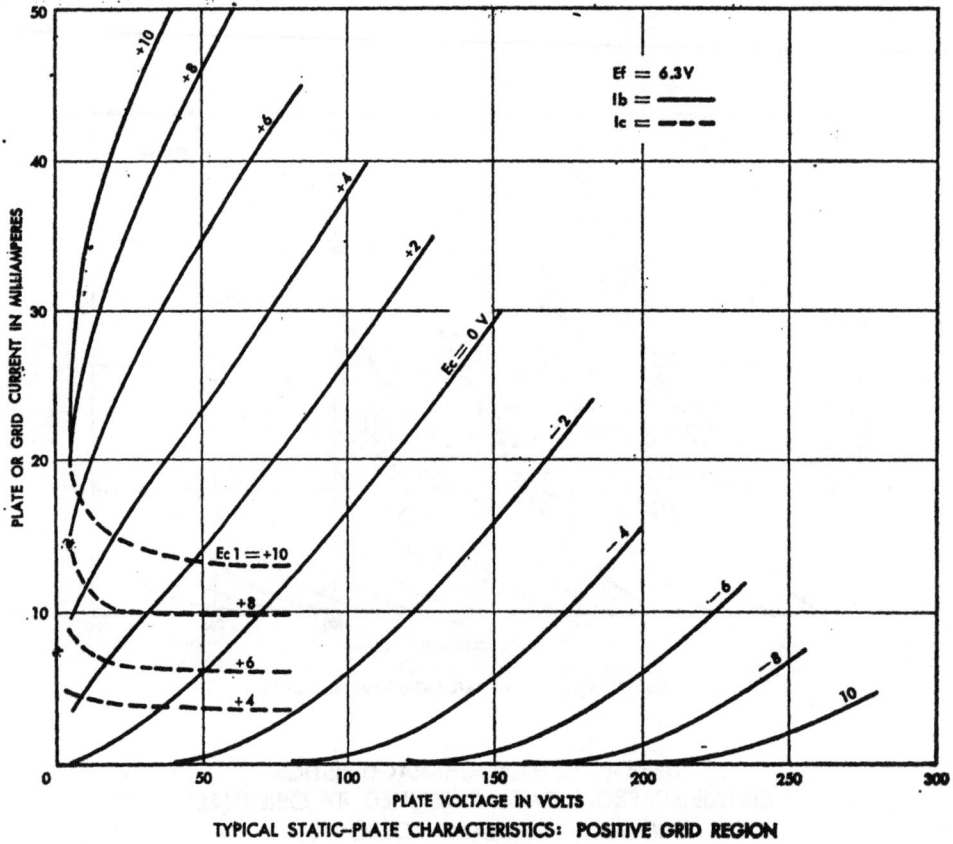

TYPICAL STATIC-PLATE CHARACTERISTICS: POSITIVE GRID REGION

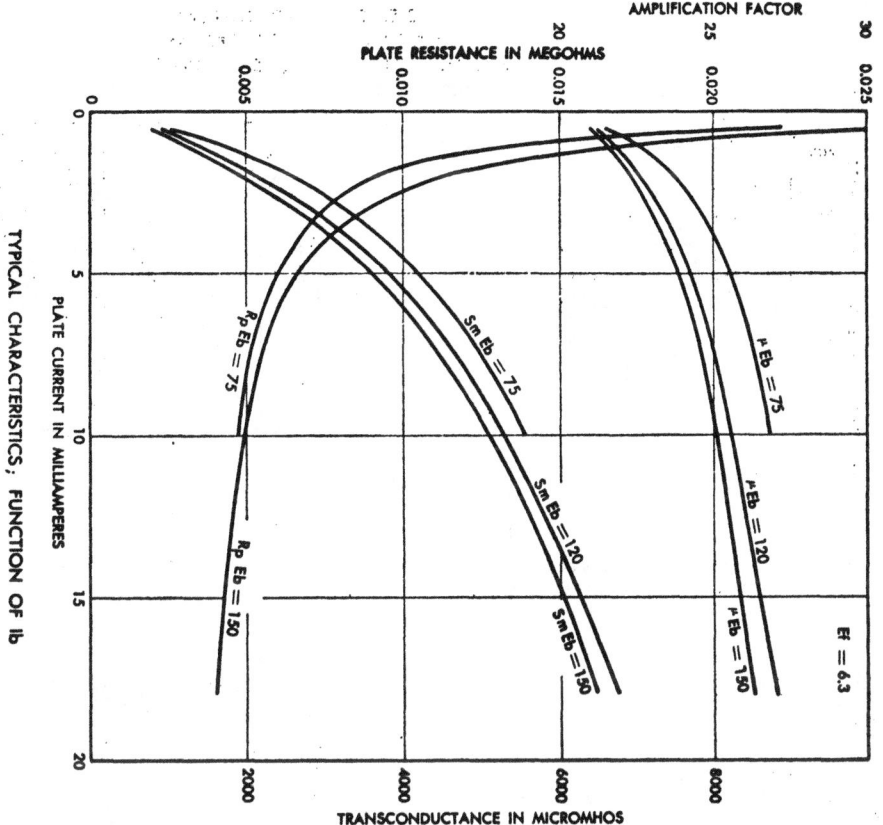

TYPICAL CHARACTERISTICS; FUNCTION OF Ib

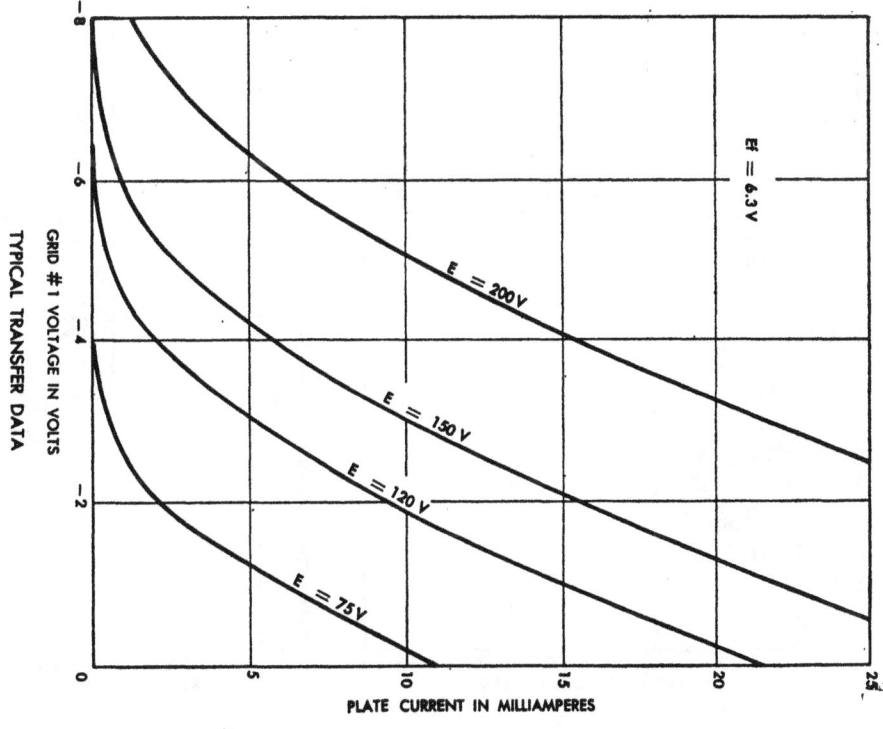

TYPICAL TRANSFER DATA

LIFE TEST PROPERTY BEHAVIOR
MIL-E-1/293C   17 SEPT. '56
PRODUCED IN 1956 BY ONE MANUFACTURER

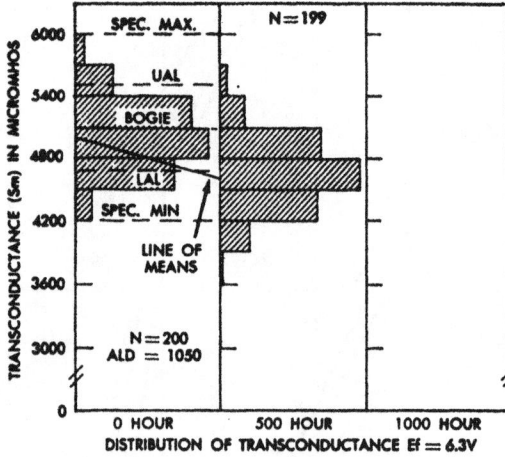

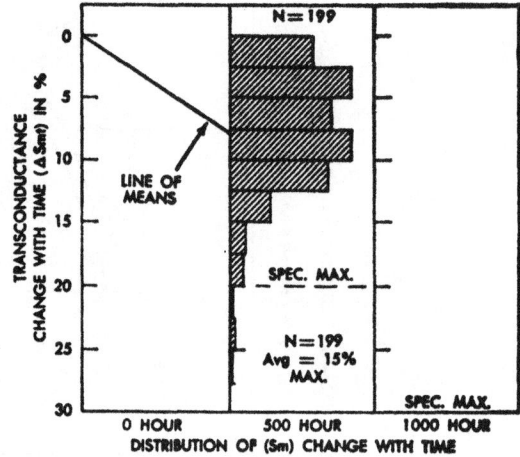

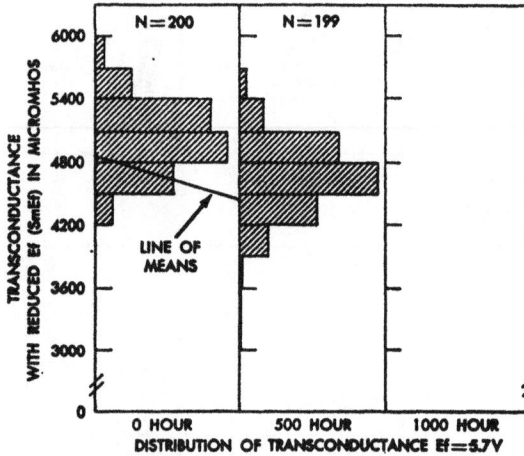

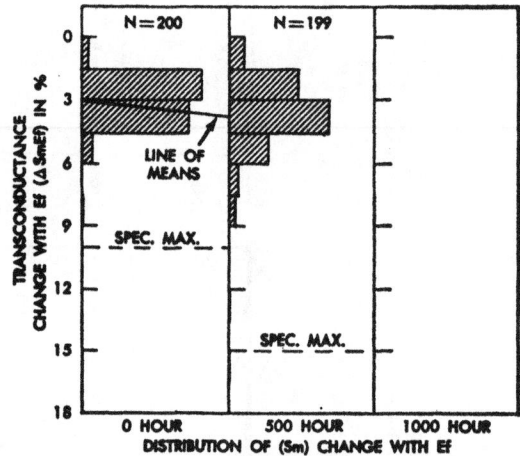

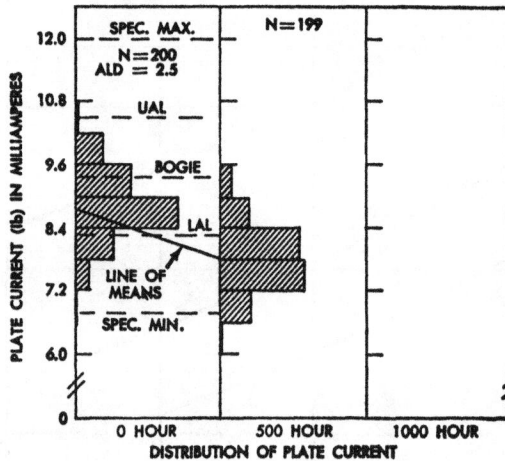

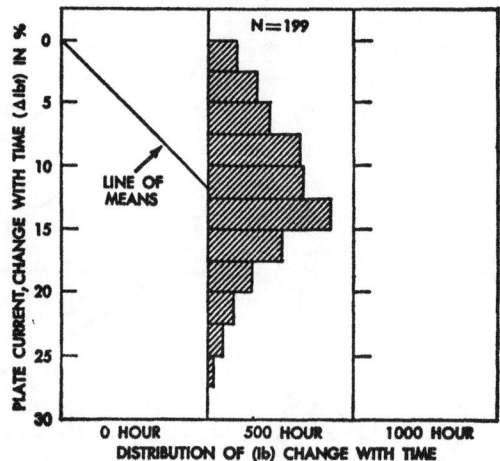

DISTRIBUTION OF TRANSCONDUCTANCE Ef = 6.3V

DISTRIBUTION OF (Sm) CHANGE WITH TIME

DISTRIBUTION OF TRANSCONDUCTANCE Ef = 5.7V

DISTRIBUTION OF (Sm) CHANGE WITH Ef

DISTRIBUTION OF PLATE CURRENT

DISTRIBUTION OF (Ib) CHANGE WITH TIME

LIFE TEST PROPERTY BEHAVIOR
MIL-E-1/293C   17 SEPT. '56
PRODUCED IN 1956 BY ONE MANUFACTURER

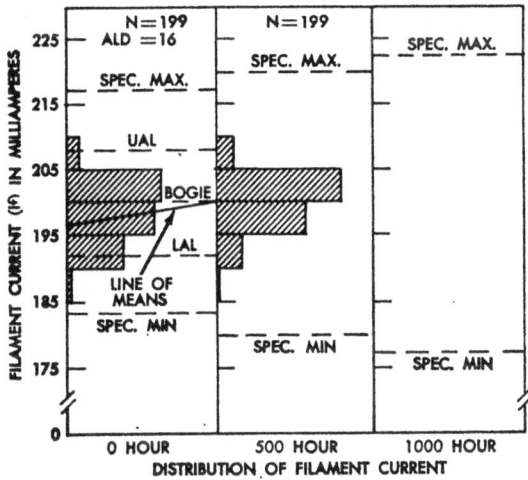

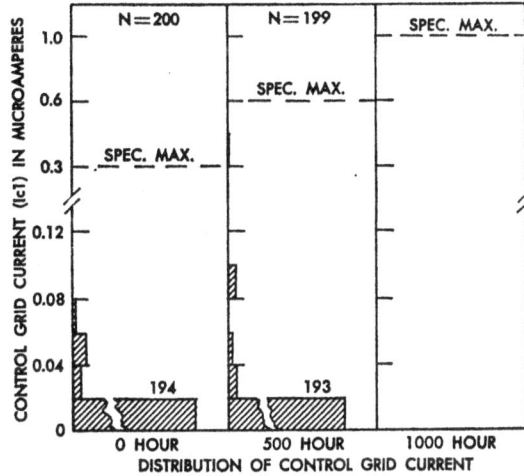

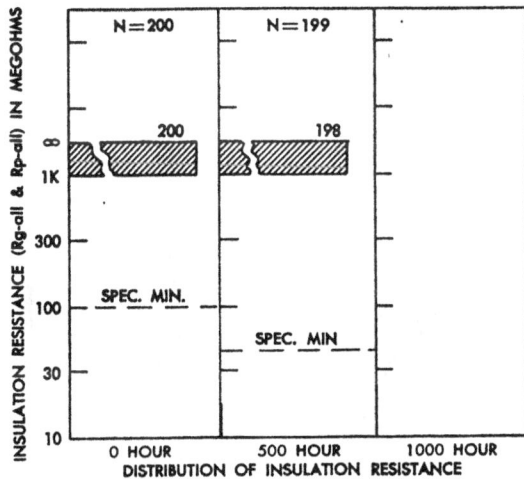

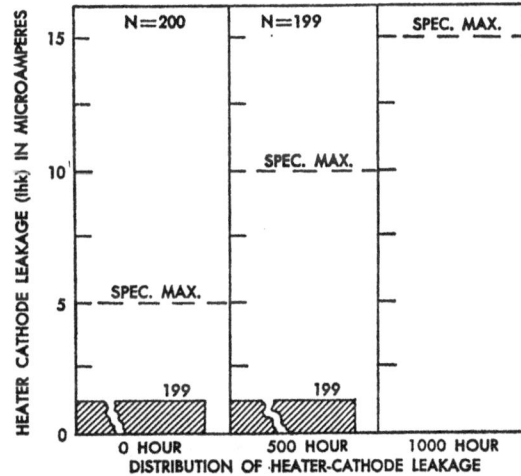

113

LIFE TEST PROPERTY BEHAVIOR
PROBABILITY OF SURVIVAL
MIL-E-1/293C   17 SEPT. '56
PRODUCED IN 1956 BY ONE MANUFACTURER

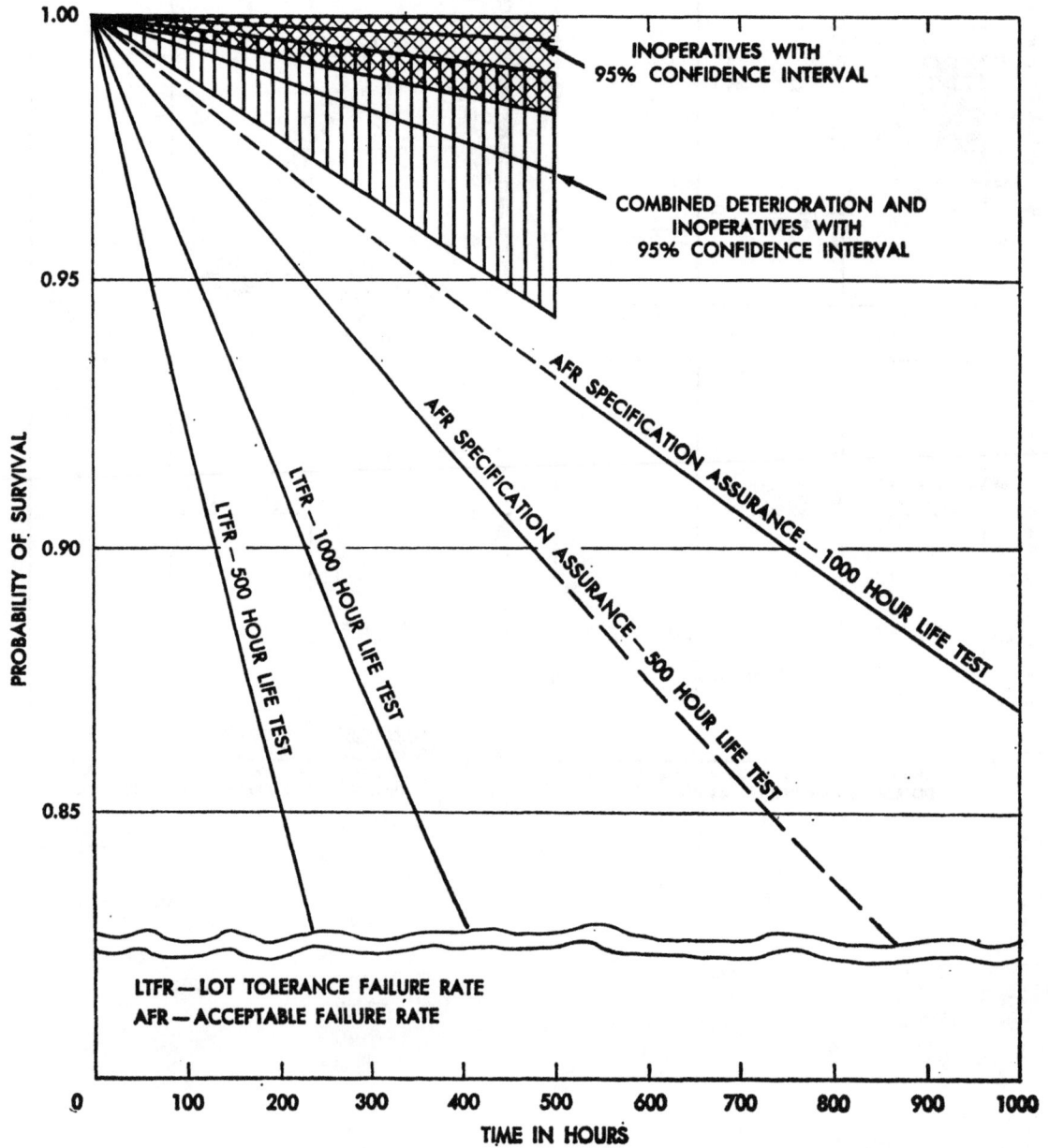

TUBE TYPE JAN-5703WB

DESCRIPTION:

The JAN-5703WB[1] is a 5 lead, pinch-press, subminiature, triode having a Mu in the range of 22.5 to 28.5 and a transconductance in the range of 4300 to 5700 micromhos. The JAN-5703WB is similar in plate characteristics to the JAN-5718 and the JAN-6111. This tube type has given satisfactory service in a variety of applications including oscillator circuits at 500 Mc.

ELECTRICAL: The electrical characteristics are as follows:
Heater Voltage...6.3 V
Heater Current...190-210 mA
Cathode...Coated Unipotential

MOUNTING: Any type mounting is adequate.

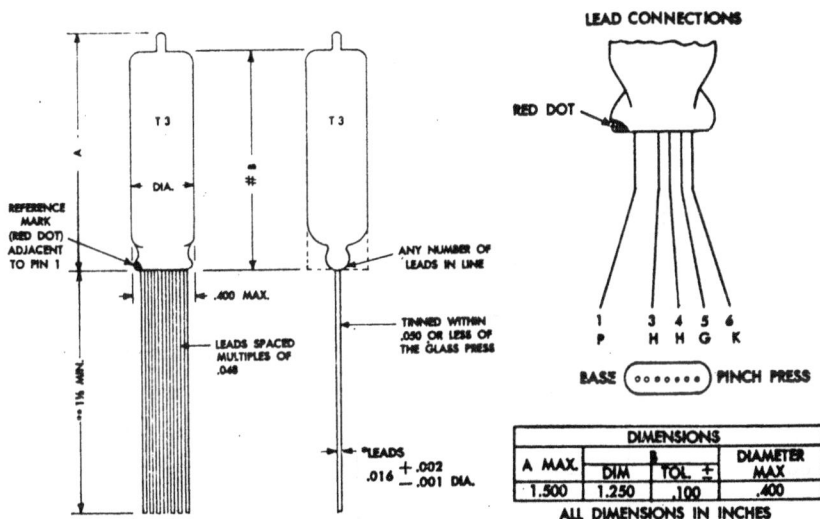

MEASURE FROM BASE SEAT TO BULB TOP-LINE AS DETERMINED BY RING GAGE OF .210 ± .001.

* LEAD DIAMETER TOLERANCE SHALL GOVERN BETWEEN .050 FROM THE GLASS TO .250 FROM THE GLASS.

** ALTERNATIVE LEAD LENGTH SHALL BE .200 ± .015 WHEN CUT LEADS ARE REQUIRED BY PROCUREMENT CONTRACT OR TSS. CUT LEADS SHALL BE ESSENTIALLY SQUARE CUT AND THE MAXIMUM BURR SHALL BE .003 INCREASE OVER THE ACTUAL LEAD DIAMETER.

RATINGS:	Ef	Eb	Ec	Ehk	Rk	Rg	Ib**	Ic*	Pp*	T En-velope	Alt
Absolute	V	Vdc	Vdc	v	ohms	Meg	mAdc	mAdc	W	°C	ft
Maximum	6.9	200	---	±200	---	1.2	15	5.5	---	220	60,000
Design Maximum	---	---	---	---	---	---	---	---	1.35	---	---
Minimum	5.7	---	---	---	---	---	---	---	---	---	---
Test Cond:	6.3	120	0	0	220	---	---	---	---	---	---

[1] The values and specification comments presented in this section are related to MIL-E-1070A dated 4 Dec 1957.

* No test at this rating exists in the specification.

** Difficulty may be encountered if this tube is operated for long periods of time with very small values of cathode current. No specification assurance of life exists under conditions of cathode current approaching the maximum.

ACCEPTANCE TEST LIMITS SUMMARY

PROPERTY		MEASUREMENT CONDITIONS	INITIAL		500 HR LIFE TEST		1000 HR LIFE TEST		UNITS
			MIN	MAX	MIN	MAX	MIN	MAX	
Heater Current	If		190	210	187	217	177	223	mA
Transconductance	Sm		4300	5700	-	-	-	-	umhos
Change in individual	Δ Smt		-	-	-	20	-	30	%
Change in average	Avg Δ Smt		-	-	-	15	-	-	%
Transconductance change with Ef	Δ SmEf	Ef=5.7V	-	5	-	15	-	-	%
Amplification Factor	Mu		22.5	28.5	-	-	-	-	-
Plate Current (1)	Ib		7.3	11.5	-	-	-	-	mAdc
Plate Current (2)	Ib	Ec1=-8.5Vdc	-	50	-	-	-	-	uAdc
Plate Current (3)	Ib	Ec=5Vdc	20	-	-	-	-	-	uAdc
Pulse Emission	is	Ef=6.0V;e(pulse)=50V;Tp=25usec;prr=200pps	300	-	-	-	-	-	mA
Power Oscillation	Po	F=500Mc;Eb=150 Vdc;Rg/Ib=20 mAdc; NOTE 2	600	-	-	-	-	-	mW
Capacitance	C gp	No shield,Ef=0	1.0	1.6	-	-	-	-	uuf
	C in		2.0	3.2	-	-	-	-	uuf
	C out		0.65	1.05	-	-	-	-	uuf
Control Grid Current	Ic1		0	-0.3	0	-0.6	-	-1.0	uAdc
Control Grid Emission	Ic	Ef=7.5V;Rg=1.0 Meg;Ec1=-10Vdc; NOTE 1	0	-0.4	-	-	-	-	uAdc
Heater Cathode Leakage	Ihk	Ehk=+100Vdc	-	5	-	10	-	15	uAdc
	Ihk	Ehk=-100Vdc	-	5	-	10	-	15	uAdc
Insulation of Electrodes	R(g1-all)	Eg1-all=-100 Vdc;Ep-all=-300Vdc;	250	-	50	-	-	-	Meg
	R(p-all)		250	-	50	-	-	-	Meg

Measurement conditions are the same as stated under Test Conditions, unless otherwise indicated.

NOTE 1: The tube shall be preheated a minimum of five minutes at test conditions for this test (except Ec1=0) prior to this test.

NOTE 2: Test per drawing 226-JAN.

116

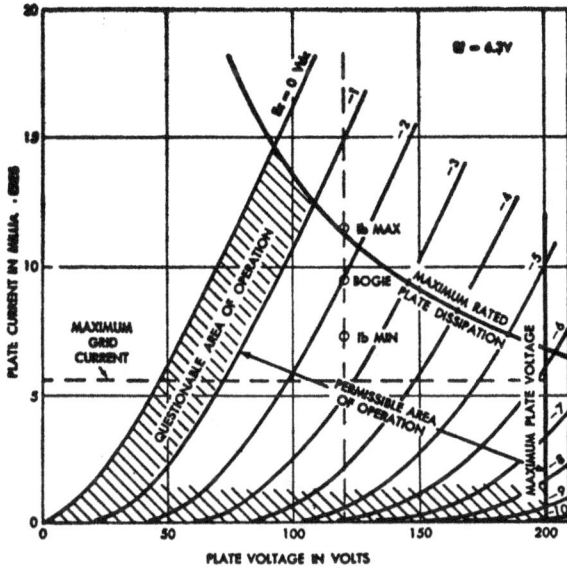

TYPICAL STATIC-PLATE CHARACTERISTICS; PERMISSIBLE AREA OF OPERATION

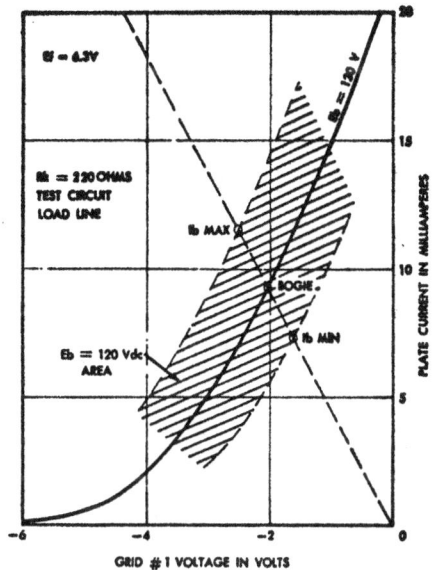

LIMIT BEHAVIOR TRANSFER DATA; VARIABILITY OF Ib

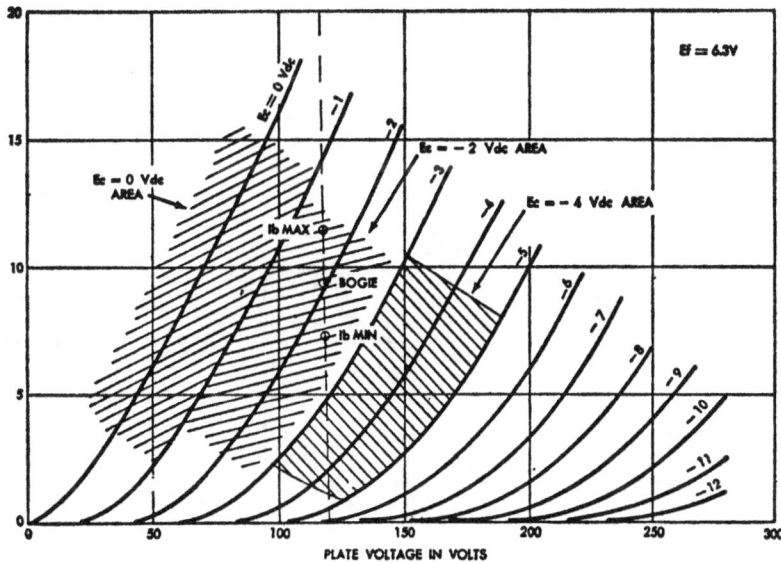

LIMIT BEHAVIOR STATIC-PLATE DATA; VARIABILITY OF Ib

DESIGN CENTER CHARACTERISTICS
OBTAINED FROM DATA PUBLISHED BY ORIGINAL
RETMA REGISTRANT FOLLOWS

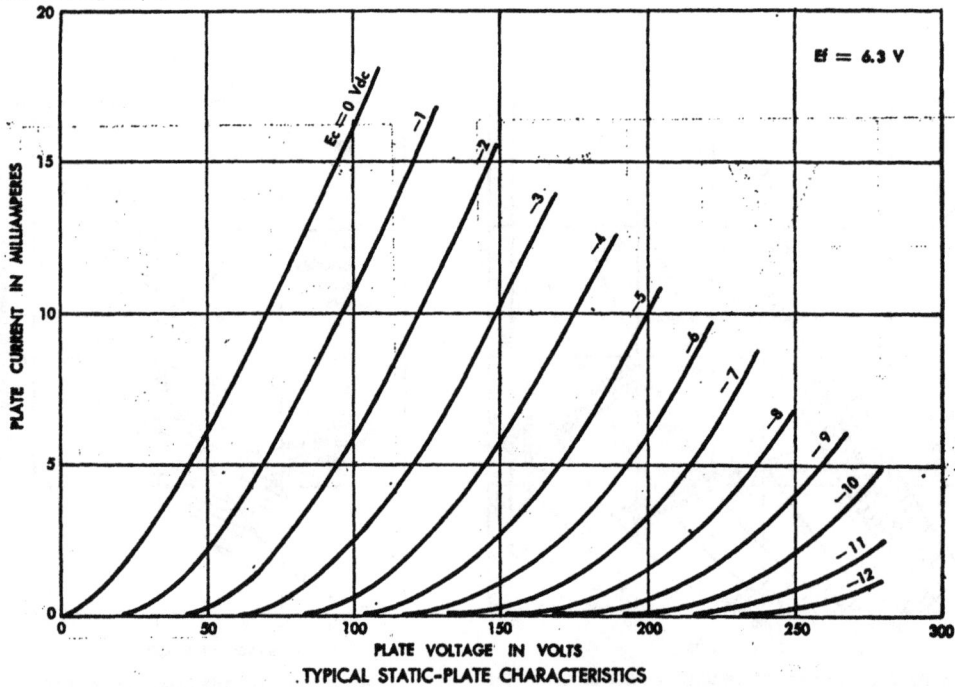

TYPICAL STATIC-PLATE CHARACTERISTICS

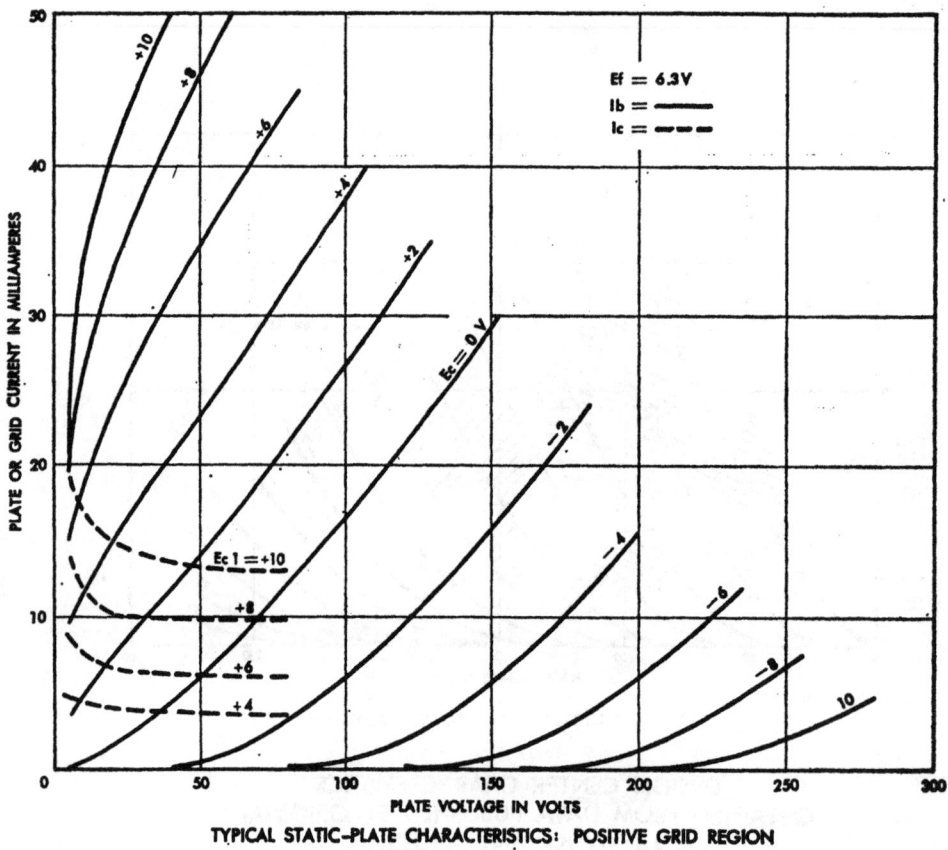

TYPICAL STATIC-PLATE CHARACTERISTICS: POSITIVE GRID REGION

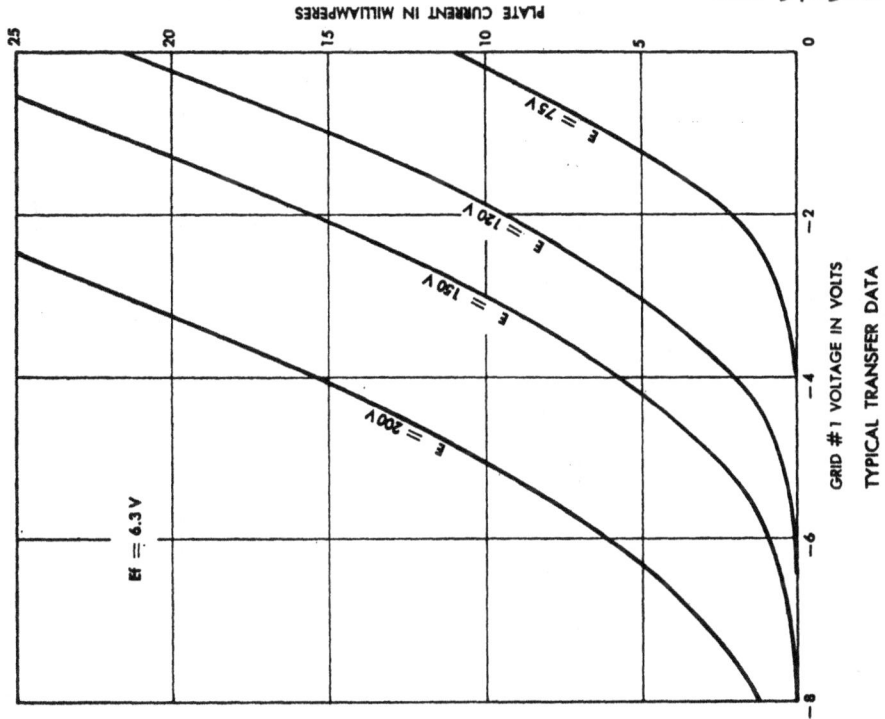

GRID #1 VOLTAGE IN VOLTS

TYPICAL TRANSFER DATA

Ef = 6.3 V

E = 75 V
E = 120 V
E = 150 V
E = 200 V

PLATE CURRENT IN MILLIAMPERES

TRANSCONDUCTANCE IN MICROMHOS

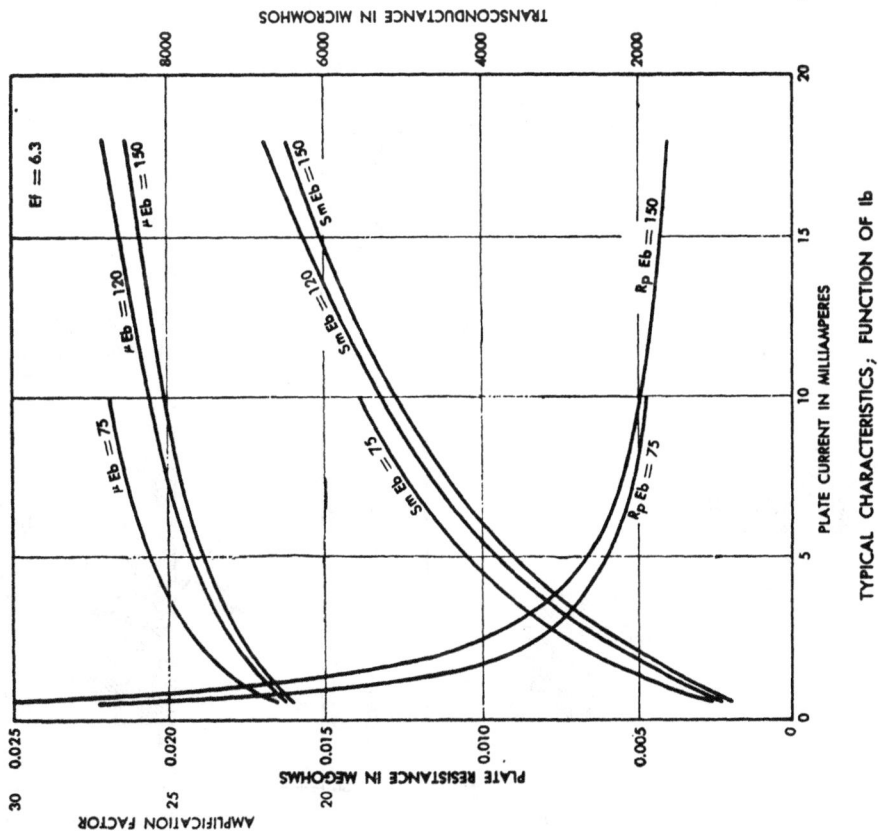

PLATE CURRENT IN MILLIAMPERES

TYPICAL CHARACTERISTICS; FUNCTION OF Ib

Ef = 6.3

μ Eb = 150
μ Eb = 120
μ Eb = 75

Sm Eb = 150
Sm Eb = 120
Sm Eb = 75

Rp Eb = 150
Rp Eb = 75

PLATE RESISTANCE IN MEGOHMS

AMPLIFICATION FACTOR

LIFE TEST PROPERTY BEHAVIOR
MIL-E-1/1070A   4 DEC. '57
PRODUCED IN 1957-'58 BY ONE MANUFACTURER

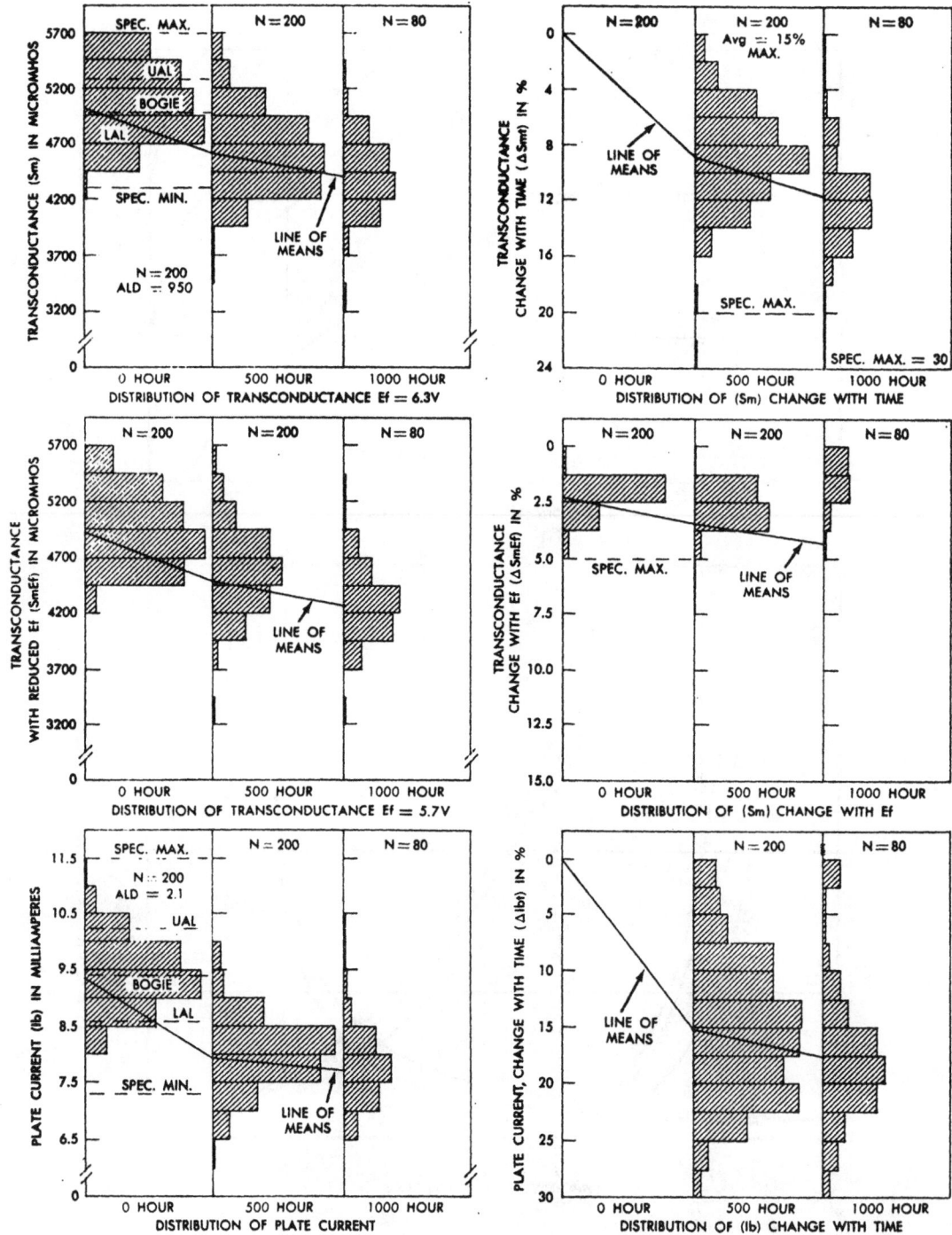

LIFE TEST PROPERTY BEHAVIOR
MIL-E-1/1070A   4 DEC. '57
PRODUCED IN 1957-'58 BY ONE MANUFACTURER

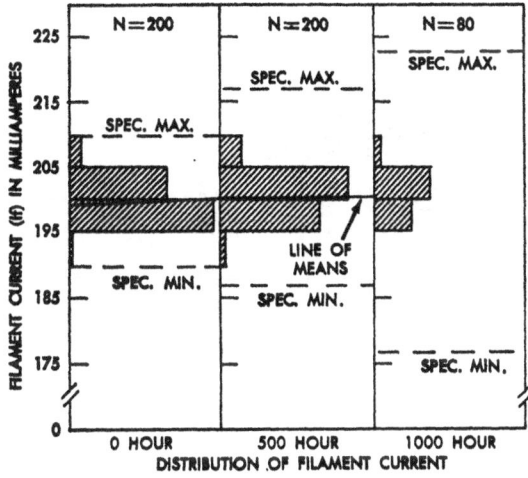

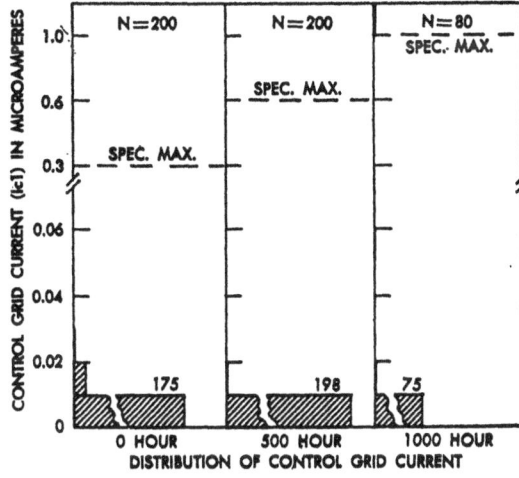

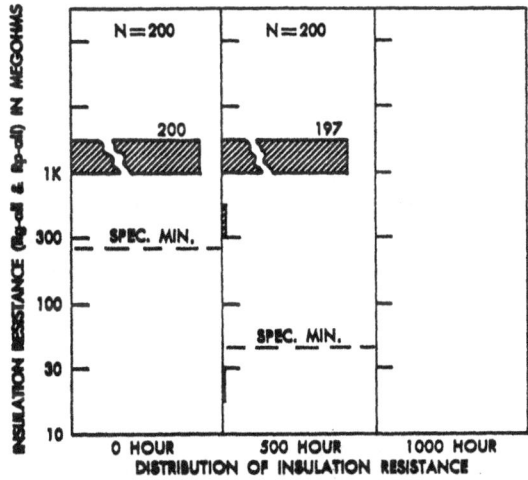

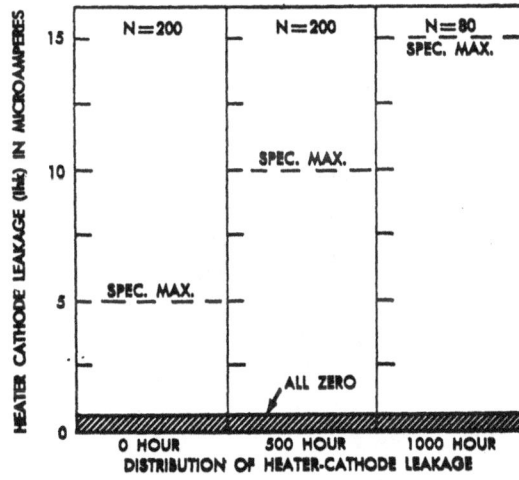

DISTRIBUTION OF FILAMENT CURRENT

DISTRIBUTION OF CONTROL GRID CURRENT

DISTRIBUTION OF INSULATION RESISTANCE

DISTRIBUTION OF HEATER-CATHODE LEAKAGE

LIFE TEST PROPERTY BEHAVIOR
PROBABILITY OF SURVIVAL
MIL-E-1/1070A   4 DEC. '57
PRODUCED IN 1957-'58 BY ONE MANUFACTURER

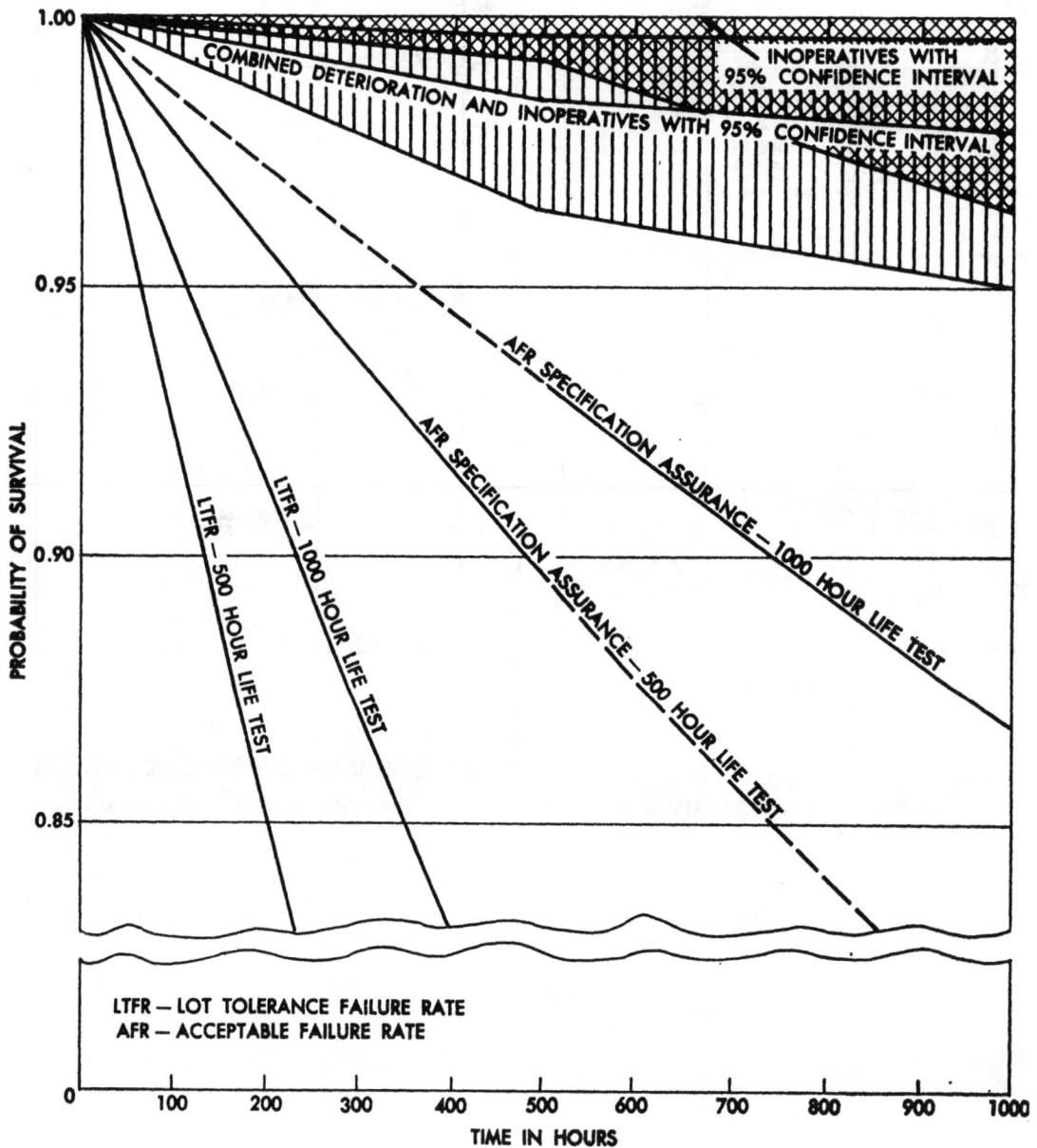

TUBE TYPE JAN-5744WA

DESCRIPTION:

The JAN-5744WA[1] is a 5 lead, flat press subminiature triode having a transconductance ranging from 3200 to 4800 micromhos and a Mu ranging from 60 to 80. The JAN-5744WA has given satisfactory service in voltage amplifier and other low current applications.

ELECTRICAL: The electrical characteristics are as follows:
Heater Voltage...6.3 V
Heater Current..183-217 mA
Cathode......................................Coated Unipotential

MOUNTING: Any type mounting is adequate.

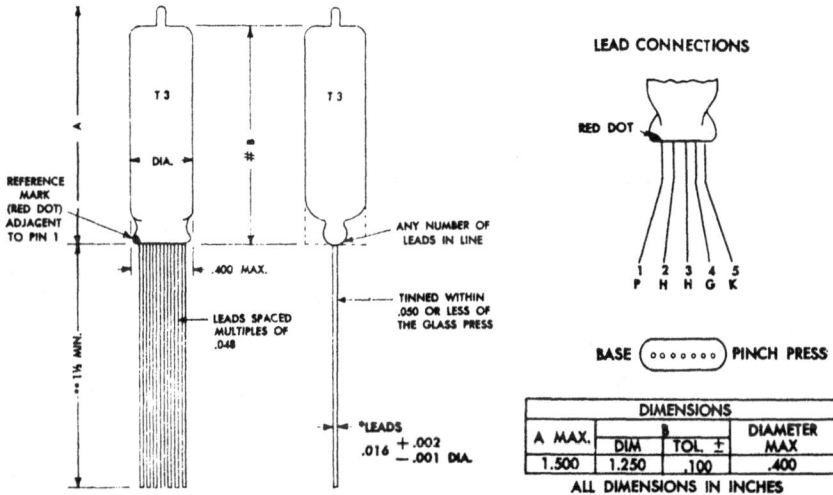

LEAD CONNECTIONS

RED DOT

1 2 3 4 5
P H H G K

BASE (oooooo) PINCH PRESS

DIMENSIONS			
A MAX.	B DIM	TOL. ±	DIAMETER MAX
1.500	1.250	.100	.400

ALL DIMENSIONS IN INCHES

\# MEASURE FROM BASE SEAT TO BULB TOP-LINE AS DETERMINED BY RING GAGE OF .210 ± .001.

* LEAD DIAMETER TOLERANCE SHALL GOVERN BETWEEN .050 FROM THE GLASS TO .250 FROM THE GLASS.

** ALTERNATIVE LEAD LENGTH SHALL BE .200 ± .015 WHEN CUT LEADS ARE REQUIRED BY PROCUREMENT CONTRACT OR TSS. CUT LEADS SHALL BE ESSENTIALLY SQUARE CUT AND THE MAXIMUM BURR SHALL BE .003 INCREASE OVER THE ACTUAL LEAD DIAMETER.

RATINGS:	Ef	Eb	Ec	Pp*	Ehk	Ib**	Rk	T Envelope	Alt
Design	V	Vdc	Vdc	W	Vdc	mAdc	ohms	°C	ft
Maximum	6.9	275	---	1.3	+200	6.5	---	265	60,000
Minimum	5.7	---	-55	---	-200	0.5	---	---	---
Normal	6.3	250	0	1.1	100	4.0	500	---	---
Test Cond:	6.3	250	0	---	0	---	500	---	---

[1] The values and specification comments presented in this section are related to MIL-E-1/84C dated 25 Jul 1956.

* No test at this rating exists in the specification.

** Difficulty may be encountered if this tube is operated for long periods of time with very small values of cathode current. No specification assurance of life exists under conditions of cathode current approaching the maximum.

ACCEPTANCE TEST LIMITS SUMMARY

PROPERTY		MEASUREMENT CONDITIONS	INITIAL		*100 HR LIFE TEST		*5000 HR LIFE TEST		UNITS
			MIN	MAX	MIN	MAX	MIN	MAX	
Heater Current	If		183	217	183	217	-	-	mA
Transconductance	Sm		3200	4800	-	-	-	-	umhos
Change in individual	Δ Smt		-	-	-	25	-	-	%
Transconductance change with Ef	Δ SmEf	Ef=5.5V	-	10	-	15	-	-	%
Amplification Factor	Mu		60	80	-	-	-	-	-
AC Amplification	Ep	Esig=200mVac; Ebb=100Vdc; Ecc=0;Rg=10 Meg;Rk=0;Rp= 0.5Meg	6.5	-	-	-	-	-	Vac
Plate Current (1)	Ib		2.8	5.7	-	-	-	-	mAdc
Plate Current (2)	Ib	Ec=-6.5Vdc	-	50	-	-	-	-	uAdc
Capacitance		0.405 in. dia shield, Ef=0							
	C gp		0.65	0.95	-	-	-	-	uuf
	C in		2.0	3.4	-	-	-	-	uuf
	C out		1.7	3.1	-	-	-	-	uuf
Control Grid Current (1)	Ic		-	-0.3	0	-0.6	-	-	uAdc
Control Grid Current (2)	Ic	Note	0	-0.3	0	-1.0	-	-	uAdc
Heater Cathode Leakage	Ihk	Ehk=+100Vdc	-	10	-	30	-	-	uAdc
	Ihk	Ehk=-100Vdc	-	10	-	30	-	-	uAdc
Insulation of Electrodes	R(g1-all)	Eg1-all=-100V	100	-	50	-	-	-	Meg
	R(p-all)	Ep-all=-300V	100	-	50	-	-	-	Meg

Measurement conditions are the same as stated under Test Conditions, unless otherwise indicated.

* Limits for 100 hours are given and 5000 hour data for information only is required in the specification.

Note: After 5 minutes at Ef=7.0V measure grid current at Ef=7.0V. Three-minute test not permitted.

LIMIT BEHAVIOR STATIC-PLATE CHARACTERISTICS;
VARIABILITY OF Ib

TYPICAL STATIC-PLATE CHARACTERISTICS;
PERMISSIBLE AREA OF OPERATION

LIMIT TRANSFER CHARACTERISTICS;
VARIABILITY OF Ib

125

DESIGN CENTER CHARACTERISTICS
OBTAINED FROM DATA PUBLISHED BY ORIGINAL
RETMA REGISTRANT

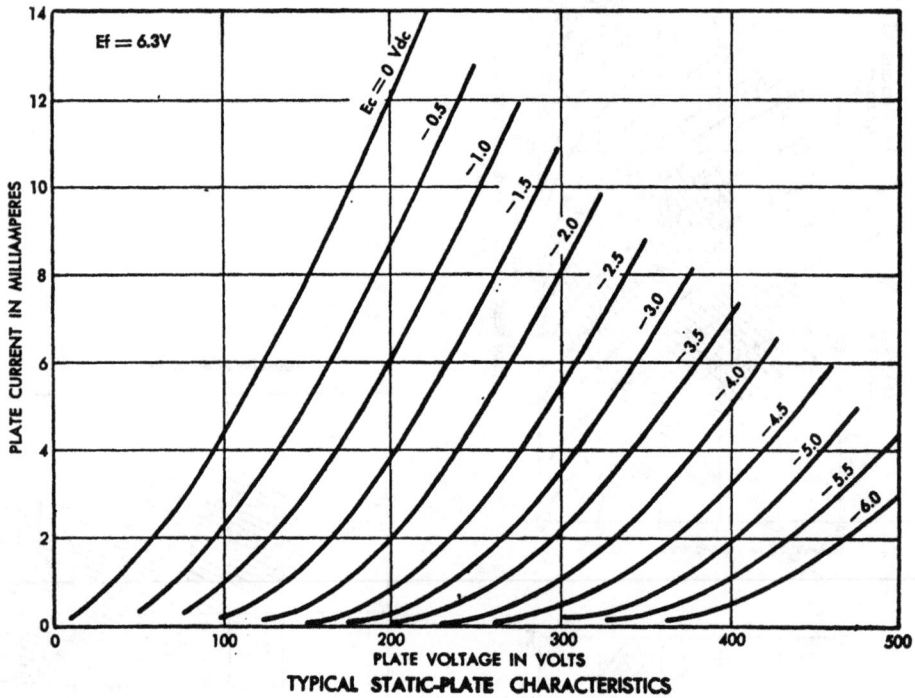

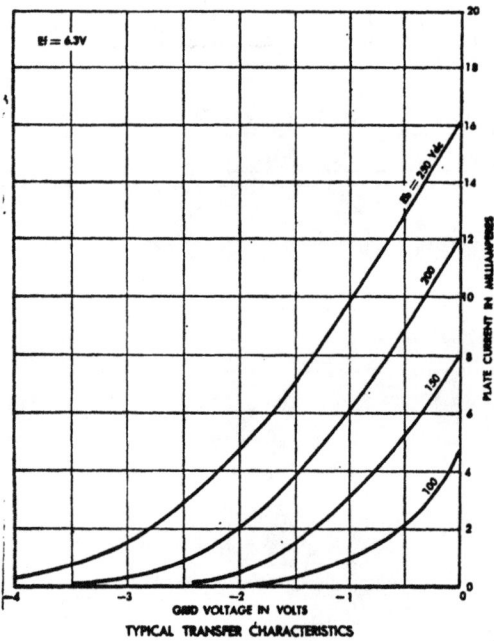

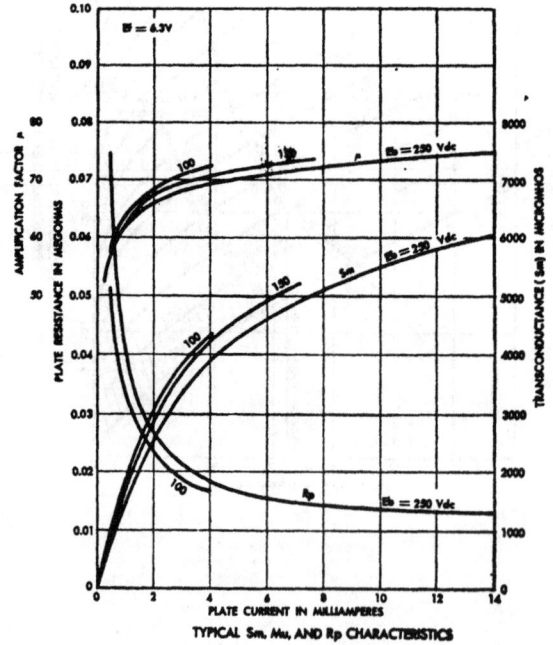

$E_f = 6.3V$

TYPICAL STATIC-PLATE CHARACTERISTICS

TYPICAL TRANSFER CHARACTERISTICS

TYPICAL S_m, M_u, AND R_p CHARACTERISTICS

126

LIFE TEST PROPERTY BEHAVIOR
MIL-E-1/84C   25 JULY '56
PRODUCED IN 1956 BY ONE MANUFACTURER

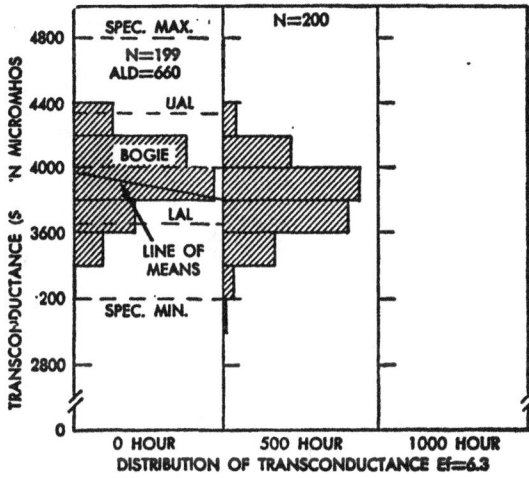

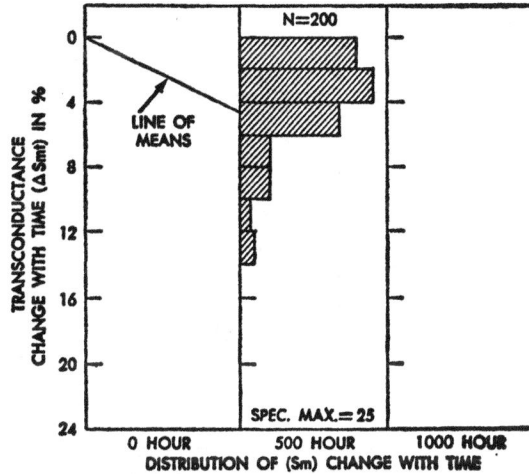

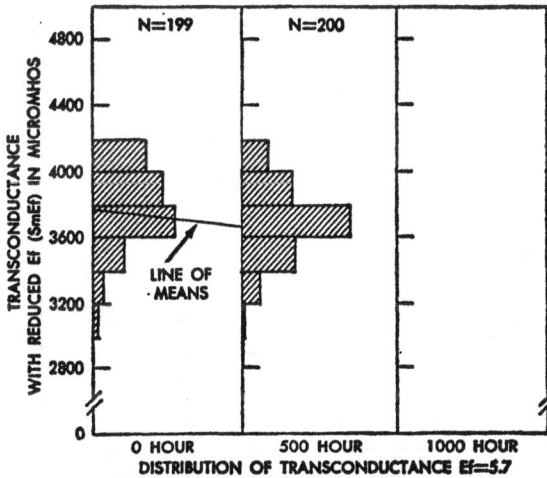

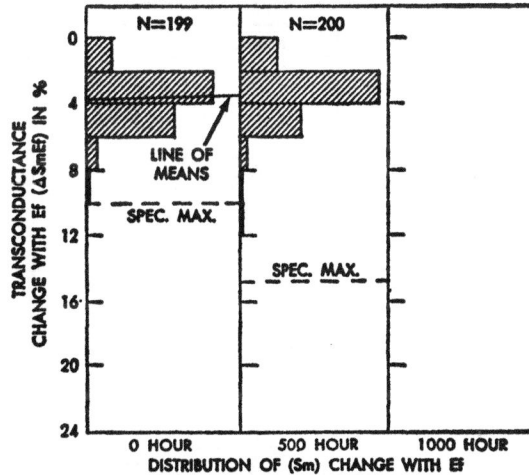

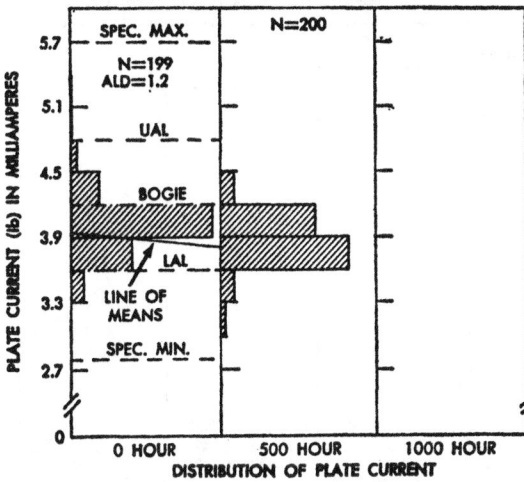

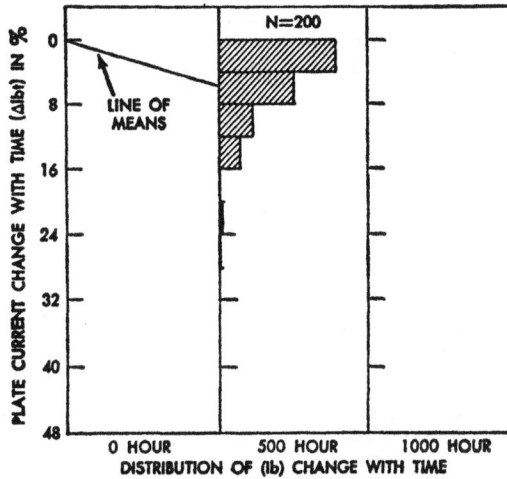

127

LIFE TEST PROPERTY BEHAVIOR
MIL-E-1/84C   25 JULY '56
PRODUCED IN 1956 BY ONE MANUFACTURER

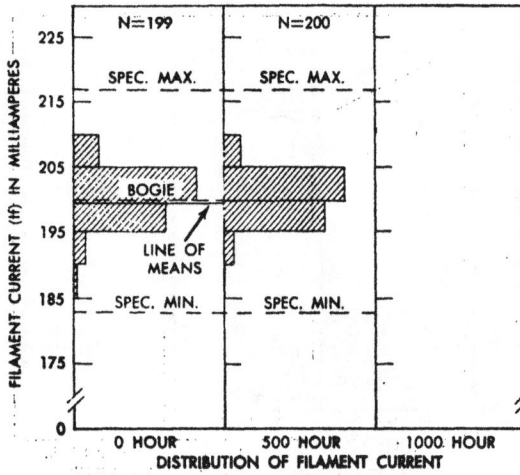

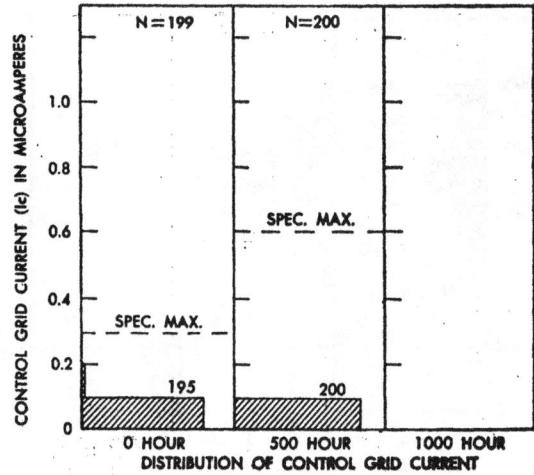

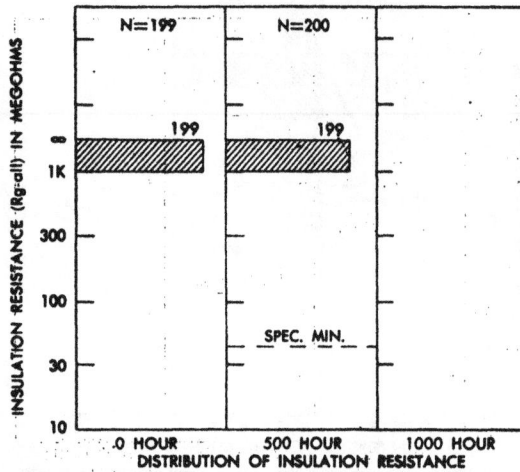

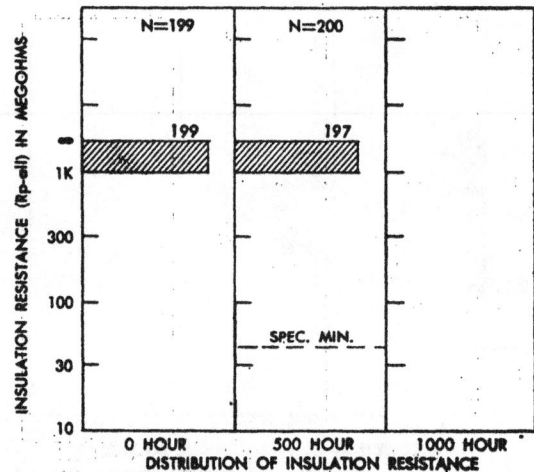

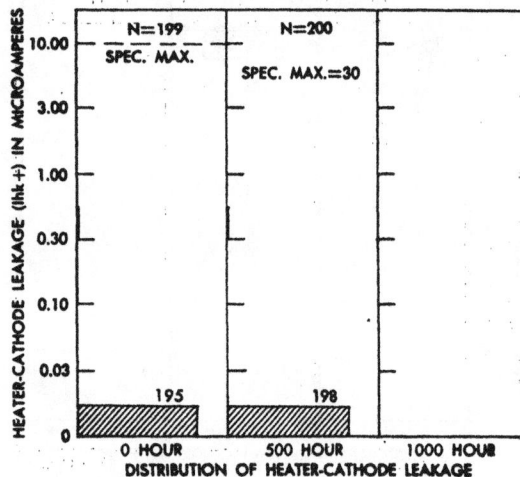

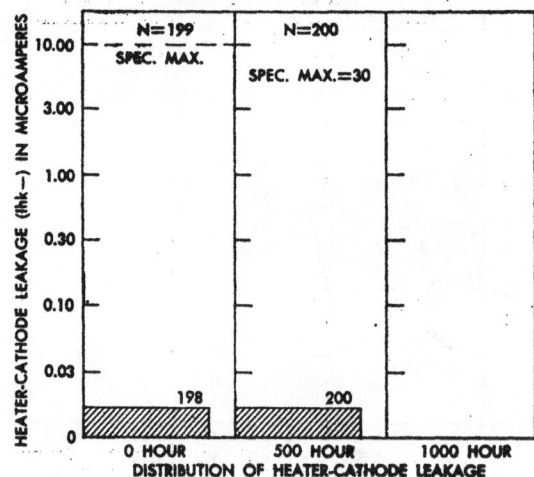

128

LIFE TEST PROPERTY, BEHAVIOR
PROBABILITY OF SURVIVAL
MIL-E-1/84C   25 JULY '56
PRODUCED IN 1956 BY ONE MANUFACTURER

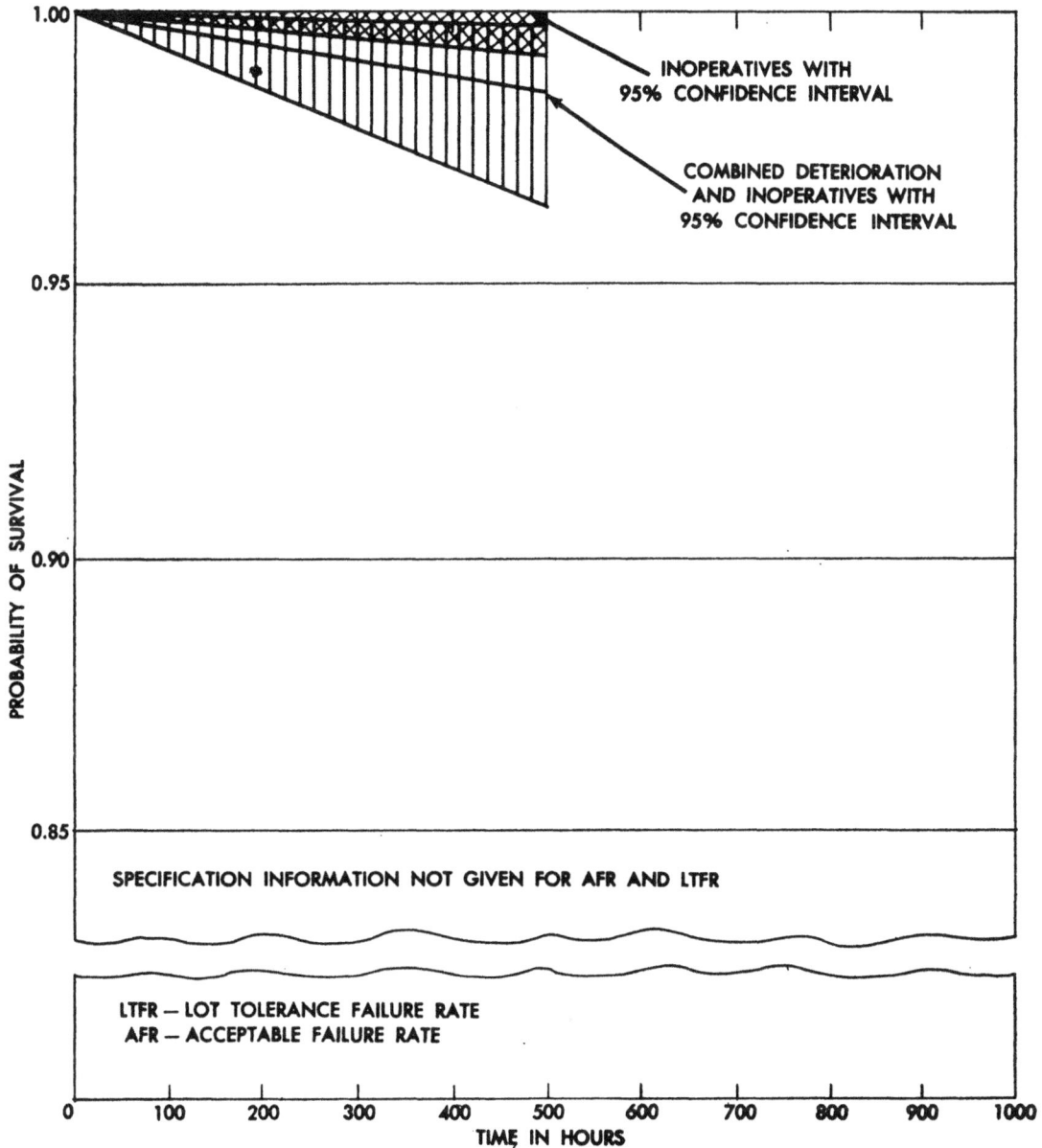

INOPERATIVES WITH
95% CONFIDENCE INTERVAL

COMBINED DETERIORATION
AND INOPERATIVES WITH
95% CONFIDENCE INTERVAL

PROBABILITY OF SURVIVAL

SPECIFICATION INFORMATION NOT GIVEN FOR AFR AND LTFR

LTFR — LOT TOLERANCE FAILURE RATE
AFR — ACCEPTABLE FAILURE RATE

TIME IN HOURS

TUBE TYPE JAN-5744WB

DESCRIPTION:

The JAN-5744WB[1] is a 5 lead flat press subminiature triode having a transconductance ranging from 3400 to 4600 micromhos, and a Mu ranging from 60 to 80. The JAN-5744WB has given satisfactory service in voltage amplifier and other low current applications.

ELECTRICAL: The electrical characteristics are as follows:
Heater Voltage...6.3 V
Heater Current...190-210 mA
Cathode...Coated Unipotential

MOUNTING: Any type mounting is adequate.

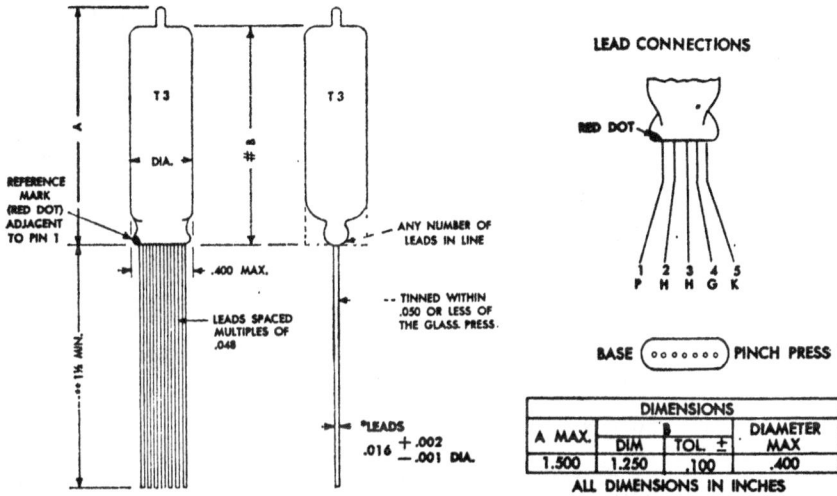

LEAD CONNECTIONS

RED DOT.

1 2 3 4 5
P H G K

BASE PINCH PRESS

DIMENSIONS			
A MAX.	B		DIAMETER MAX
	DIM	TOL ±	
1.500	1.250	.100	.400

ALL DIMENSIONS IN INCHES

REFERENCE MARK (RED DOT) ADJACENT TO PIN 1

.400 MAX.

LEADS SPACED MULTIPLES OF .048

ANY NUMBER OF LEADS IN LINE

-- TINNED WITHIN .050 OR LESS OF THE GLASS PRESS.

*LEADS .016 +.002 −.001 DIA.

MEASURE FROM BASE SEAT TO BULB TOP-LINE AS DETERMINED BY RING GAGE OF .210 ± .001.

* LEAD DIAMETER TOLERANCE SHALL GOVERN BETWEEN .050 FROM THE GLASS TO .250 FROM THE GLASS.

** ALTERNATIVE LEAD LENGTH SHALL BE .200 ± .015 WHEN CUT LEADS ARE REQUIRED BY PROCUREMENT CONTRACT OR TSS. CUT LEADS SHALL BE ESSENTIALLY SQUARE CUT AND THE MAXIMUM BURR SHALL BE .003 INCREASE OVER THE ACTUAL LEAD DIAMETER.

RATINGS:	Ef	Eb	Ec	Ehk	Rk	Rg	Ib**	Ic*	Pp*	T Envelope	Alt
Absolute	V	Vdc	Vdc	v	ohms	Meg	mAdc	mAdc	W	°C	ft
Maximum	6.9	275	---	±200	---	1.2	6.5	1.0	---	220	60,000
Design Maximum	---	---	---	---	---	---	---	---	1.3	---	---
Minimum	5.7	---	-55	---	---	---	0.5	---	---	---	---
Test Cond:	6.3	250	0	0	500	---	---	---	---	---	---

[1] The values and specification comments presented in this section are related to MIL-E-1/1073A dated 4 Dec 1957.

* No test at this rating exists in the specification.

** Difficulty may be encountered if this tube is operated for long periods of time with very small values of cathode current. No specification assurance of life exists under conditions of cathode current approaching the maximum.

ACCEPTANCE TEST LIMITS SUMMARY

PROPERTY		MEASUREMENT CONDITIONS	INITIAL		500 HR LIFE TEST		1000 HR LIFE TEST		UNITS
			MIN	MAX	MIN	MAX	MIN	MAX	
Heater Current	If		190	210	187	217	177	223	mA
Transconductance	Sm		3400	4600	-	-	-	-	umhos
Change in individual	Δ Smt		-	-	-	20	-	30	%
Change in average	Avg Δ Smt		-	-	-	15	-	-	%
Transconductance change with Ef	Δ SmEf	Ef=5.7V	-	5	-	15	-	-	%
Amplification Factor	Mu		60	80	-	-	-	-	-
AC Amplification	Ep	Esig=0.2Vac; Ebb=100Vdc; Ecc=0;Rg1=10 Meg;Rp=0.5Meg; Rk=0	6.5	-	-	-	-	-	Vac
Plate Current (1)	Ib		3.2	5.2	-	-	-	-	mAdc
Plate Current (2)	Ib	Ec1=-6.5Vdc	-	50	-	-	-	-	uAdc
Plate Current (3)	Ib	Ec1=-4.0Vdc	5	-	-	-	-	-	uAdc
Capacitance		0.405 in. dia. shield, Ef=0							
	C gp		0.65	0.95	-	-	-	-	uuf
	C in		2.0	3.4	-	-	-	-	uuf
	C out		1.6	3.0	-	-	-	-	uuf
Control Grid Current	Ic		-	-0.3	-	-0.6	-	-1.0	uAdc
Control Grid Emission	Ic	Ef=7.5V;Ec1= -10Vdc. Note	-	-0.4	-	-	-	-	uAdc
Heater Cathode Leakage	Ihk	Ehk=+100Vdc	-	5	-	10	-	15	uAdc
	Ihk	Ehk=-100Vdc	-	5	-	10	-	15	uAdc
Insulation of Electrodes	R(g1-all)	Eg1-all=-100V	100	-	50	-	-	-	Meg
	R(p-all)	Ep-all=-300V	100	-	50	-	-	-	Meg

Measurement conditions are the same as stated under Test Conditions, unless otherwise indicated.

Note: The tube shall be preheated a minimum of five minutes at test conditions for this test (except Ec1=0; Rg=1.0 Meg) prior to this test.

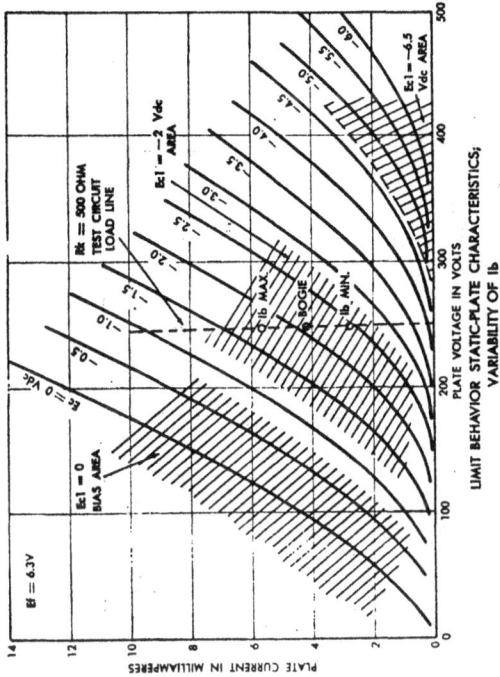

LIMIT BEHAVIOR STATIC-PLATE CHARACTERISTICS;
VARIABILITY OF Ib

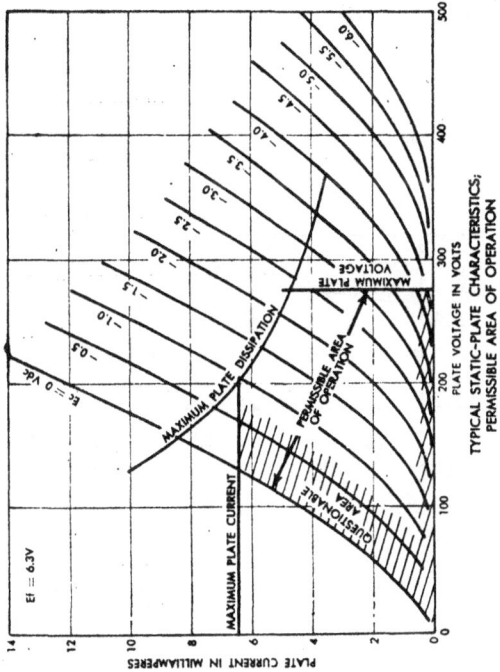

TYPICAL STATIC-PLATE CHARACTERISTICS;
PERMISSIBLE AREA OF OPERATION

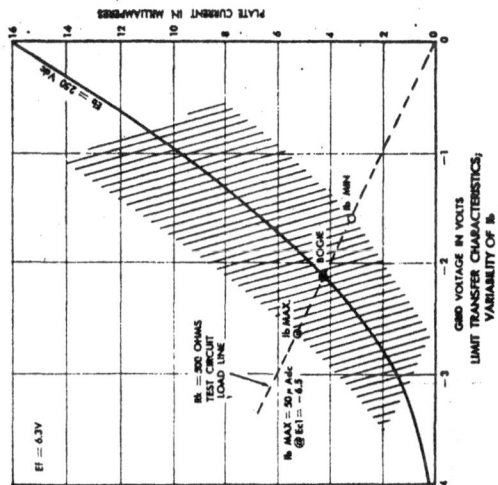

LIMIT TRANSFER CHARACTERISTICS;
VARIABILITY OF Ib

133

DESIGN CENTER CHARACTERISTICS
OBTAINED FROM DATA PUBLISHED BY ORIGINAL
RETMA REGISTRANT

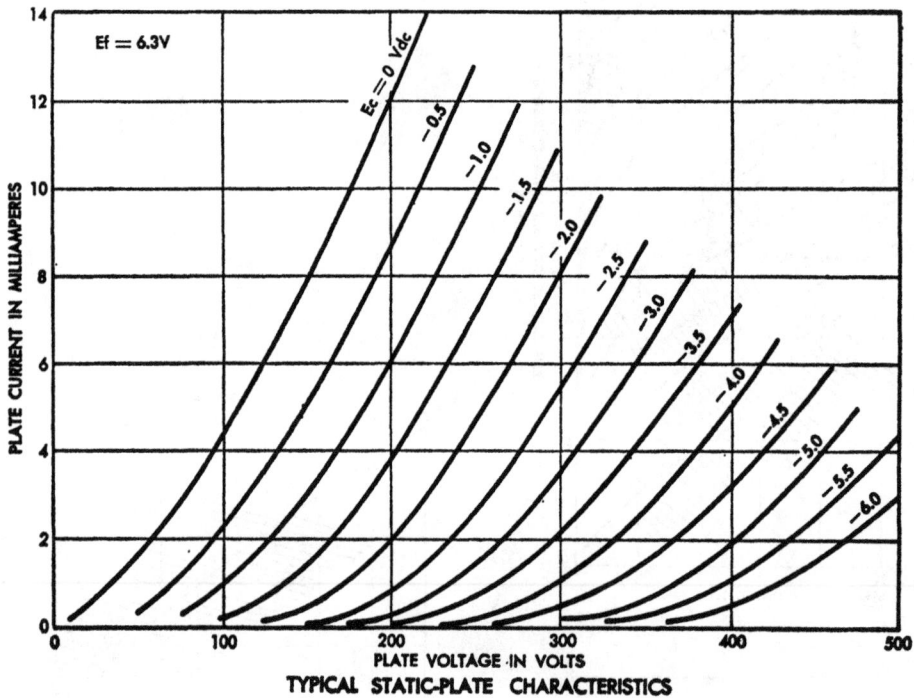

TYPICAL STATIC-PLATE CHARACTERISTICS

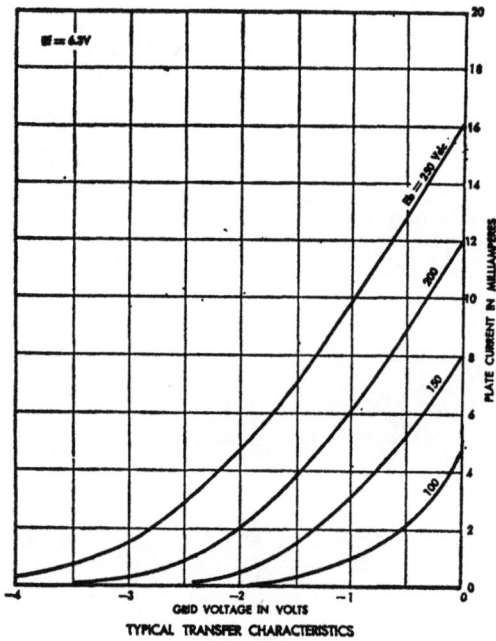

TYPICAL TRANSFER CHARACTERISTICS

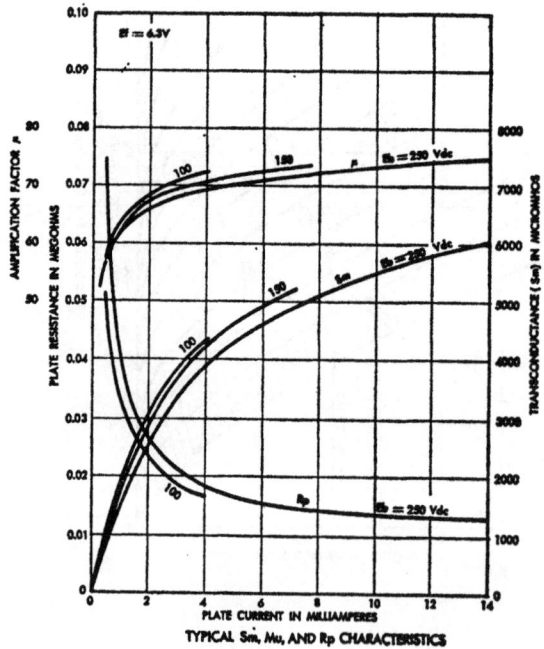

TYPICAL Sm, Mu, AND Rp CHARACTERISTICS

LIFE TEST PROPERTY BEHAVIOR
MIL-E-1/1073A   4 DEC. '57
PRODUCED IN 1958-'59 BY ONE MANUFACTURER

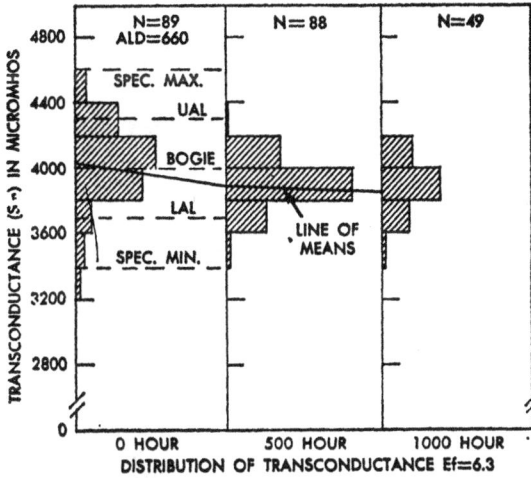

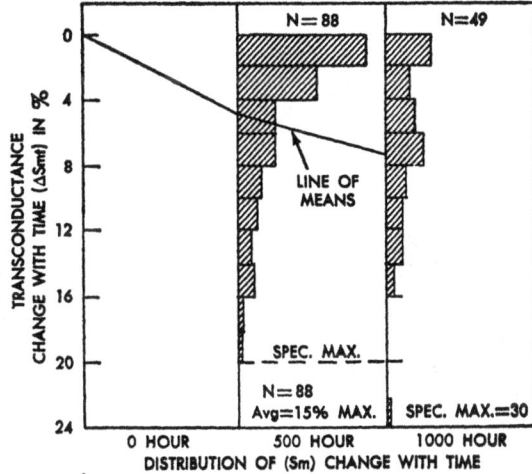

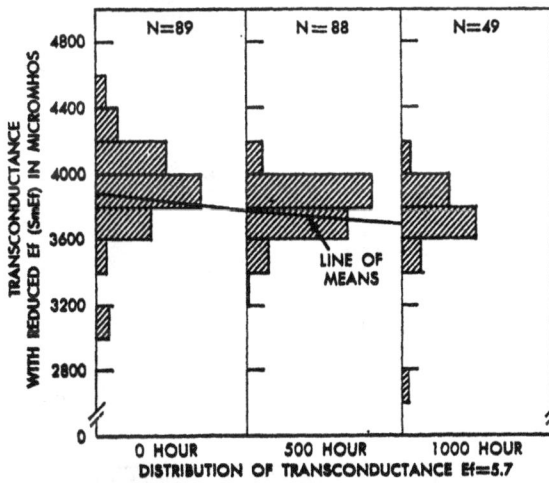

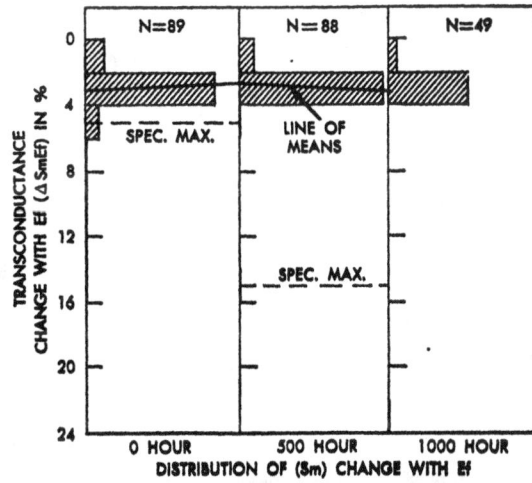

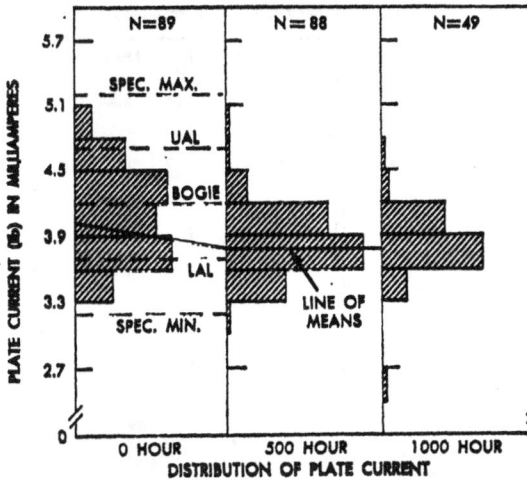

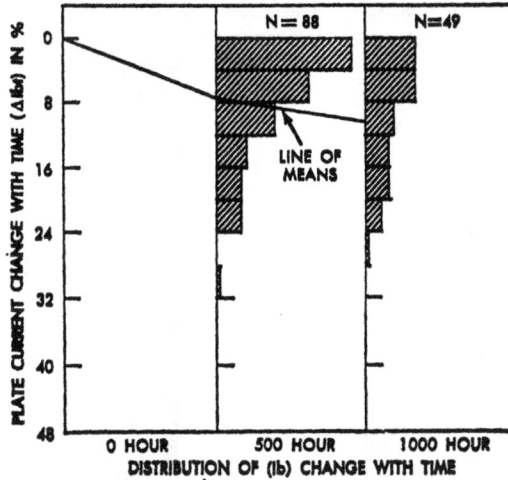

LIFE TEST PROPERTY BEHAVIOR
MIL-E-1/1073A   4 DEC. '57
PRODUCED IN 1957-'59 BY ONE MANUFACTURER

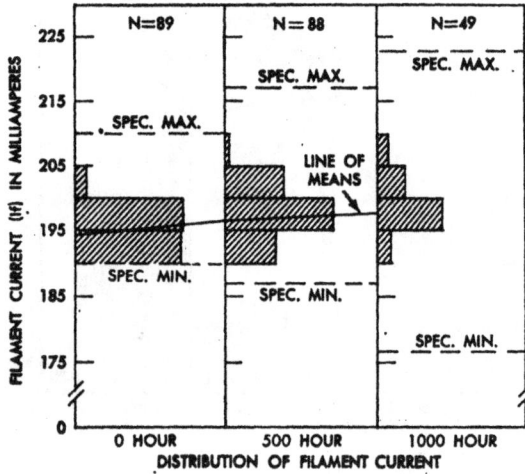

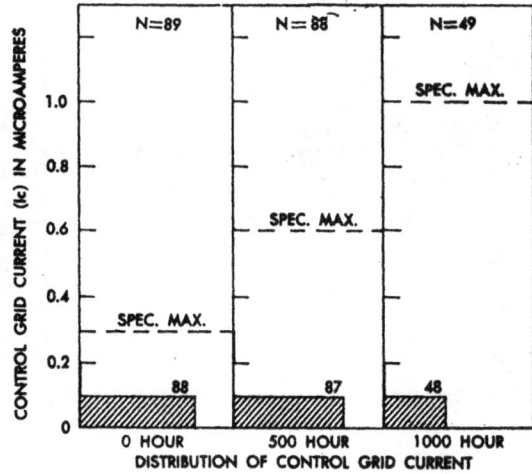

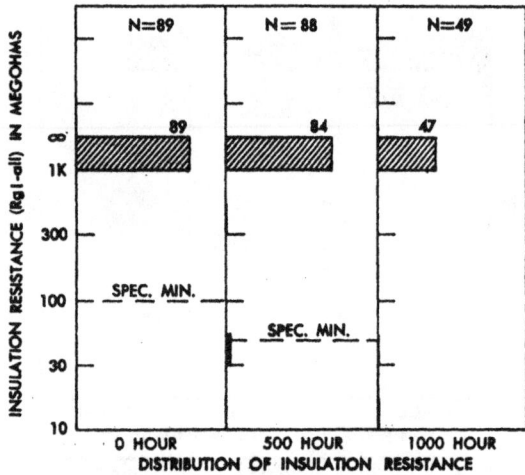

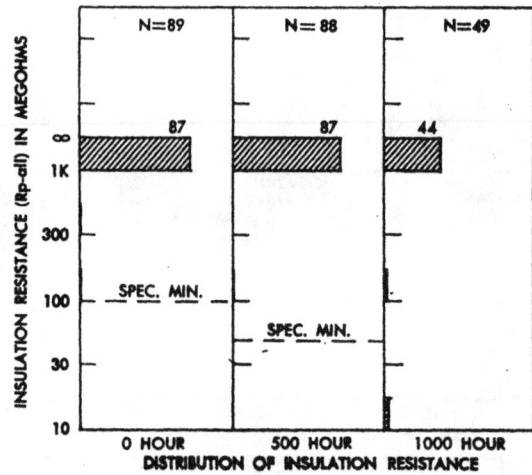

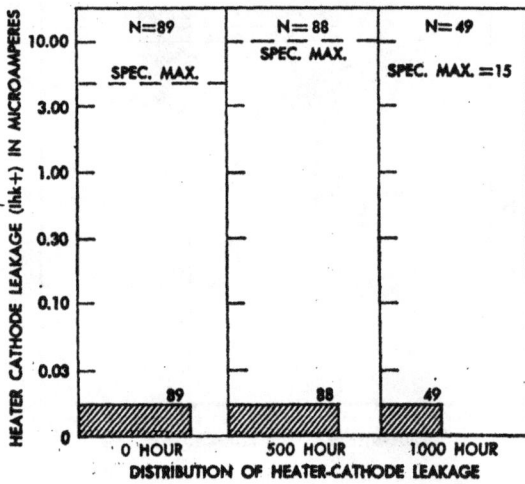

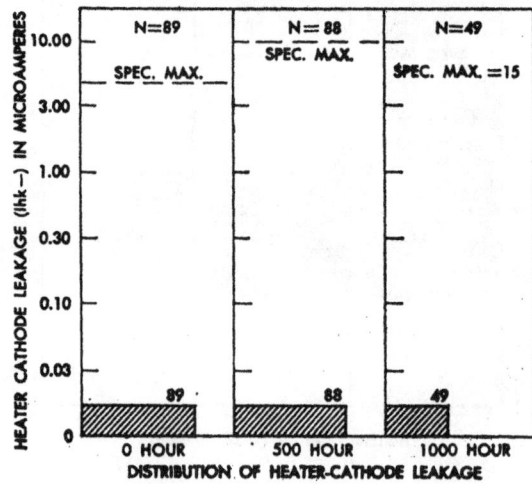

TUBE TYPE JAN-5755

DESCRIPTION:

   The JAN-5755[1] is a 9 pin miniature, twin triode having a Mu in the range
of 63 to 77 and transconductance in the range of 1100 to 2000 micromhos. Each
triode is electrically independent, although the two heaters have a common
connection. A "mechanical stability" test for balance exists in which the tube
is tapped at shock levels of 400 and 600 G.

ELECTRICAL: The electrical characteristics are as follows:

	Series	Parallel
Heater Voltage	12.6 V	6.3 V
Heater Current	165-195 mA	330-390 mA
Cathode	Coated	Unipotential

MOUNTING: Any type mounting is adequate.

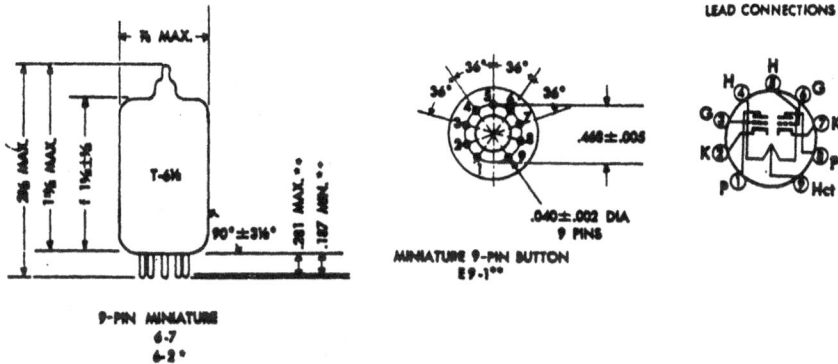

*REFERS TO JETEC PUBLICATION JO-G2-2, MARCH 1955 SUPERSEDED BY JO-G2-2, MARCH 1958

**REFERS TO JETEC PUBLICATION JO-G3-1, MARCH 1955 SUPERSEDED BY JO-G3-2, MARCH 1958

f MEASURE FROM BASE SEAT TO BULB TOP-LINE AS DETERMINED BY RING GAGE OF 7/16 I.D.

ALL DIMENSIONS IN INCHES

RATINGS:	Ef	Eb	Ec	Ehk	Pp/p*	Alt*
Absolute	V	Vdc	Vdc	v	W	ft
Maximum	6.3-12.6 ± 10%	250	---	75	1.0	10,000
Test Cond:	6.3	180	0	---	---	---

[1] The values and specification comments presented in this section are
     related to MIL-E-1/167 dated 20 May 1953.

*   No test at this rating exists in the specification.

    Difficulty may be encountered if this tube is operated for long periods
    of time with very small values of cathode current. No specification
    assurance of life exists under conditions of cathode current approaching
    the maximum.

137

ACCEPTANCE TEST LIMITS SUMMARY

PROPERTY		MEASUREMENT CONDITIONS	INITIAL		500 HR LIFE TEST		1000 HR LIFE TEST		UNITS
			MIN	MAX	MIN	MAX	MIN	MAX	
Heater Current	If		330	390	-	-	-	-	mA
Transconductance	Sm		1100	2000	1000	-	-	-	umhos
Transconductance Change with Ef	Δ SmEf	Ef=5.7V	-	15	-	-	-	-	%
Amplification Factor	Mu	Eb=110Vdc;Ec/ Ib=0.150mAdc	63	77	-	-	-	-	-
Plate Current	Ib		1.3	2.9	-	-	-	-	mAdc
Control Grid Current	Ic	Eb=110Vdc;Ec/ Ib=0.150mAdc	-	-.001 +200	-	-	-	-	uAdc uuAdc
Heater Cathode Leakage	Ihk Ihk	Ehk=+100Vdc Ehk=-100Vdc	- -	20 20	- -	- -	- -	- -	uAdc uAdc
*Drift Test	Δ Ec		-	5	-	-	-	-	mVdc
*Electrical Stability	Δ Ec		-	2	-	-	-	-	mVdc
*Mechanical Stability	Δ Ec		-	25	-	-	-	-	mVdc
*Balance	Δ E1cE2c		-	±0.3	-	-	-	-	Vdc

Measurement conditions are the same as stated under Test Conditions, unless otherwise indicated.

* See specification for additional information.

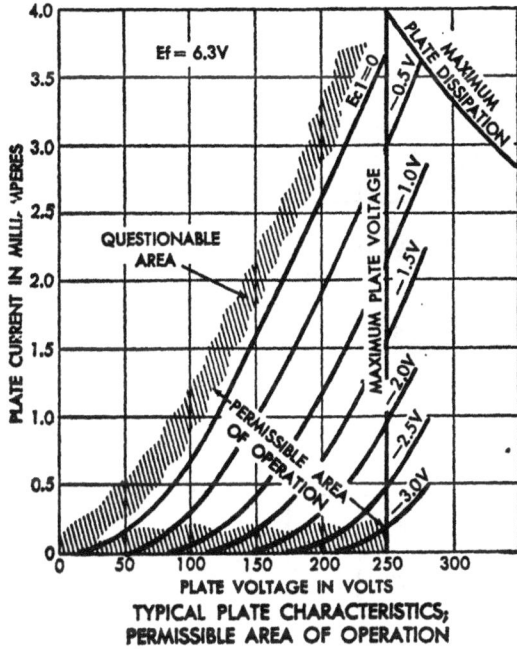

TYPICAL PLATE CHARACTERISTICS;
PERMISSIBLE AREA OF OPERATION

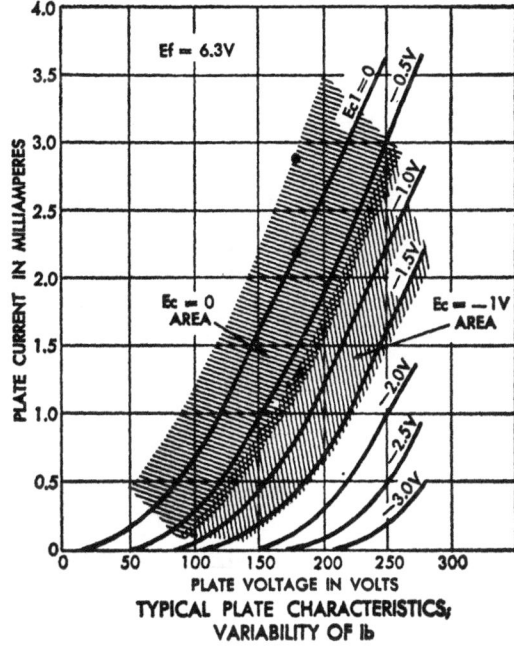

TYPICAL PLATE CHARACTERISTICS;
VARIABILITY OF Ib

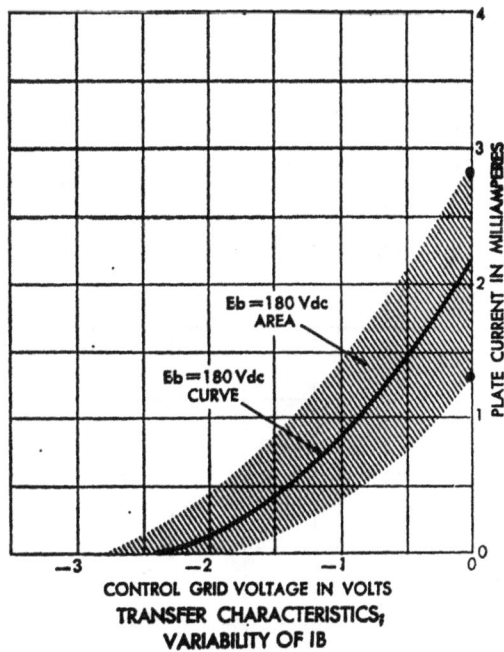

TRANSFER CHARACTERISTICS;
VARIABILITY OF IB

139

**DESIGN CENTER CHARACTERISTICS
OBTAINED FROM DATA PUBLISHED BY ORIGINAL
RETMA REGISTRANT**

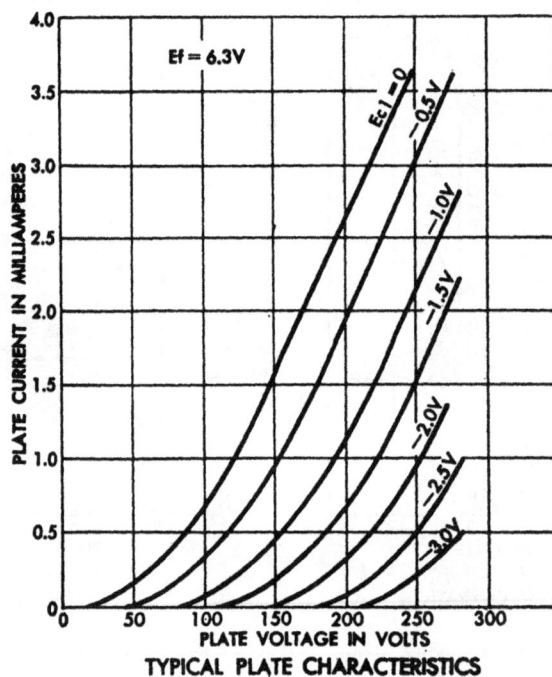

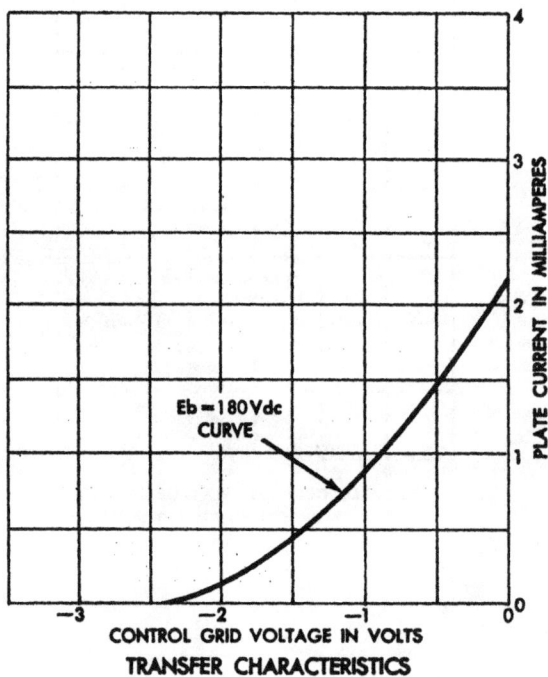

TYPICAL PLATE CHARACTERISTICS

TRANSFER CHARACTERISTICS

LIFE TEST PROPERTY BEHAVIOR
MIL-E-1/167   20 MAY '53
PRODUCED IN 1958 BY ONE MANUFACTURER

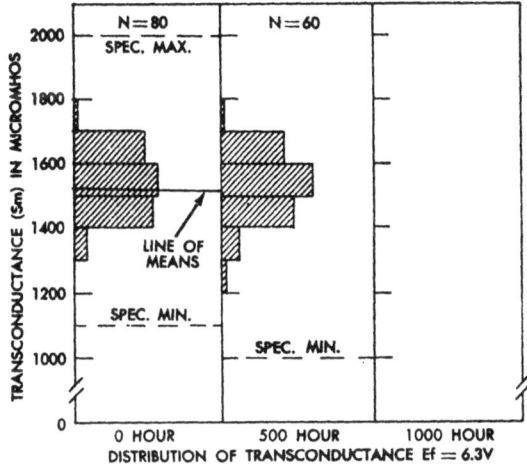

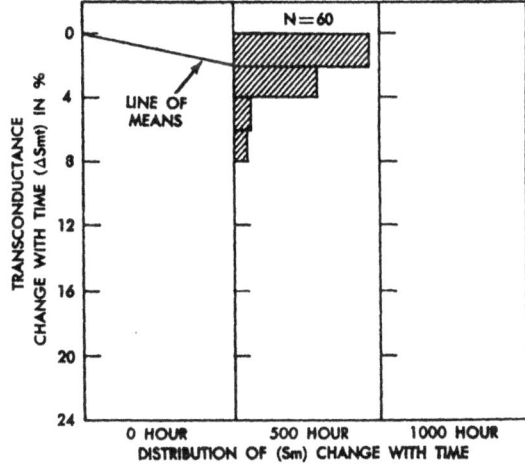

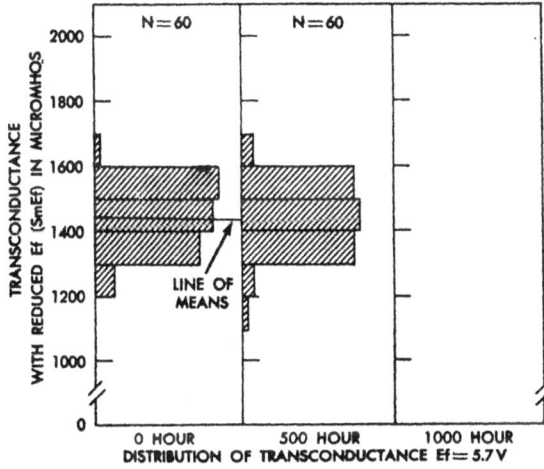

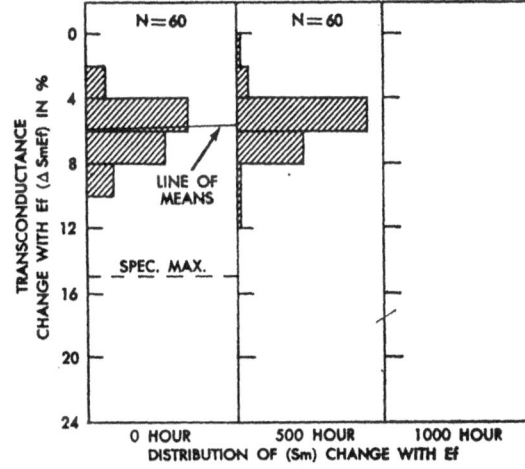

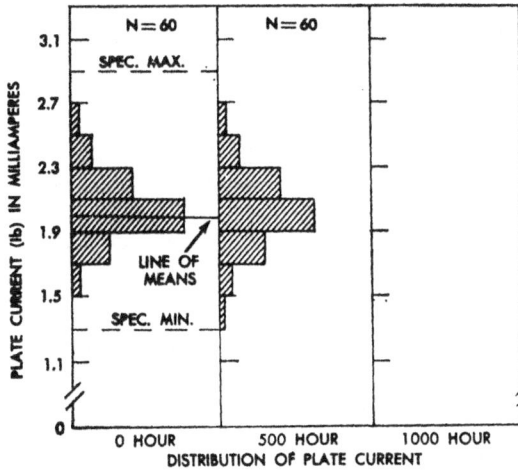

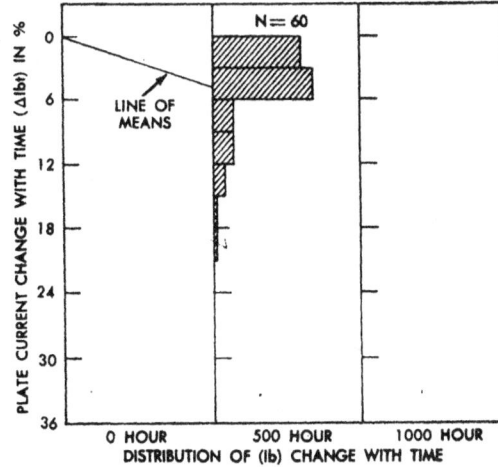

141

TUBE TYPE JAN-5784WA

DESCRIPTION:

The JAN-5784WA[1] is a flat press, seven lead subminiature, dual control, pentode having a transconductance in the range 2650 to 3950 micromhos.

ELECTRICAL: The electrical characteristics are as follows:
Heater Voltage..6.3 V
Heater Current...183-217 mA
Cathode..Coated Unipotential

MOUNTING: Any type mounting is adequate.

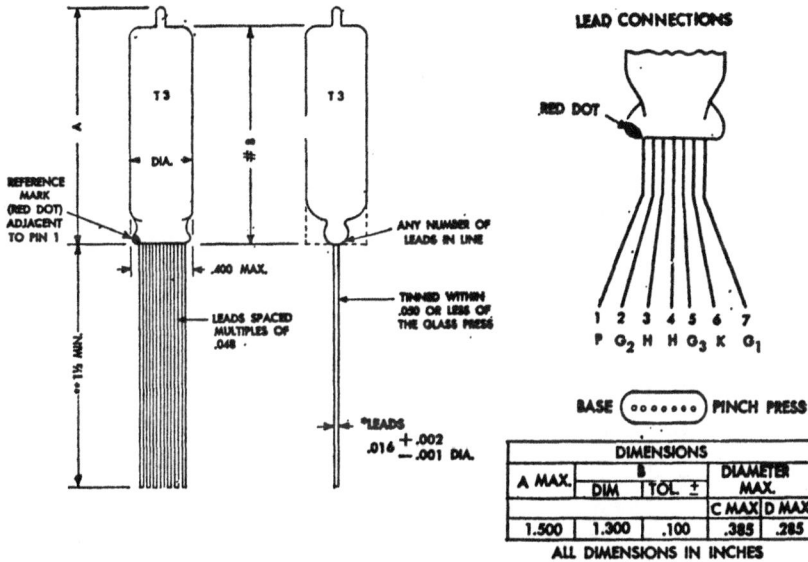

MEASURE FROM BASE SEAT TO BULB TOP-LINE AS DETERMINED BY RING GAGE OF .210 ± .001.

* LEAD DIAMETER TOLERANCE SHALL GOVERN BETWEEN .050 FROM THE GLASS TO .250 FROM THE GLASS.

** ALTERNATIVE LEAD LENGTH SHALL BE .200 ± .015 WHEN CUT LEADS ARE REQUIRED BY PROCUREMENT CONTRACT OR TSS. CUT LEADS SHALL BE ESSENTIALLY SQUARE CUT AND THE MAXIMUM BURR SHALL BE .003 INCREASE OVER THE ACTUAL LEAD DIAMETER.

RATINGS:	E_f	E_b	E_{c1}	E_{c2}	E_{c3}	E_{hk}	I_{c1}	R_k	R_{g1}	I_k**	P_p*	P_{g2}*	I_{c3}	T Envelope	Alt
Design	V	Vdc	Vdc	Vdc	Vdc	v	mAdc	ohms	Meg	mAdc	W	W	mAdc	°C	ft
Maximum	6.9	165	0	155		30	200	1.0	---	1.2	16.5	0.79	0.6	0.2	220	60,000
Minimum	5.7	---	-55	---	-55	---	---	---	---	---	---	---	---	---	---
Test Cond:	6.3	120	0	120	0	0	---	230	---	---	---	---	---	---	---

[1] The values and specification comments presented in this section are related to MIL-E-1/88D dated 26 Dec 1956.

* No test at this rating exists in the specification.

** Difficulty may be encountered if this tube is operated for long periods of time with very small values of cathode current. No specification assurance of life exists under conditions of cathode current approaching the maximum.

ACCEPTANCE TEST LIMITS SUMMARY

PROPERTY	MEASUREMENT CONDITIONS	INITIAL		500 HR LIFE TEST		1000 HR LIFE TEST		UNITS
		MIN	MAX	MIN	MAX	MIN	MAX	
Heater Current If		183	217	180	220	177	223	mA
Transconductance (1) Sm		2650	3950	–	–	–	–	umhos
Change in individual Δ Smt		–	–	–	20	–	30	%
Change in average Avg Δ Smt		–	–	–	15	–	–	%
Transconductance change with Ef Δ SmEf	Ef=5.7V	–	15	–	15	–	–	%
Transconductance(3)Sg3-p	Ec3=-1.0Vdc;*	400	1100	–	–	–	–	umhos
Transconductance(4)Sg3-p	Ec3=+22Vdc; *	–	25	–	–	–	–	umhos
Plate Current (1) Ib		3.9	7.1	–	–	–	–	mAdc
Plate Current (2) Ib	Ec3=-10Vdc*	–	200	–	–	–	–	uAdc
Plate Current (3) Ib	Ec3=-6Vdc*	5	–	–	–	–	–	uAdc
Plate Current (4) Ib	Ec1=-9Vdc	–	200	–	–	–	–	uAdc
Plate Current (5) Ib	Ec1=-5Vdc	5	–	–	–	–	–	uAdc
Screen Current Ic2		2.8	5.4	–	–	–	–	mAdc
Capacitance C g1p	0.405 in. dia shield, Ef=0	–	0.030	–	–	–	–	uuf
C in		3.5	5.5	–	–	–	–	uuf
C out		2.8	4.4	–	–	–	–	uuf
Control Grid Current (1) Ic1		–	-0.3	0	-0.9	0	-1.0	uAdc
Control Grid Current (2) Ic1	Ef=7.0V Note	0	-0.5	–	–	–	–	uAdc
Heater Cathode Leakage Ihk	Ehk=+100Vdc	–	5	–	10	–	15	uAdc
Ihk	Ehk=-100Vdc	–	5	–	10	–	15	uAdc
Insulation of Electrodes R(g1-all)	Eg1-all=-100V	100	–	50	–	–	–	Meg
R(g3-all)	Eg3-all=-100V	100	–	50	–	–	–	Meg
R(p-all)	Ep-all=-300V	100	–	50	–	–	–	Meg

Measurement conditions are the same as stated under Test Conditions, unless otherwise indicated.

Note: The tube shall be preheated a minimum of five minutes at test conditions for this test (except Rg1=1.0 Meg) prior to this test.

*   The reference point for Ec3, on this test, shall be the negative side of the cathode resistor.

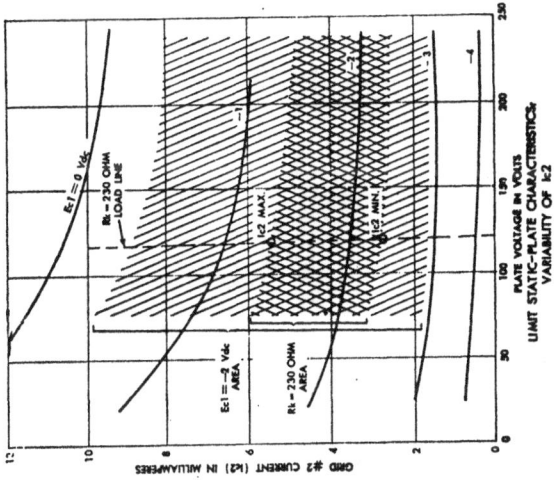

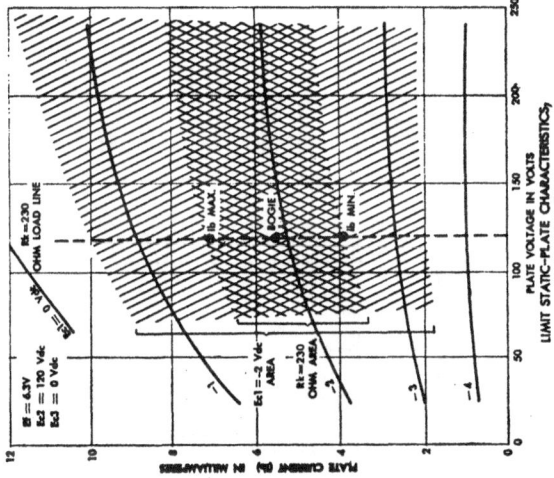

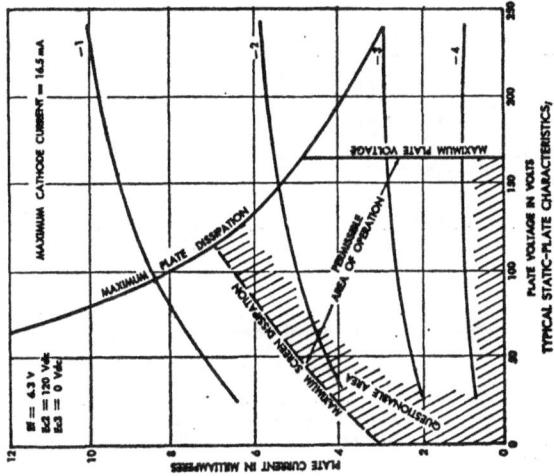

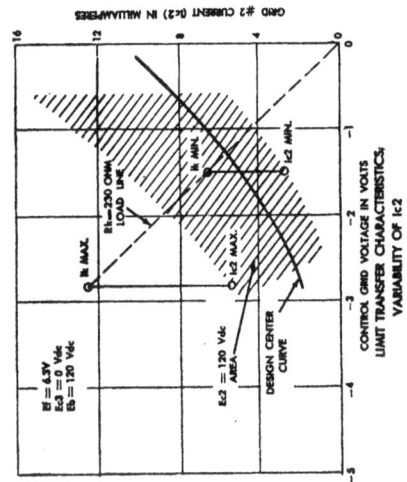

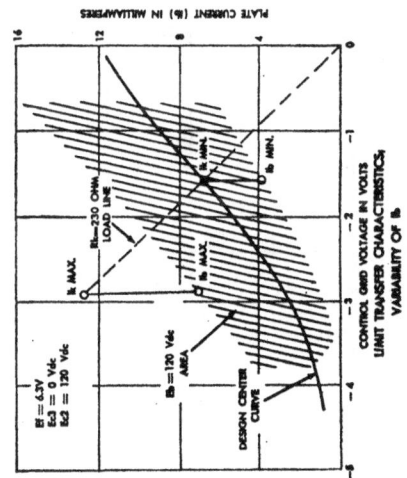

145

DESIGN CENTER CHARACTERISTICS
OBTAINED FROM DATA PUBLISHED BY ORIGINAL
RETMA REGISTRANT

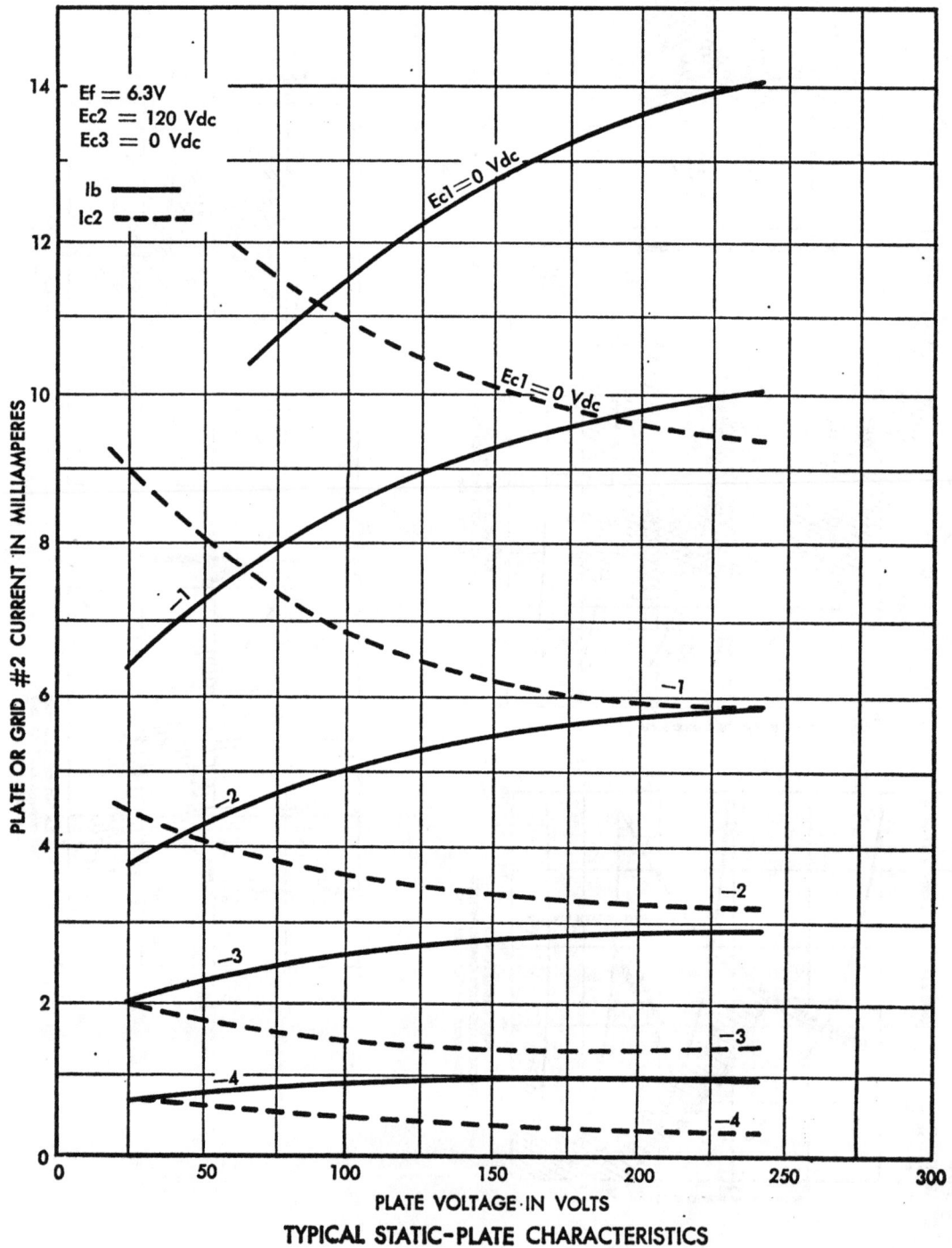

Ef = 6.3V
Ec2 = 120 Vdc
Ec3 = 0 Vdc

Ib ———
Ic2 ‐ ‐ ‐ ‐

Ec1 = 0 Vdc

PLATE OR GRID #2 CURRENT IN MILLIAMPERES

PLATE VOLTAGE IN VOLTS

TYPICAL STATIC-PLATE CHARACTERISTICS

146

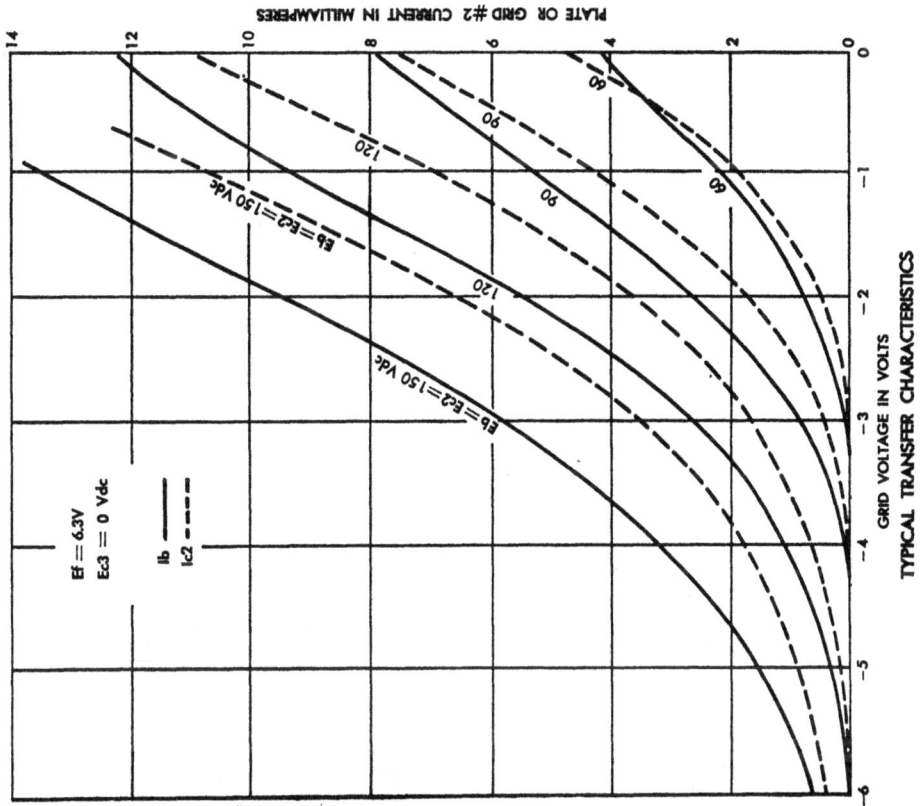

PLATE OR GRID #2 CURRENT IN MILLIAMPERES

GRID VOLTAGE IN VOLTS
TYPICAL TRANSFER CHARACTERISTICS

Ef = 6.3V
Ec3 = 0 Vdc

Ib
Ic2

Eb=Ec2=150 Vdc

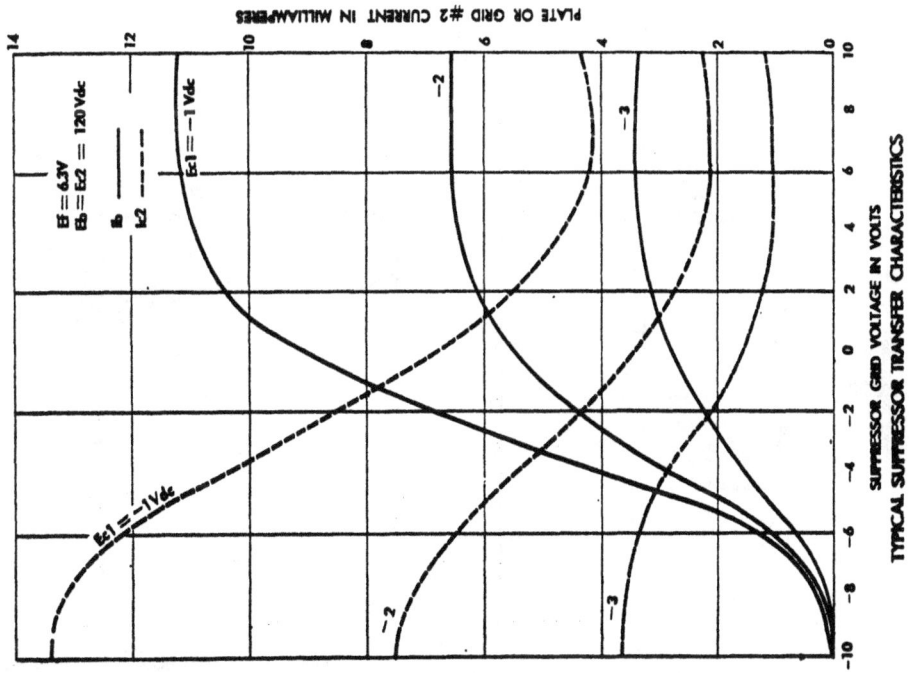

PLATE OR GRID #2 CURRENT IN MILLIAMPERES

SUPPRESSOR GRID VOLTAGE IN VOLTS
TYPICAL SUPPRESSOR TRANSFER CHARACTERISTICS

Ef = 6.3V
Eb = Ec2 = 120 Vdc

Ib
Ic2

Ec1 = −1 Vdc

147

LIFE TEST PROPERTY BEHAVIOR
MIL-E-1/88D   26 DEC '56
PRODUCED IN 1958-'59 BY ONE MANUFACTURER

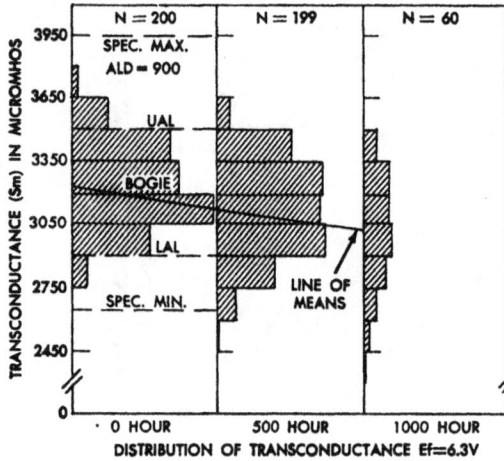

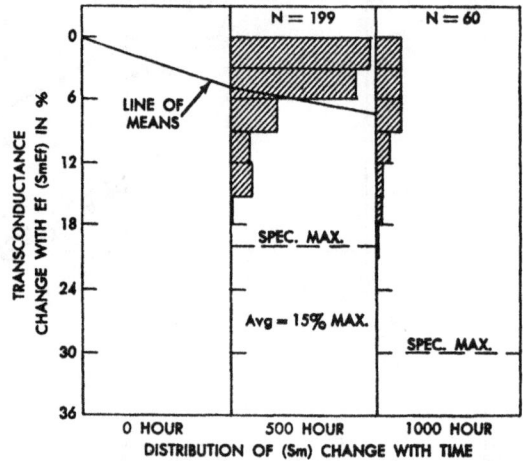

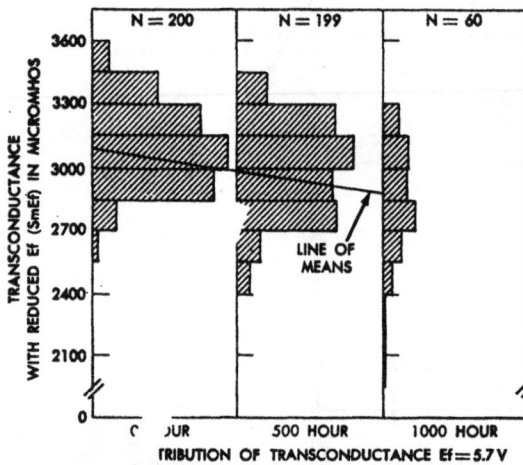

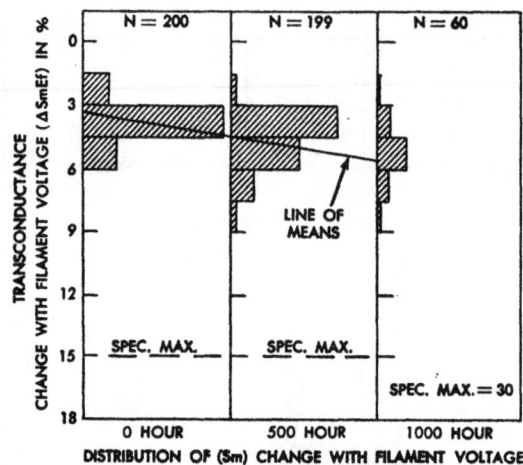

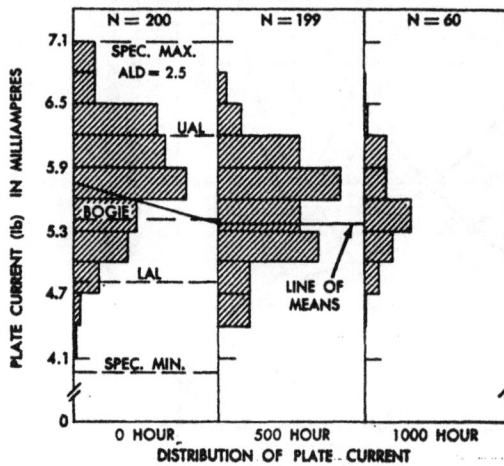

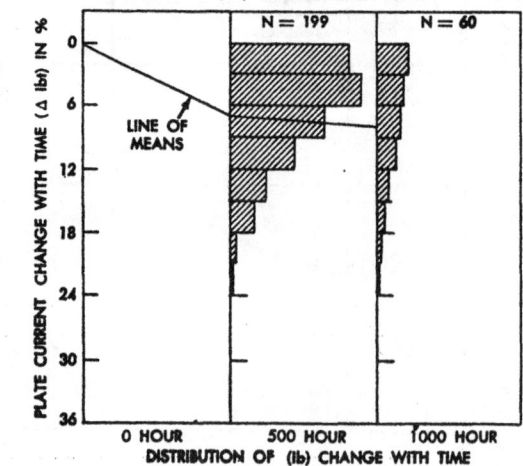

DISTRIBUTION OF TRANSCONDUCTANCE Ef=6.3V

DISTRIBUTION OF (Sm) CHANGE WITH TIME

DISTRIBUTION OF TRANSCONDUCTANCE Ef=5.7 V

DISTRIBUTION OF (Sm) CHANGE WITH FILAMENT VOLTAGE

DISTRIBUTION OF PLATE CURRENT

DISTRIBUTION OF (Ib) CHANGE WITH TIME

148

LIFE TEST PROPERTY BEHAVIOR
MIL-E-1/88D   26 DEC '56
PRODUCED IN 1958-'59 BY ONE MANUFACTURER

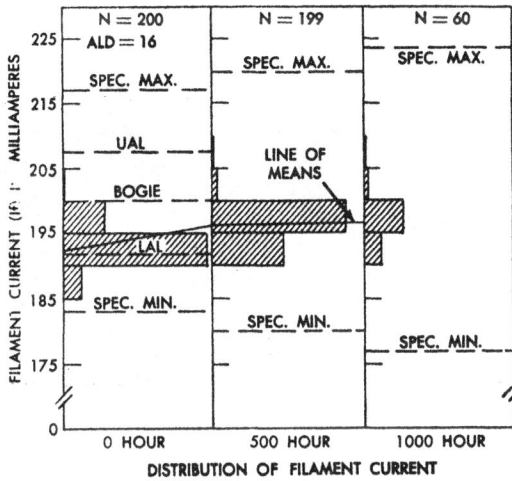

DISTRIBUTION OF FILAMENT CURRENT

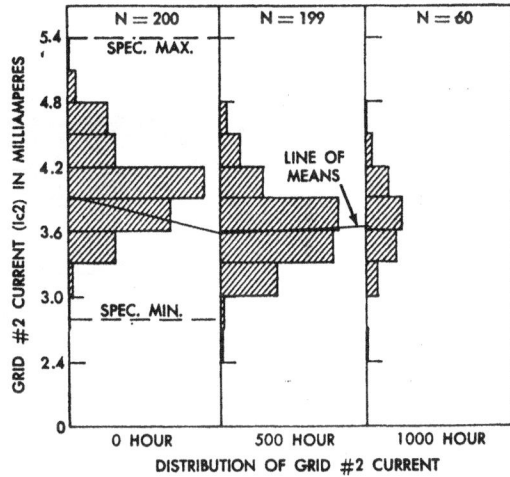

DISTRIBUTION OF GRID #2 CURRENT

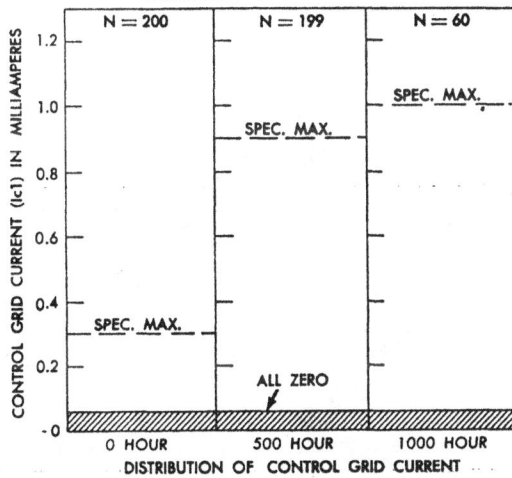

DISTRIBUTION OF CONTROL GRID CURRENT

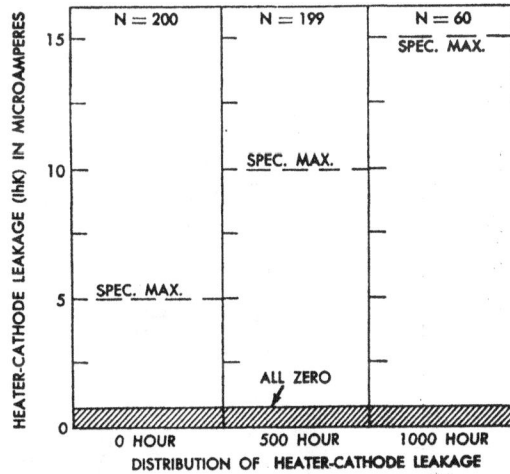

DISTRIBUTION OF HEATER-CATHODE LEAKAGE

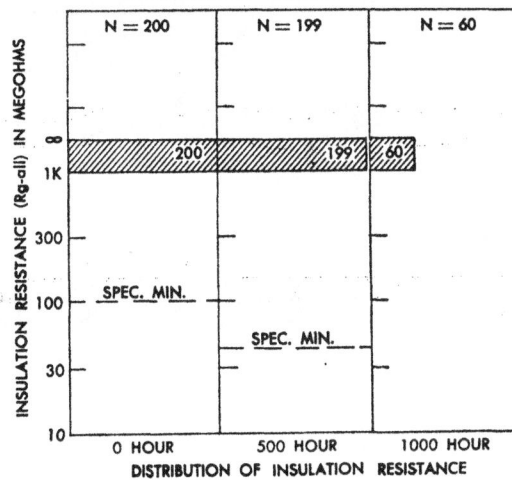

DISTRIBUTION OF INSULATION RESISTANCE

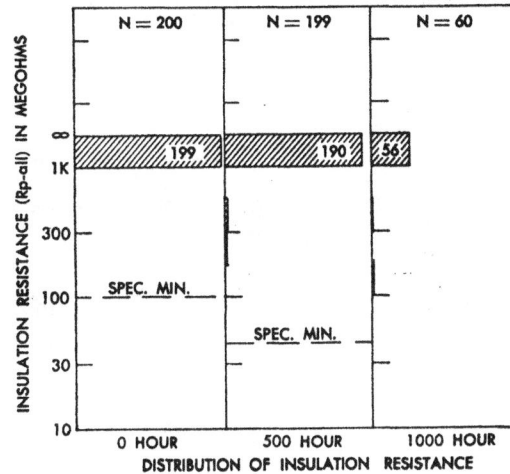

DISTRIBUTION OF INSULATION RESISTANCE

149

LIFE TEST PROPERTY BEHAVIOR
PROBABILITY OF SURVIVAL
MIL-E-1/88D   26 DEC. '56
PRODUCED IN 1958-'59 BY ONE MANUFACTURER

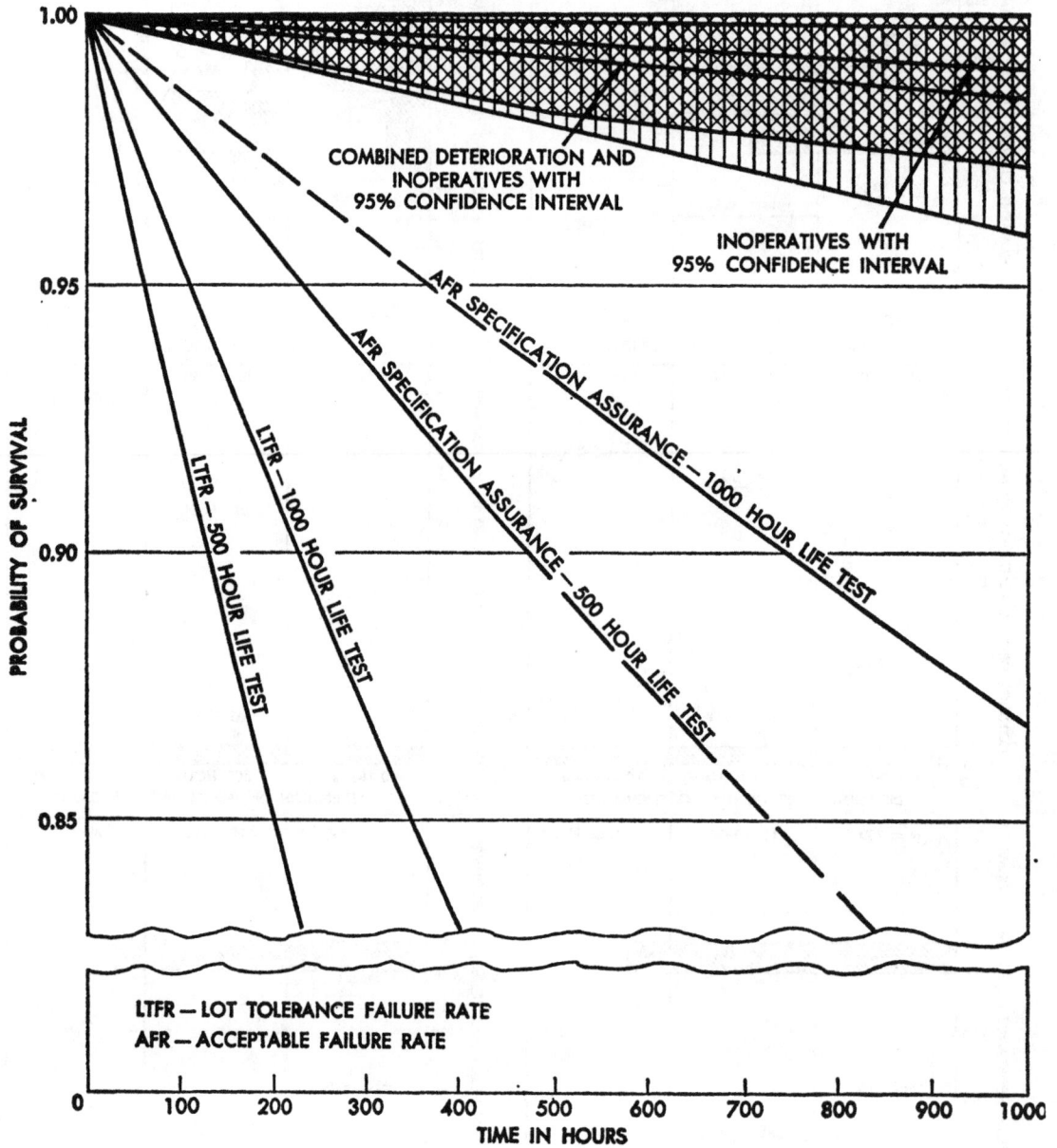

COMBINED DETERIORATION AND
INOPERATIVES WITH
95% CONFIDENCE INTERVAL

INOPERATIVES WITH
95% CONFIDENCE INTERVAL

AFR SPECIFICATION ASSURANCE — 1000 HOUR LIFE TEST

AFR SPECIFICATION ASSURANCE — 500 HOUR LIFE TEST

LTFR — 500 HOUR LIFE TEST

LTFR — 1000 HOUR LIFE TEST

LTFR — LOT TOLERANCE FAILURE RATE

AFR — ACCEPTABLE FAILURE RATE

PROBABILITY OF SURVIVAL

TIME IN HOURS

TUBE TYPE JAN-5814A

DESCRIPTION:

The JAN-5814A[1/] is a 9 pin miniature medium-mu twin triode having separate cathode connections. The heater may be connected for either series or parallel operations. The JAN-5814A has a Mu in the range 15.5 to 18.5 and a transconductance ranging from 1750 to 2650 micromhos.

ELECTRICAL: The electrical characteristics are as follows:
Heater Voltage...12.6 or 6.3 V
Heater Current...175 or 350 mA
Cathode...Coated Unipotential

MOUNTING: Any type mounting is adequate.

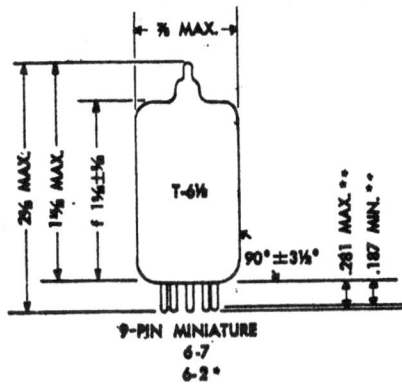

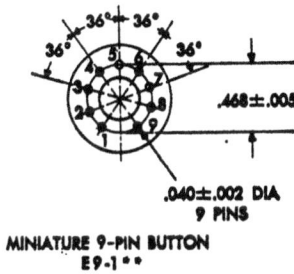

9-PIN MINIATURE
6-7
6-2*

MINIATURE 9-PIN BUTTON
E9-1**

LEAD CONNECTIONS

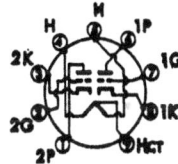

*REFERS TO JETEC PUBLICATION JO-G2-2, MARCH 1955 SUPERSEDED BY JO-G2-2, MARCH 1958

**REFERS TO JETEC PUBLICATION JO-G3-1, MARCH 1955 SUPERSEDED BY JO-G2-2, MARCH 1958

f MEASURE FROM BASE SEAT TO BULB TOP-LINE AS DETERMINED BY RING GAGE OF 7/16 I.D.

ALL DIMENSIONS IN INCHES

RATINGS:	Ef	Eb	Ec	Ehk	Rg/g	Ik/k**	Ic/g*	Pp/p	T Envelope	Alt
Absolute	V	Vdc	Vdc	v	Meg	mAdc	mAdc	W	°C	ft
Maximum	13.9 6.9	330	0	100	0.5	22	5.0	3.0	+165	60,000 Note
Minimum	11.3 5.7	---	-55	---	---	---	---	---	---	---
Test Cond:	12.6	250	-8.5	0	---	---	---	---	---	---

1/ The values and specification comments presented in this section are related to MIL-E-1/12A dated 23 Dec 1955.

* No test at this rating exists in the specification.

** Difficulty may be encountered if this tube is operated for long periods of time with very small values of cathode current. No specification assurance of life exists under conditions of cathode current approaching the maximum.

Note: If altitude rating is exceeded, reduction of instantaneous voltages (Ef excluded) may be required.

ACCEPTANCE TEST LIMIT SUMMARY

PROPERTY		MEASUREMENT CONDITIONS	INITIAL		500 HR LIFE TEST		1000 HR LIFE TEST		UNITS
			MIN	MAX	MIN	MAX	MIN	MAX	
Heater Current	If		160	190	160	193	160	196	mA
Transconductance(1)	Sm		1750	2650	-	-	-	-	umhos
Change in individual	ΔSmt		-	-	-	15	-	15	%
Change in average	Avg ΔSmt		-	-	-	10	-	-	%
Transconductance change with Ef	ΔSmEf	Ef=11.4V	-	15	-	15	-	-	%
Transconductance(3)	Sm	Eb=100Vdc; Ec=0	2500	4000	-	-	-	-	umhos
Amplification Factor	Mu		15.5	18.5	-	-	-	-	-
Plate Current (1)	Ib		6.5	14.5	-	-	-	-	mAdc
Plate Current (2)	Ib	Ec=-30Vdc; Rp=0.1Meg	-	20	-	-	-	-	uAdc
Plate Current (3)	Ib	Ec=-18Vdc	5	-	-	-	-	-	uAdc
Plate Current (1) difference between sections	Ib		-	3.5	-	-	-	-	mAdc
Capacitance		No shield, Ef=0							
C	gp		1.20	1.80	-	-	-	-	uuf
C	in		1.25	1.95	-	-	-	-	uuf
C	out 1		0.30	0.70	-	-	-	-	uuf
C	out 2		0.20	0.60	-	-	-	-	uuf
Control Grid Current	Ic	Rg=0.5Meg	0	-0.5	0	-0.5	0	-0.5	uAdc
Control Grid Emission	Isc	Ef=15.0V Ec=-30Vdc Rg=0.5Meg Note	-	-1.5	-	-	-	-	uAdc
Heater Cathode Leakage	Ihk	Ehk=+100Vdc	-	7	-	7	-	7	uAdc
	Ihk	Ehk=-100Vdc	-	7	-	7	-	7	uAdc
Insulation of Electrodes	R(g1-all)	Eg1-all=-100V	500	-	250	-	-	-	Meg
	R(p-all)	Ep-all=-300V	500	-	250	-	-	-	Meg

Measurement conditions are the same as stated under Test Conditions, unless otherwise indicated.

Note: **The tube shall be preheated a minimum of five minutes at test conditions for this test (except Ef=15 V; Ec=-8.5 Vdc; Eb=200 Vdc; Rg=0.5 Meg) prior to this test.**

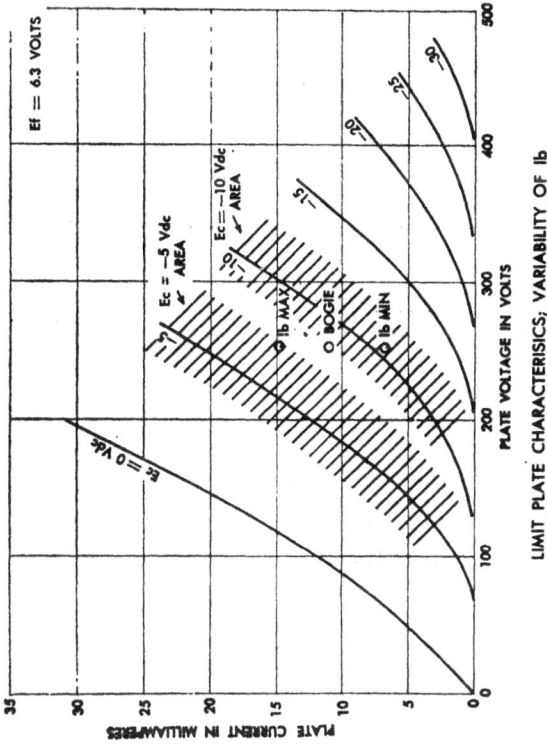

LIMIT PLATE CHARACTERISTICS; VARIABILITY OF Ib

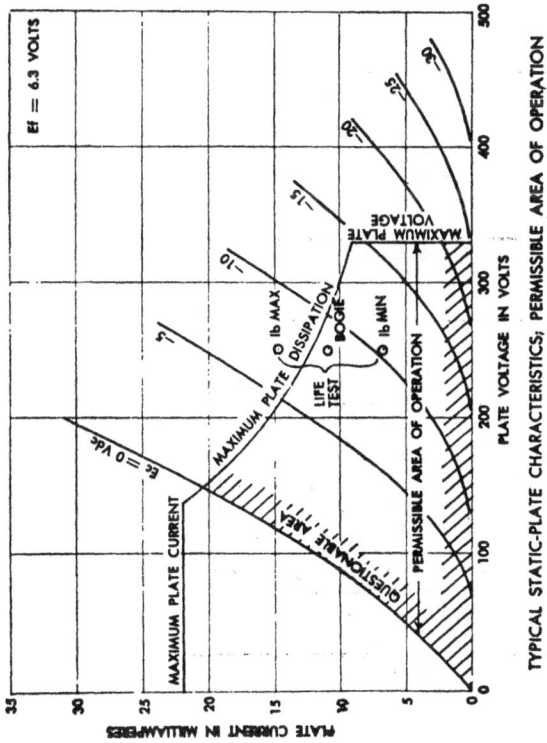

TYPICAL STATIC-PLATE CHARACTERISTICS; PERMISSIBLE AREA OF OPERATION

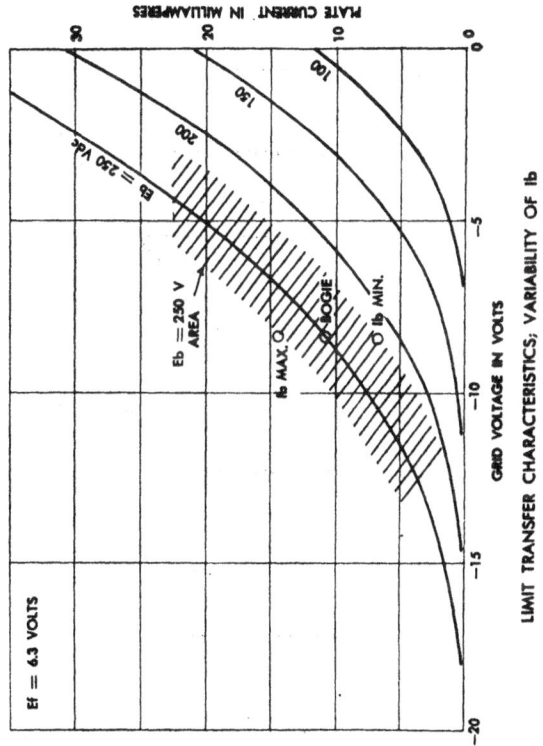

LIMIT TRANSFER CHARACTERISTICS; VARIABILITY OF Ib

153

DESIGN CENTER CHARACTERISTICS
OBTAINED FROM DATA PUBLISHED BY ORIGINAL
RETMA REGISTRANT

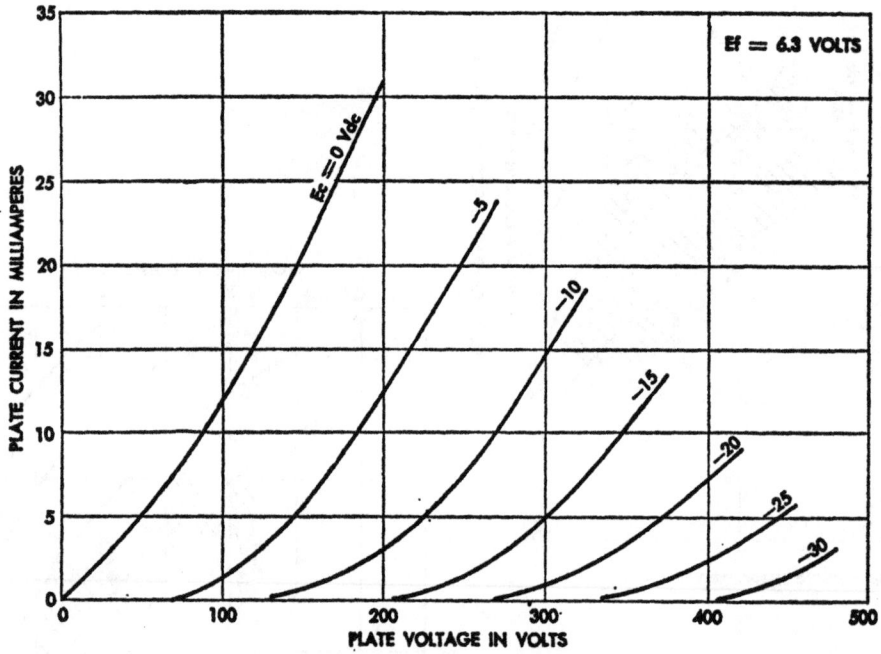

TYPICAL STATIC-PLATE CHARACTERISTICS

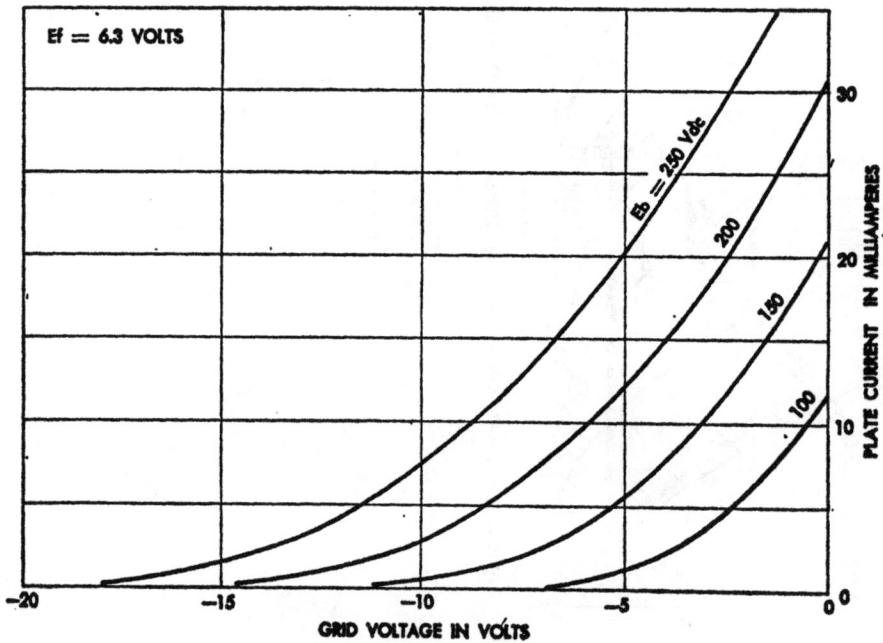

TYPICAL TRANSFER CHARACTERISTICS

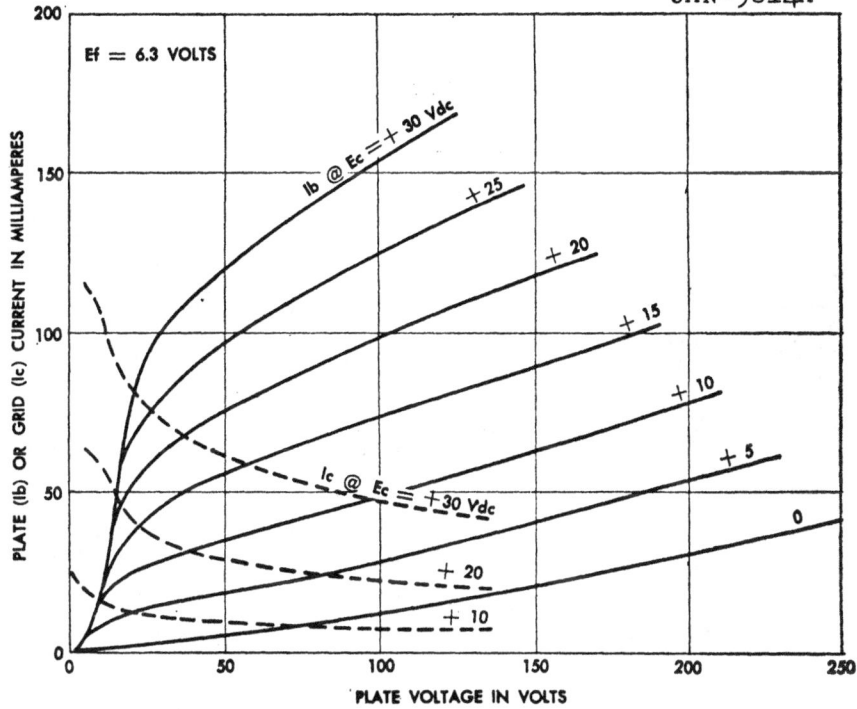

Ef = 6.3 VOLTS

Ib @ Ec = + 30 Vdc

+ 25

+ 20

+ 15

+ 10

+ 5

0

Ic @ Ec = + 30 Vdc

+ 20

+ 10

TYPICAL STATIC-PLATE AND GRID CHARACTERISTICS;
POSITIVE GRID REGION

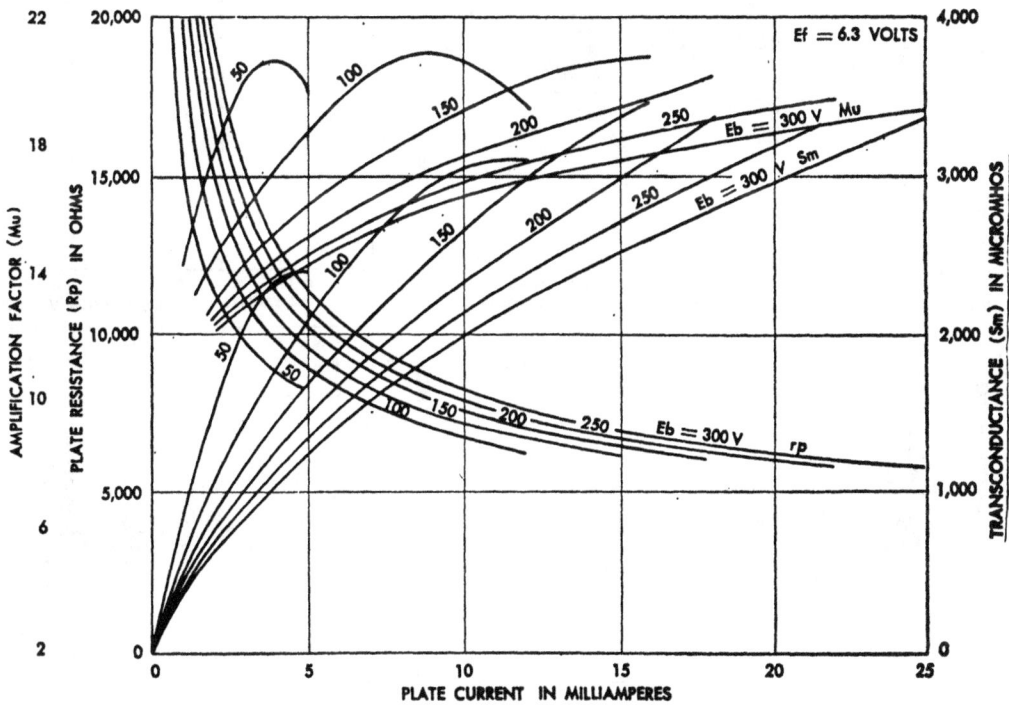

Ef = 6.3 VOLTS

TYPICAL Sm, Mu, AND Rp CHARACTERISTICS

LIFE TEST PROPERTY BEHAVIOR
MIL-E-1/12A   23 DEC. '55
PRODUCED IN 1956-'59 BY FOUR MANUFACTURERS

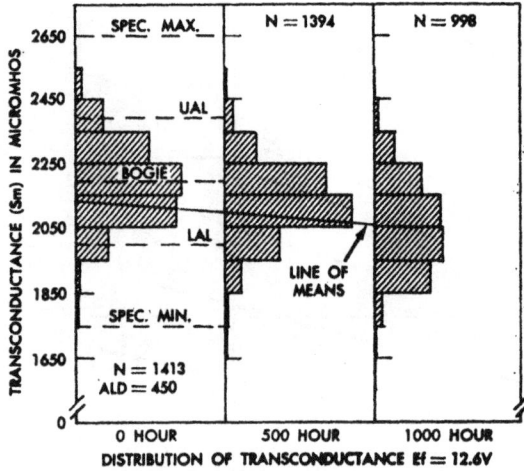

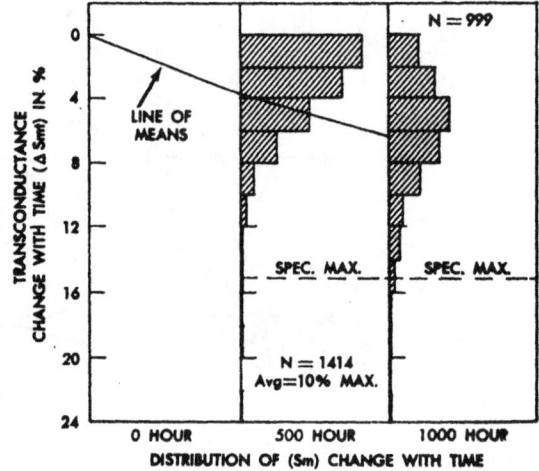

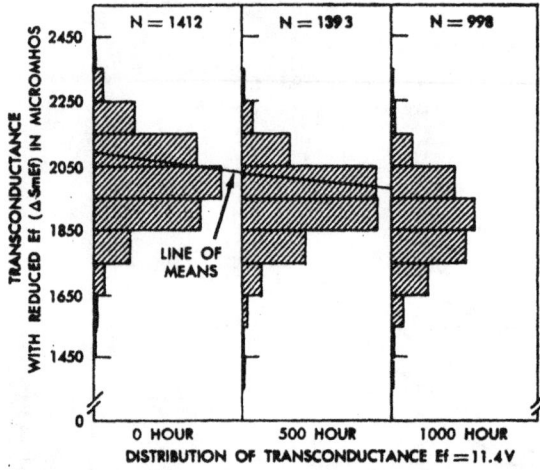

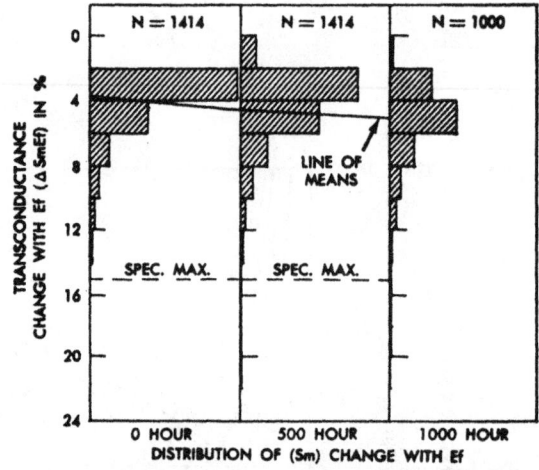

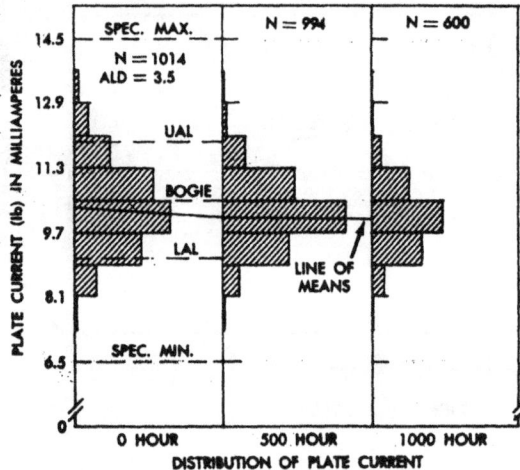

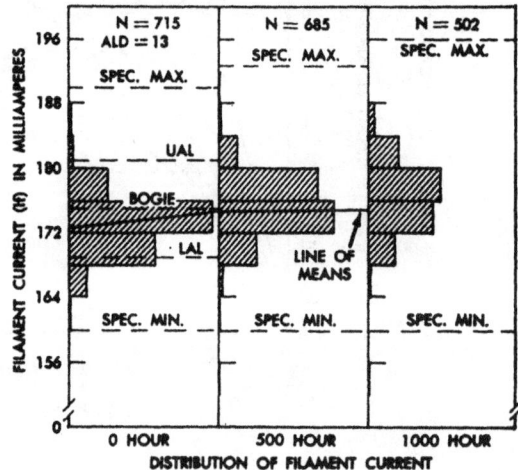

Distribution of Transconductance Ef = 12.6V

Distribution of (Sm) Change with Time

Distribution of Transconductance Ef = 11.4V

Distribution of (Sm) Change with Ef

Distribution of Plate Current

Distribution of Filament Current

LIFE TEST PROPERTY BEHAVIOR
MIL-E-1/12A   23 DEC. '55
PRODUCED IN 1956-'59 BY FOUR MANUFACTURERS

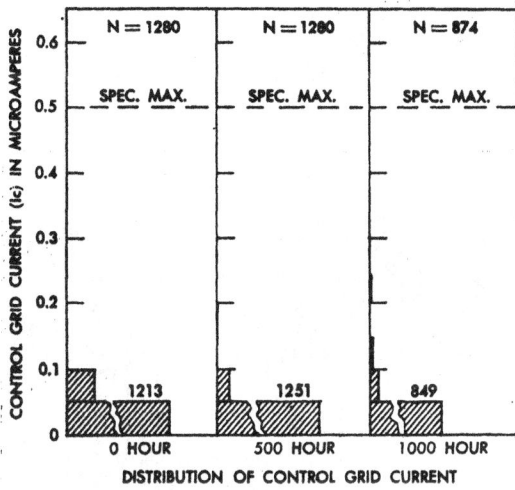

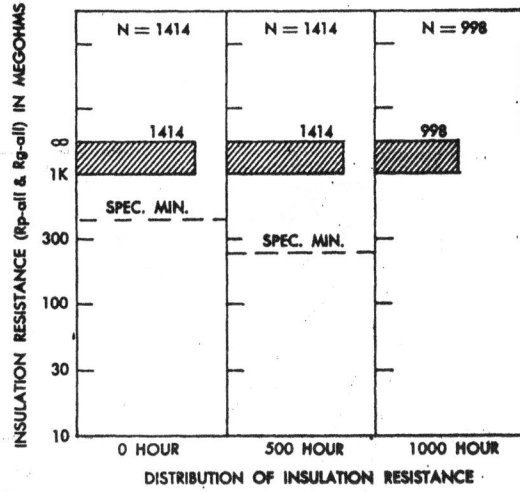

DISTRIBUTION OF CONTROL GRID CURRENT

DISTRIBUTION OF INSULATION RESISTANCE

LIFE TEST PROPERTY BEHAVIOR
PROBABILITY OF SURVIVAL
MIL-E-1/12A   23 DEC. '55
PRODUCED IN 1956-'59 BY FOUR MANUFACTURERS

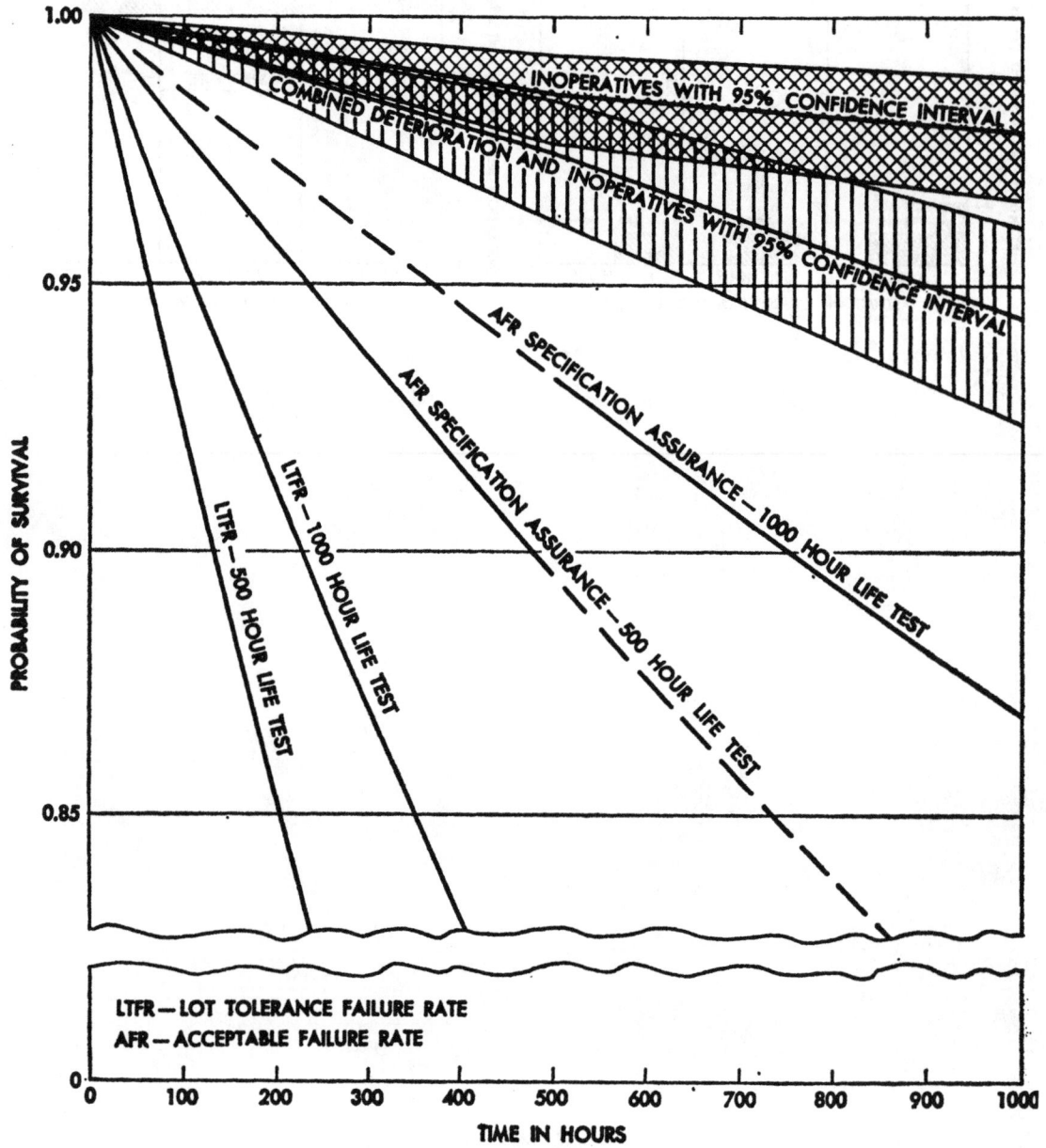

TUBE TYPE JAN-5840

DESCRIPTION:

The JAN-5840[1] is an 8 pin subminiature, sharp cutoff pentode having a transconductance in the range 4200 to 5800. The JAN-5840 is identical to JAN-6205 except grid No. 3 and the cathode are internally connected and is similar in plate characteristics to JAN-5702WA and the miniature type JAN-5654/6AK5W.

ELECTRICAL: The electrical characteristics are as follows:
Heater Voltage..6.3 V
Heater Current:...140-160 mA
Cathode...Coated Unipotential

MOUNTING: Any type mounting is adequate.

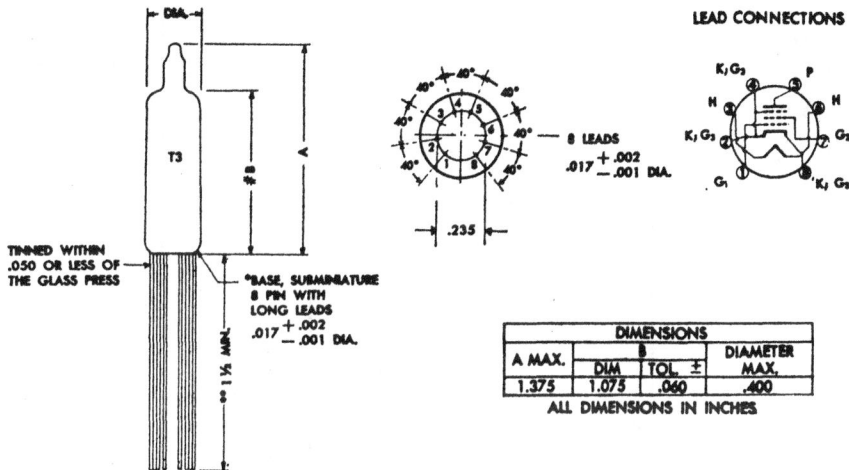

LEAD CONNECTIONS

8 LEADS
.017 +.002 -.001 DIA.

.235

TINNED WITHIN .050 OR LESS OF THE GLASS PRESS

*BASE, SUBMINIATURE 8 PIN WITH LONG LEADS
.017 +.002 -.001 DIA.

DIMENSIONS			
A MAX.	B		DIAMETER MAX.
	DIM	TOL ±	
1.375	1.075	.060	.400

ALL DIMENSIONS IN INCHES

\# MEASURE FROM BASE SEAT TO BULB TOP-LINE AS DETERMINED BY RING GAGE OF .210 ± .001.

* LEAD DIAMETER TOLERANCE SHALL GOVERN BETWEEN .050 FROM THE GLASS TO .250 FROM THE GLASS.

** ALTERNATIVE LEAD LENGTH SHALL BE .200 ± .015 WHEN CUT LEADS ARE REQUIRED BY PROCUREMENT CONTRACT OR TSS. CUT LEADS SHALL BE ESSENTIALLY SQUARE CUT AND THE MAXIMUM BURR SHALL BE .003 INCREASE OVER THE ACTUAL LEAD DIAMETER.

RATINGS:	Ef	Eb	Ec1	Ec2	Ec3	Ehk	Rk	Rg1	Ik*	Pp	Pg2	T Envelope	Alt
Absolute	V	Vdc	Vdc	Vdc	Vdc	v	ohms	Meg	mAdc	W	W	°C	ft
Maximum	6.6	165	0	155	22	200	---	1.1	16.5	0.80	0.35	+220	60,000 Note
Minimum	6.0	---	-55	---	---	---	---	---	---	---	---	---	---
Test Cond:	6.3	100	0	100	0	0	150	---	---	---	---	---	---

1/ The value and specification comments presented in this section are related to MIL-E-1/140B dated 5 Aug 1955.

* Difficulty may be encountered if this tube is operated for long periods of time with very small values of cathode current. No specification assurance of life exists under conditions of cathode current approaching the maximum.

Note: If altitude rating is exceeded, reduction of instantaneous voltages (Ef excluded) may be required.

ACCEPTANCE TEST LIMITS SUMMARY

PROPERTY		MEASUREMENT CONDITIONS	INITIAL		500 HR LIFE TEST		1000 HR LIFE TEST		UNITS
			MIN	MAX	MIN	MAX	MIN	MAX	
Heater Current	If		140	160	138	164	-	-	mA
Transconductance	Sm		4200	5800	-	-	-	-	umhos
Change in individual	ΔSmt		-	-	-	20	-	-	%
Change in average	Avg ΔSmt		-	-	-	15	-	-	%
Transconductance Change with Ef	ΔSmEf	Ef=5.7V	-	10	-	15	-	-	%
Plate Resistance	rp		0.175	-	-	-	-	-	Meg
Plate Current (1)	Ib		5.5	9.5	-	-	-	-	mAdc
Plate Current (2)	Ib	Ec1=-9.0Vdc; Rk=0	-	50	-	-	-	-	uAdc
Screen Grid Current	Ic2		1.5	3.3	-	-	-	-	mAdc
Capacitance		0.405 in. dia. shield, Ef=0							
	C g1p		-	0.015	-	-	-	-	uuf
	C in		3.5	4.9	-	-	-	-	uuf
	C out		2.9	3.9	-	-	-	-	uuf
Control Grid Current	Ic1	Rg1=1.0Meg	0	-0.3	0	-0.8	-	-	uAdc
Control Grid Emission	Ic1	Ef=7.5V;Ec1= -9.0Vdc;Rg1= 1.0Meg;Rk=0 Note	0	-0.5	-	-	-	-	uAdc
Heater Cathode Leakage	Ihk	Ehk=+100Vdc	-	5.0	-	10	-	-	uAdc
	Ihk	Ehk=-100Vdc	-	5.0	-	10	-	-	uAdc
Insulation of Electrodes	R(g1-all)	Eg1-all=-100V	100	-	50	-	-	-	Meg
	R(p-all)	Ep-all=-300V	100	-	50	-	-	-	Meg

Measurement conditions are the same as stated under Test Conditions, unless otherwise indicated.

Note: The tube shall be preheated a minimum of five minutes at test conditions for this test (except Ec1=0; Rk=150 ohms) prior to this test.

TYPICAL STATIC-PLATE CHARACTERISTICS; PERMISSIBLE AREA OF OPERATION

LIMIT BEHAVIOR STATIC-PLATE CHARACTERISTICS, VARIABILITY OF Ib

LIMIT BEHAVIOR STATIC-PLATE CHARACTERISTICS, VARIABILITY OF Ic2

VARIABILITY OF Ib

VARIABILITY OF Ic2

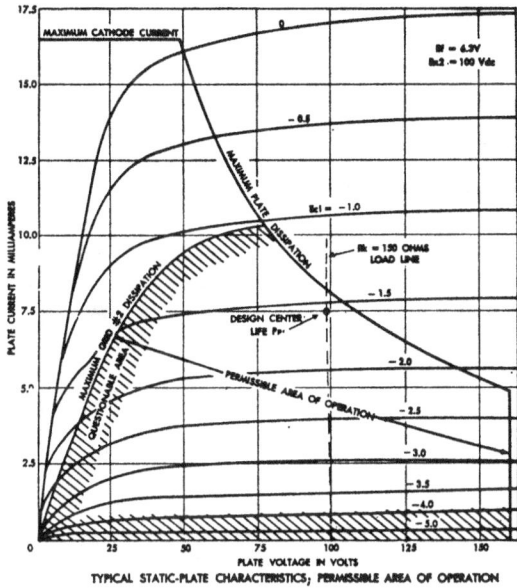

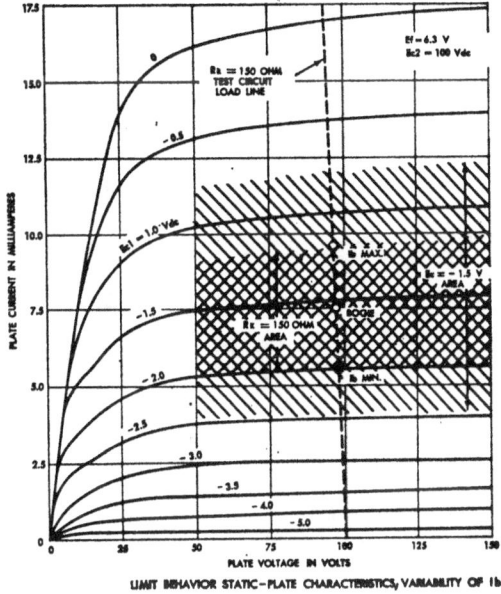

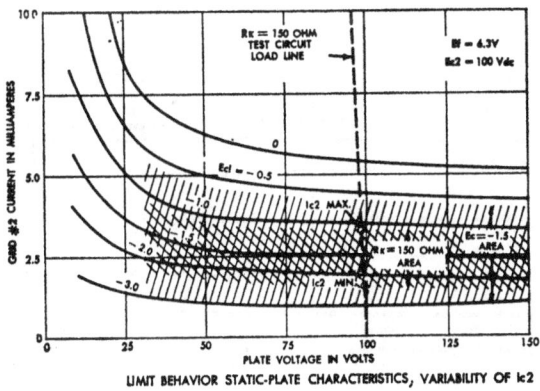

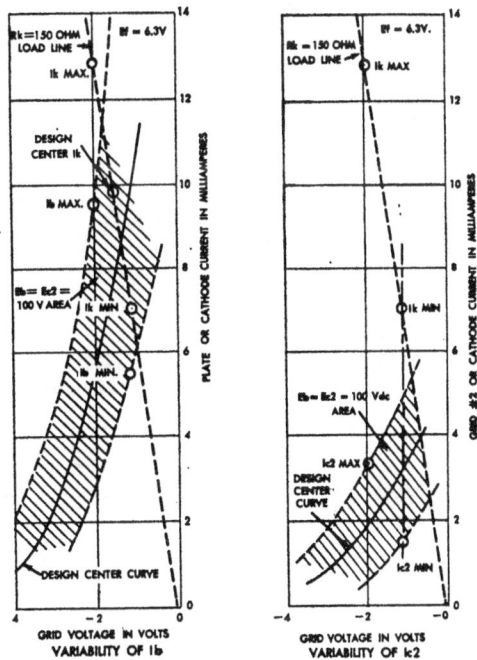

LIMIT BEHAVIOR TRANSFER CHARACTERISTICS

161

DESIGN CENTER CHARACTERISTICS
OBTAINED FROM DATA PUBLISHED BY ORIGINAL
RETMA REGISTRANT

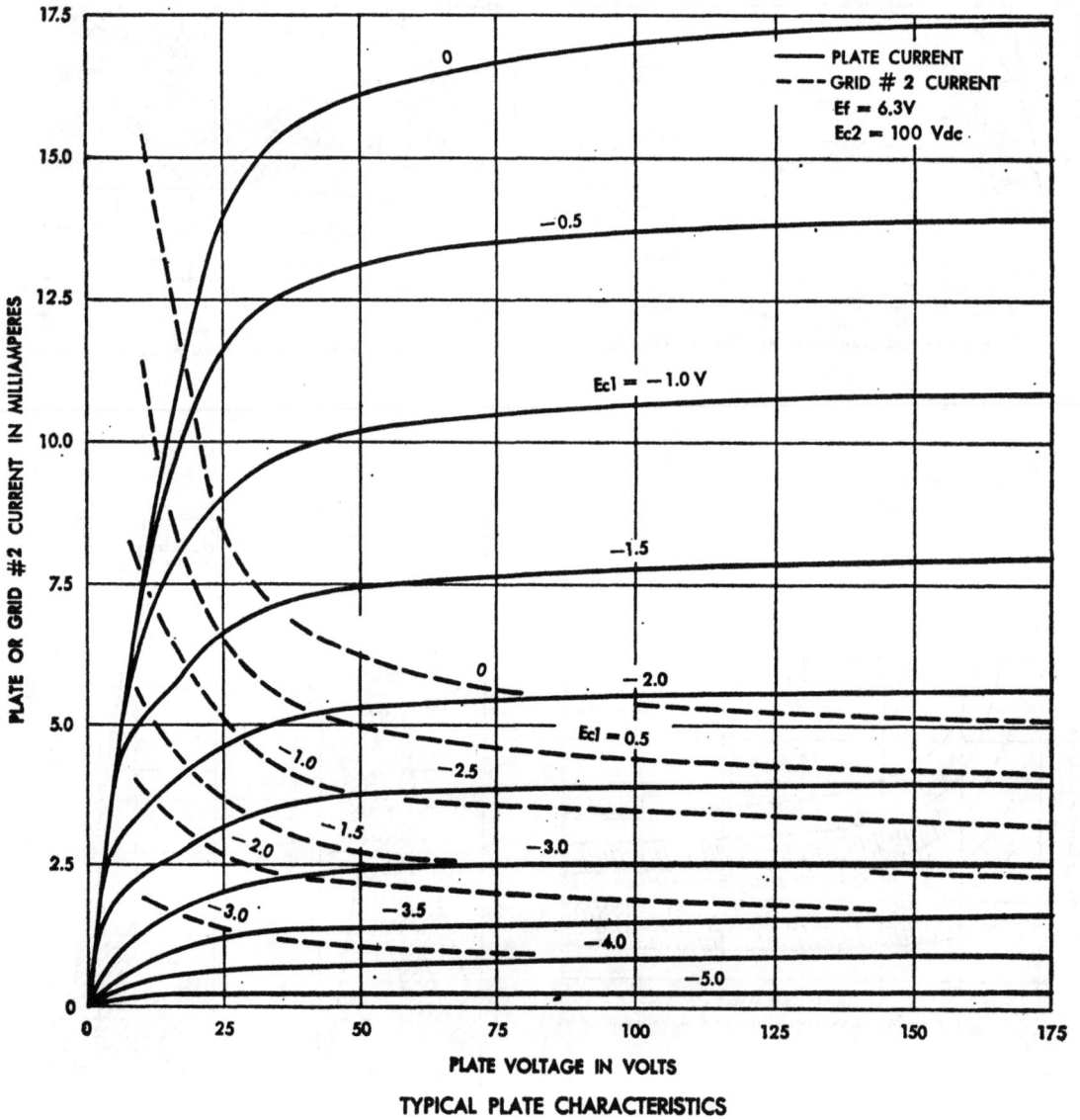

TYPICAL PLATE CHARACTERISTICS

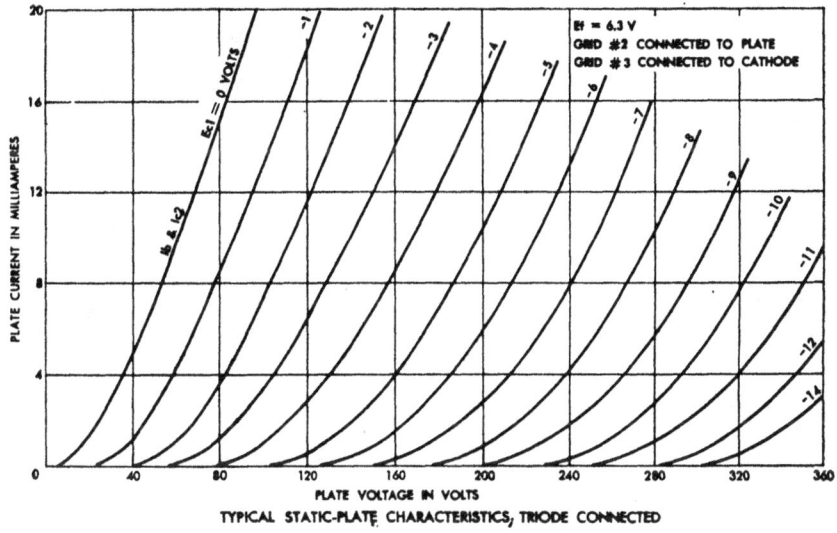

TYPICAL STATIC-PLATE CHARACTERISTICS, TRIODE CONNECTED

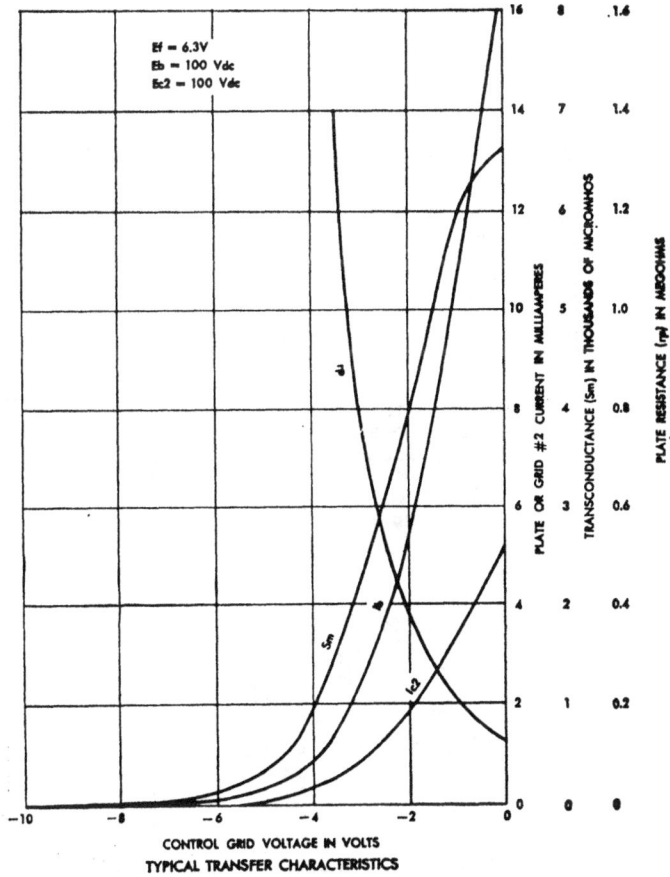

TYPICAL TRANSFER CHARACTERISTICS

163

LIFE TEST PROPERTY BEHAVIOR
MIL-E-1/140 B   5 AUG. '55
PRODUCED IN 1956-'57 BY FOUR MANUFACTURERS

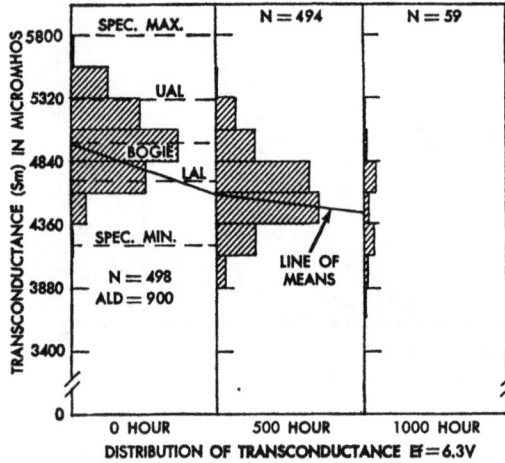

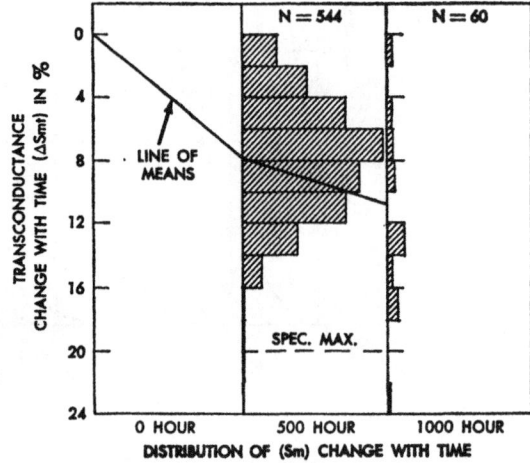

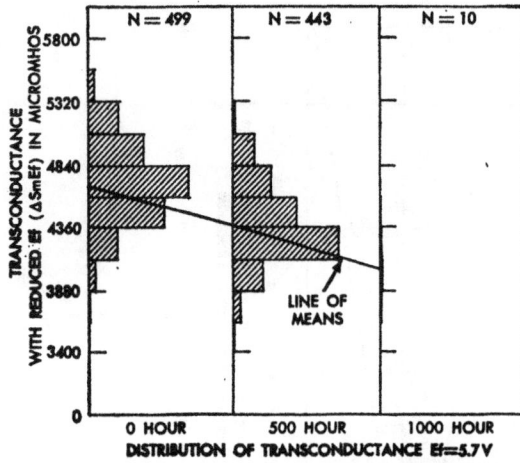

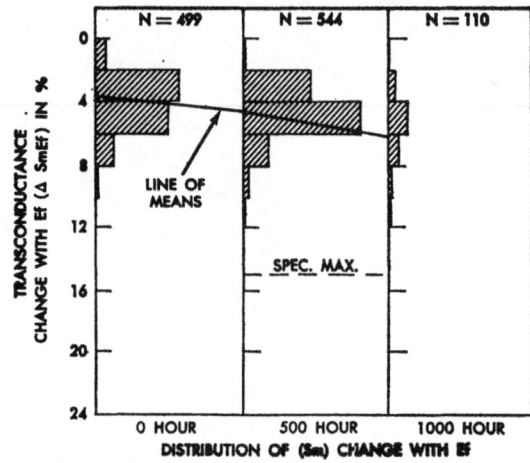

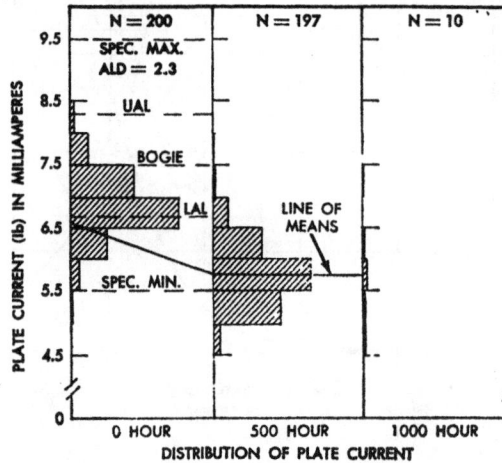

164

LIFE TEST PROPERTY BEHAVIOR
MIL-E-1/140B   5 AUG. '55
PRODUCED IN 1956-'57 BY FOUR MANUFACTURERS

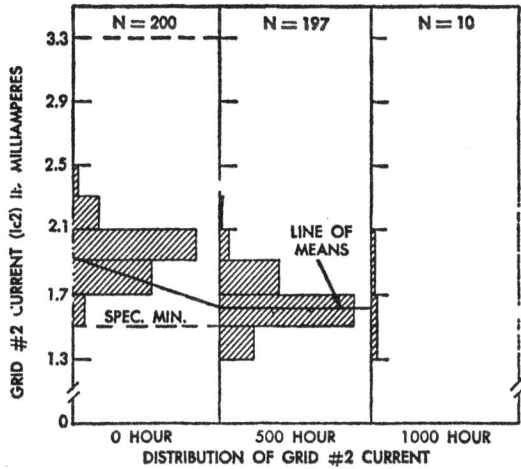

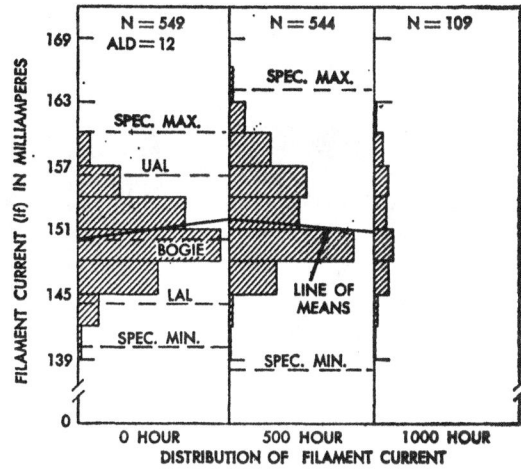

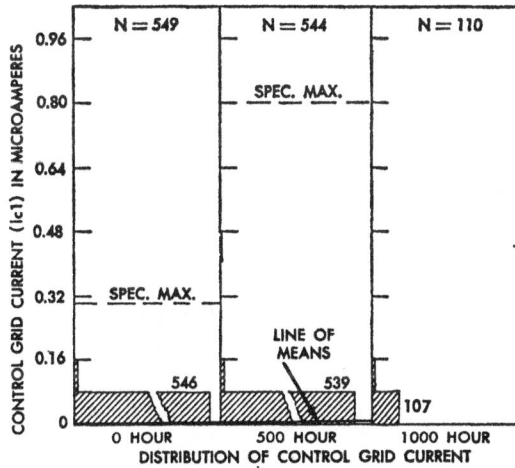

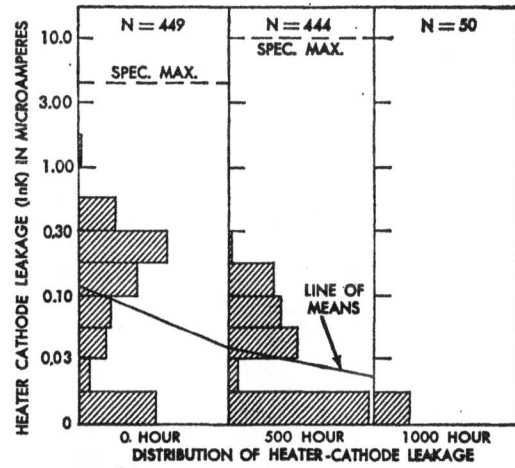

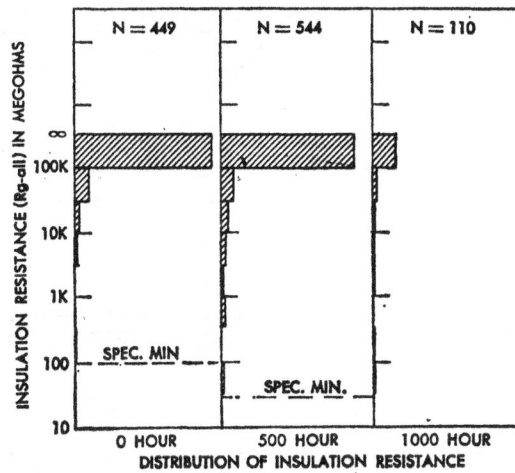

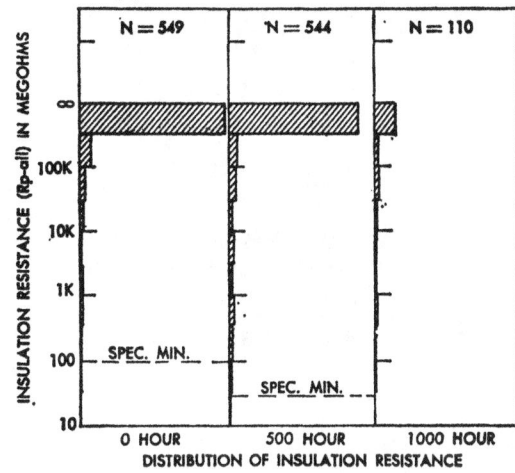

LIFE TEST PROPERTY BEHAVIOR
PROBABILITY OF SURVIVAL
MIL-E-1/140B   5 AUG. '55
PRODUCED IN 1956-'57 BY THREE MANUFACTURERS

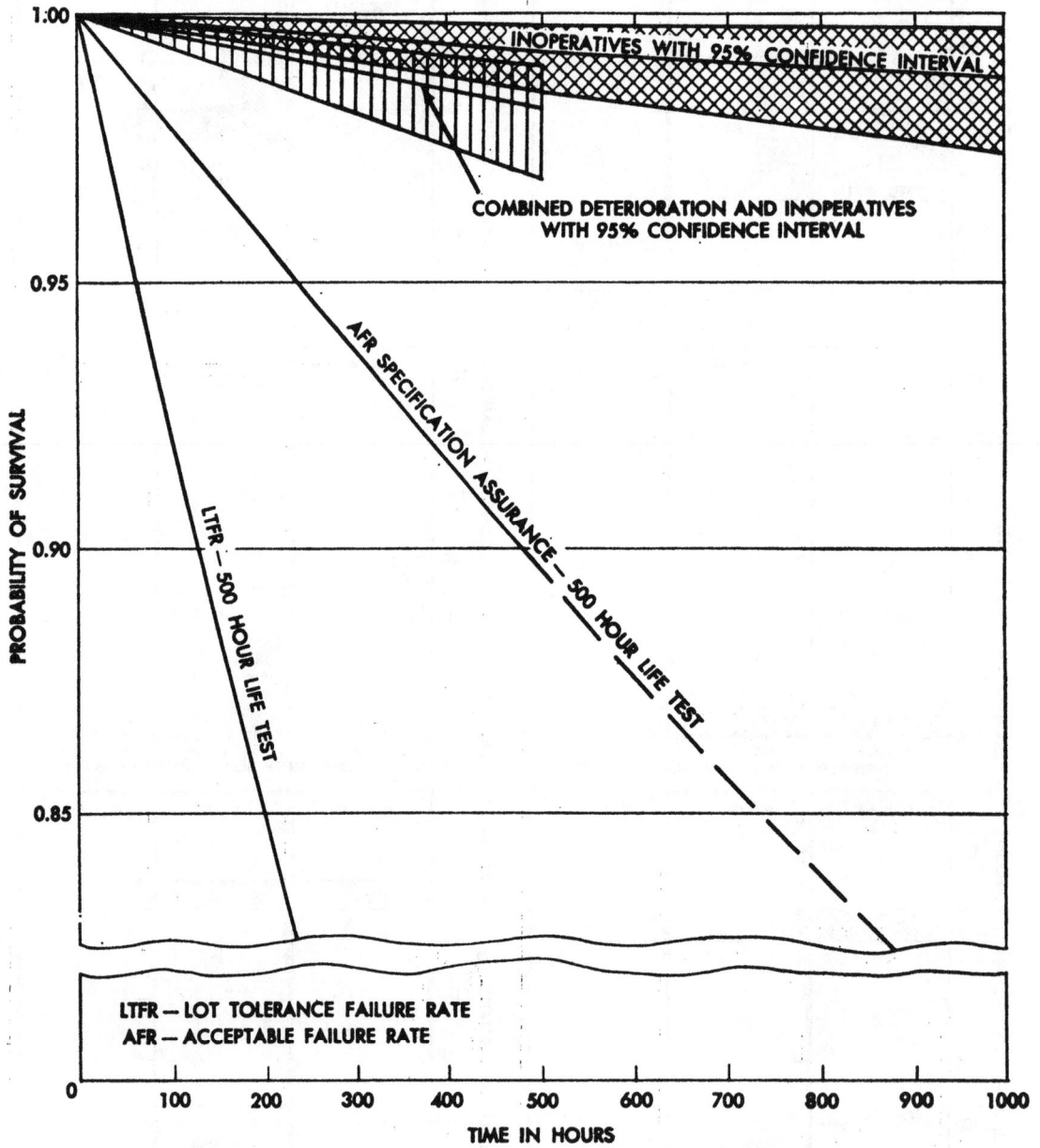

TUBE TYPE JAN-5899

DESCRIPTION:

The JAN-5899[1] is an 8 lead, button base, subminiature, semi-remote cut-off pentode having a transconductance in the range of 3800 to 5200. The JAN-5899 is identical to Type 6206 except suppressor grid is internally connected.

ELECTRICAL: The electrical characteristics are as follows:
Heater Voltage...6.3 V
Heater Current..140-160 A
Cathode...Coated Unipotential

MOUNTING: Any type mounting is adequate.

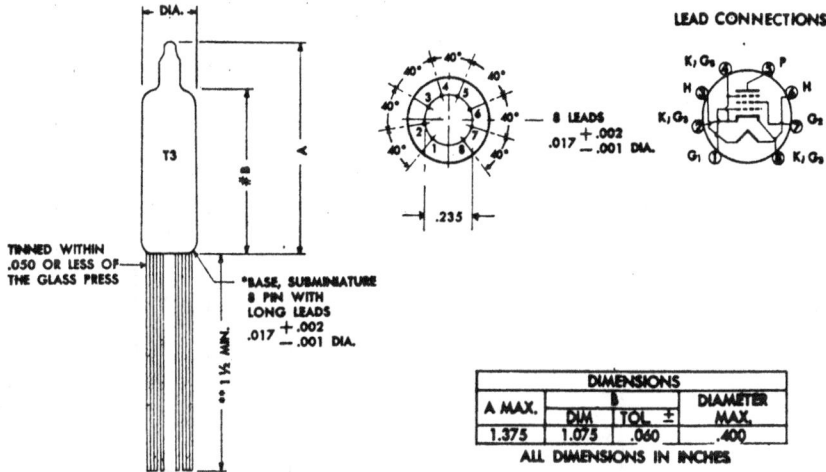

	DIMENSIONS		
A MAX.	DIM	TOL ±	DIAMETER MAX.
1.375	1.075	.060	.400

ALL DIMENSIONS IN INCHES

\# MEASURE FROM BASE SEAT TO BULB TOP-LINE AS DETERMINED BY RING GAGE OF .210 ± .001.

* LEAD DIAMETER TOLERANCE SHALL GOVERN BETWEEN .050 FROM THE GLASS TO .250 FROM THE GLASS.

** ALTERNATIVE LEAD LENGTH SHALL BE .200 ± .015 WHEN CUT LEADS ARE REQUIRED BY PROCUREMENT CONTRACT OR TSS. CUT LEADS SHALL BE ESSENTIALLY SQUARE CUT AND THE MAXIMUM BURR SHALL BE .003 INCREASE OVER THE ACTUAL LEAD DIAMETER.

RATINGS:	Ef	Eb	Ec1	Ec2	Ec3	Ehk	Rk	Rg1	Ik*	Pp	Pg2	T Envelope	Alt
Absolute	V	Vdc	Vdc	Vdc	Vdc	v	ohms	Meg	mAdc	W	W	°C	ft
Maximum	6.6	165	0	155	22	200	---	1.1	16.5	---	---	+220	60,000
Design Maximum	---	---	---	---	---	---	---	---	---	0.85	0.25	---	---
Minimum	6.0	---	-55	---	---	---	---	---	---	---	---	---	---
Test Cond:	6.3	100	0	100	0	0	120	---	---	---	---	---	---

[1] The values and specification comments presented in this section are related to MIL-E-1/97D dated 22 Oct 1957.

* Difficulty may be encountered if this tube is operated for long periods of time with very small values of cathode current. No specification assurance of life exists under conditions of cathode current approaching the maximum.

ACCEPTANCE TEST LIMIT SUMMARY

PROPERTY		MEASUREMENT CONDITIONS	INITIAL		500 HR LIFE TEST		1000 HR LIFE TEST		UNITS
			MIN	MAX	MIN	MAX	MIN	MAX	
Heater Current	If		140	160	138	164	–	–	mA
Transconductance(1)	Sm		3800	5200	–	–	–	–	umhos
Change in individual	ΔSmt		–	–	–	20	–	–	%
Change in average	Avg ΔSmt		–	–	–	15	–	–	%
Transconductance change with Ef	ΔSmEf	Ef=5.7V	–	10	–	15	–	–	%
Transconductance(3)	Sm	Ec1=-14Vdc; Rk=0	1.0	75	–	–	–	–	umhos
Plate Resistance	rp		0.175	–	–	–	–	–	Meg
Plate Current	Ib		5.2	9.2	–	–	–	–	mAdc
Screen Grid Current	Ic2		1.0	3.0	–	–	–	–	mAdc
Capacitance		0.405 in.dia. shield, Ef=0							
C	g1p		–	0.015	–	–	–	–	uuf
C	in		3.5	4.5	–	–	–	–	uuf
C	out		2.9	3.9	–	–	–	–	uuf
Control Grid Current	Ic1	Rg1=1.0 Meg	0	-0.3	0	-0.8	–	–	uAdc
Control Grid Emission	Ic1	Ef=7.5V;Ec1= -14Vdc;Rg1= 1.0Meg;Rk=0; NOTE	0	-0.5	–	–	–	–	uAdc
Heater Cathode Leakage	Ihk	Ehk=+100Vdc	–	5.0	–	10	–	–	uAdc
	Ihk	Ehk=-100Vdc	–	5.0	–	10	–	–	uAdc
Insulation of	R(g1-all)	Eg1-all=-100V	100	–	50	–	–	–	Meg
Electrodes	R(p-all)	Ep-all=-300V	100	–	50	–	–	–	Meg

Measurement conditions are the same as stated under Test Conditions, unless otherwise indicated.

NOTE: The tube shall be preheated a minimum of five minutes at test conditions for this test (except Ec1=0; Rk=120 ohms) prior to this test.

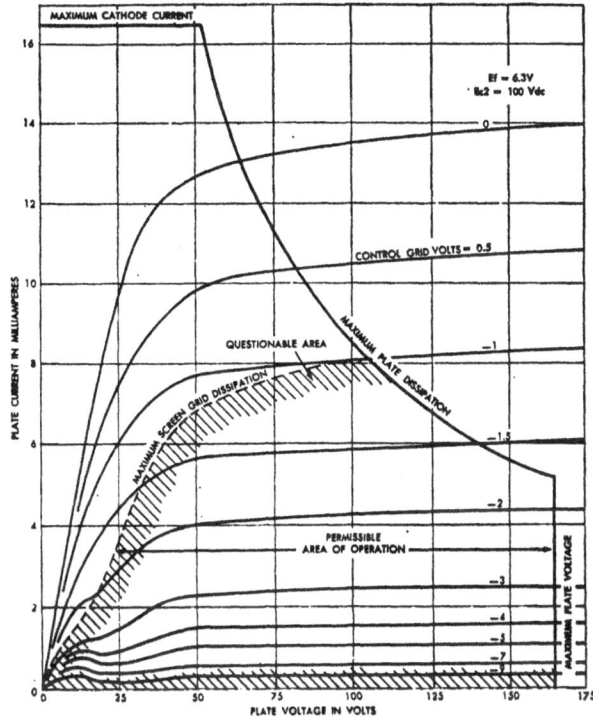

TYPICAL STATIC-PLATE CHARACTERISTICS; PERMISSIBLE AREA OF OPERATION

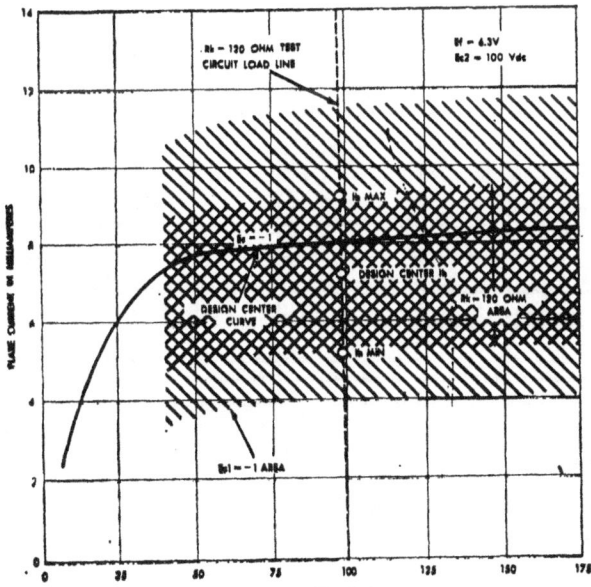

LIMIT BEHAVIOR STATIC - PLATE DATA; VARIABILITY OF Ib

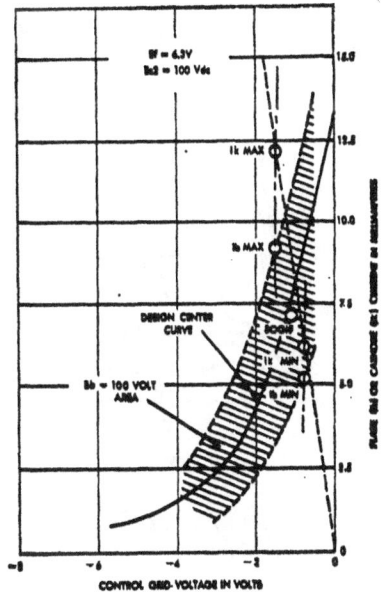

LIMIT TRANSFER DATA; VARIABILITY OF Ib

169

DESIGN CENTER CHARACTERISTICS
OBTAINED FROM DATA PUBLISHED BY ORIGINAL
RETMA REGISTRANT

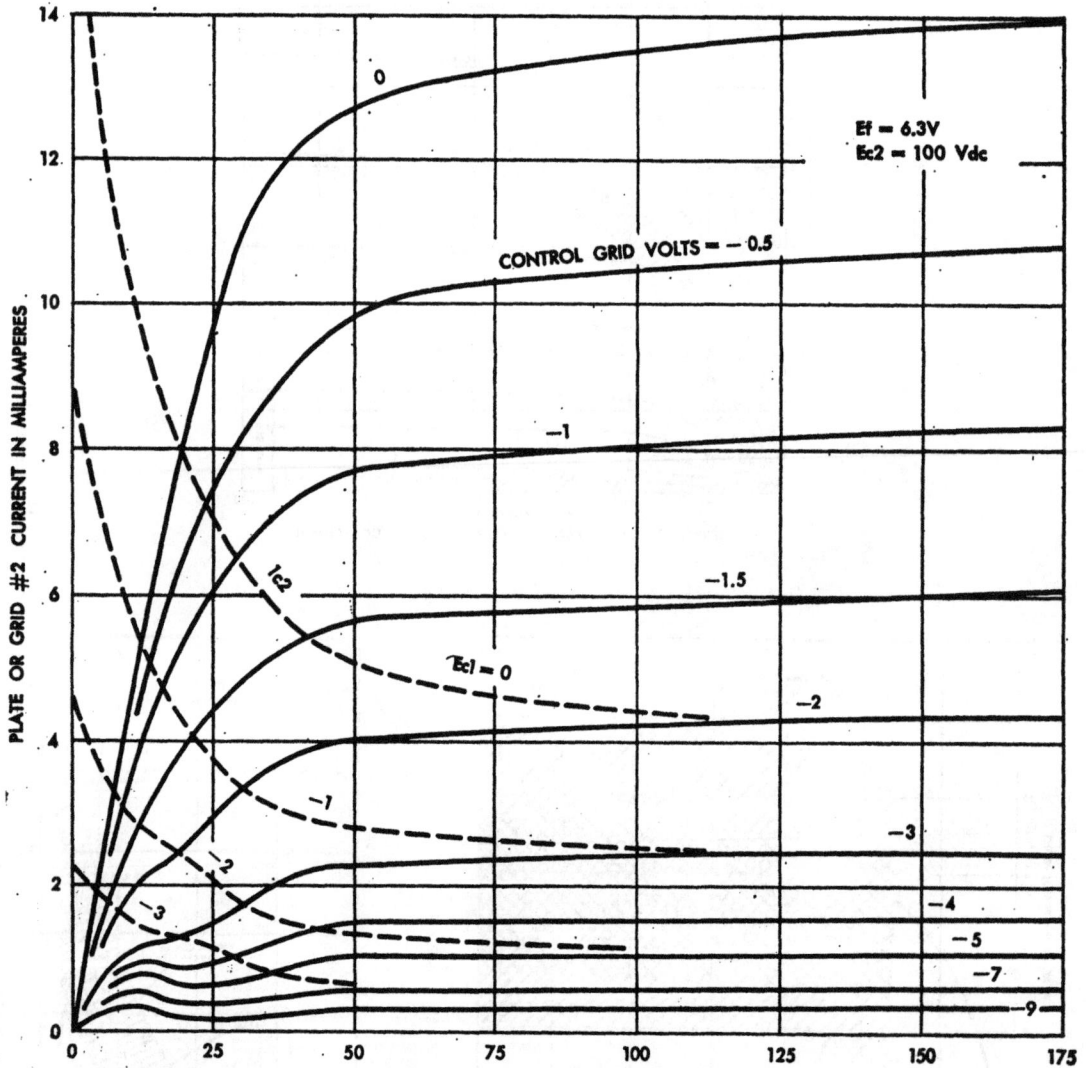

TYPICAL STATIC — PLATE CHARACTERISTICS

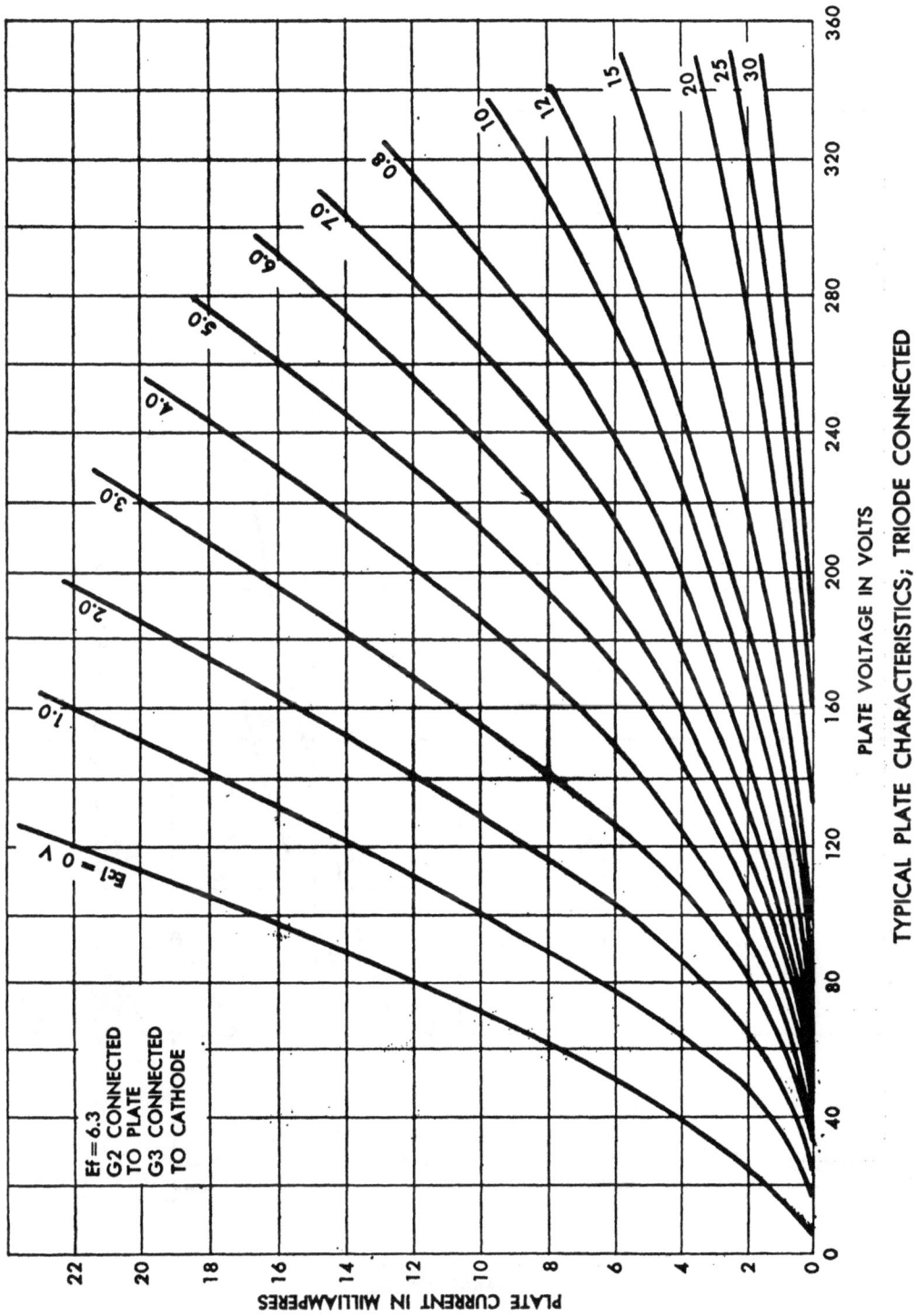

TYPICAL PLATE CHARACTERISTICS; TRIODE CONNECTED

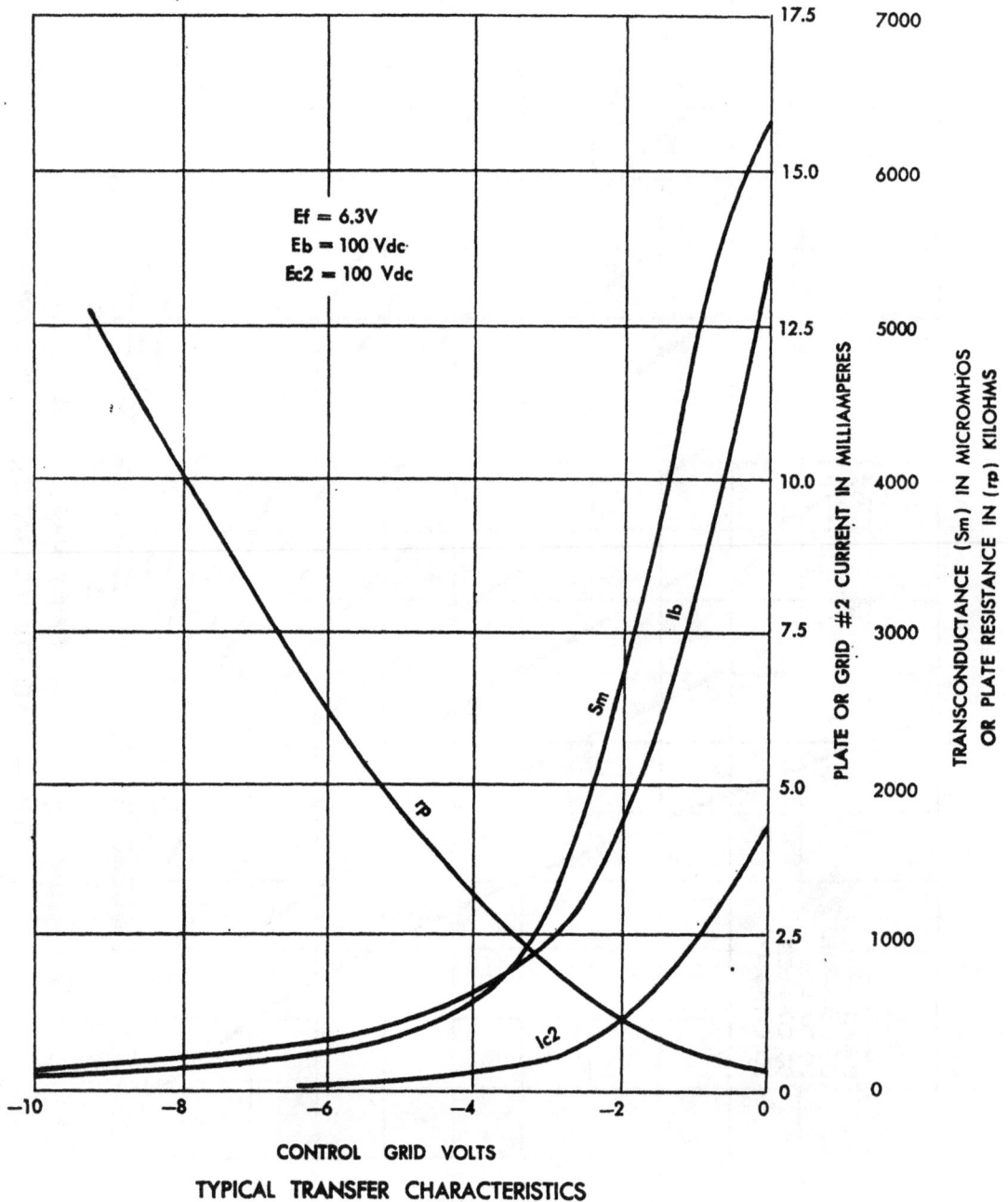

Ef = 6.3V
Eb = 100 Vdc
Ec2 = 100 Vdc

CONTROL GRID VOLTS

TYPICAL TRANSFER CHARACTERISTICS

LIFE TEST PROPERTY BEHAVIOR
MIL-E-1/97D   23 JUNE '55
PRODUCED IN 1955-'58 BY THREE MANUFACTURERS

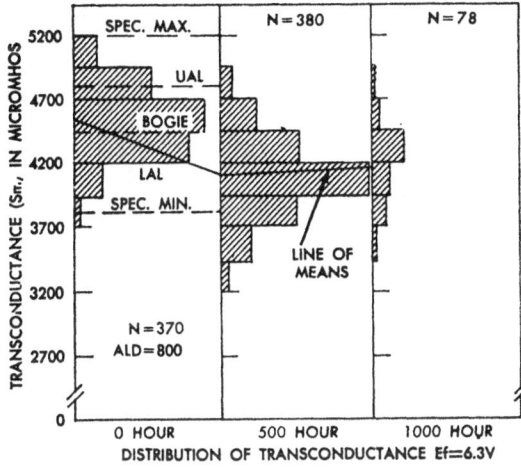

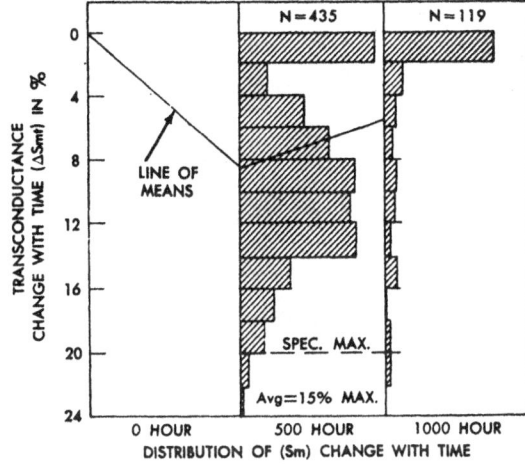

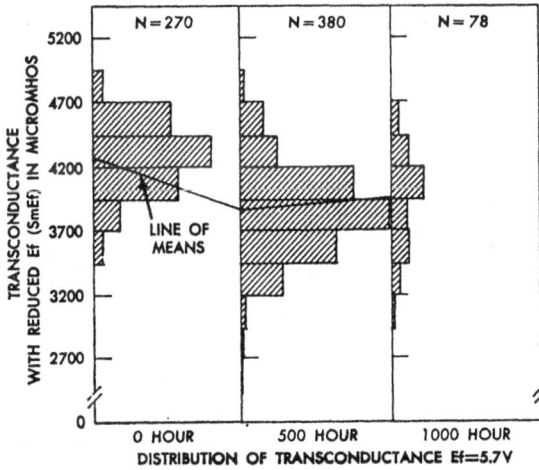

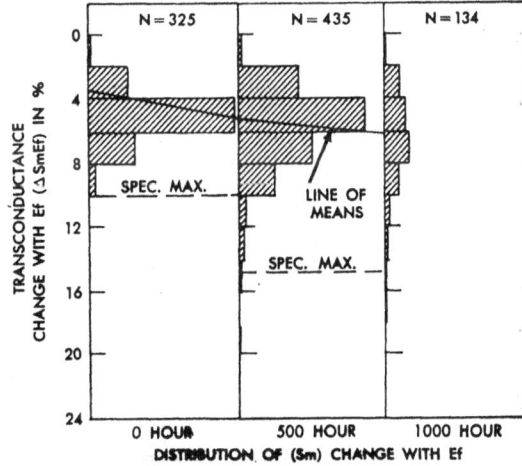

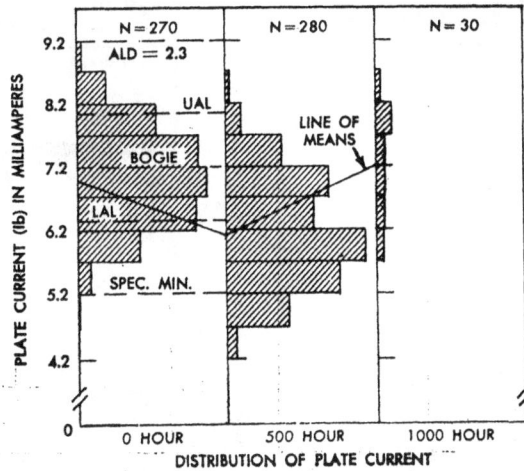

LIFE TEST PROPERTY BEHAVIOR
MIL-E-1/97D   23 JUNE '55
PRODUCED IN 1955-'58 BY THREE MANUFACTURERS

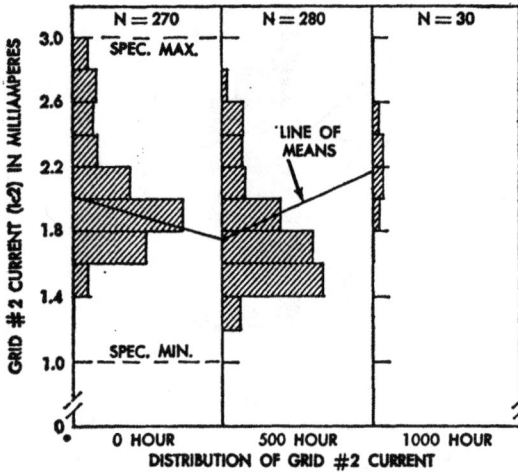

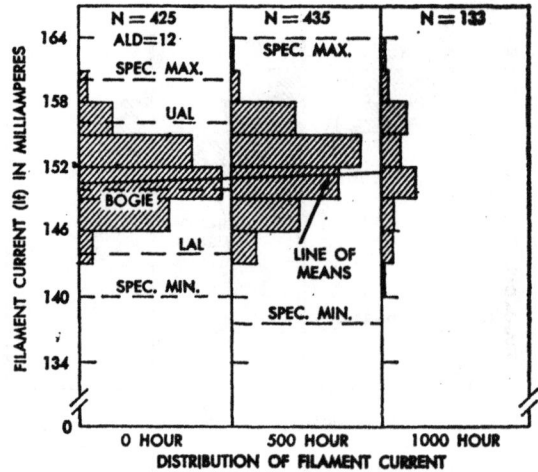

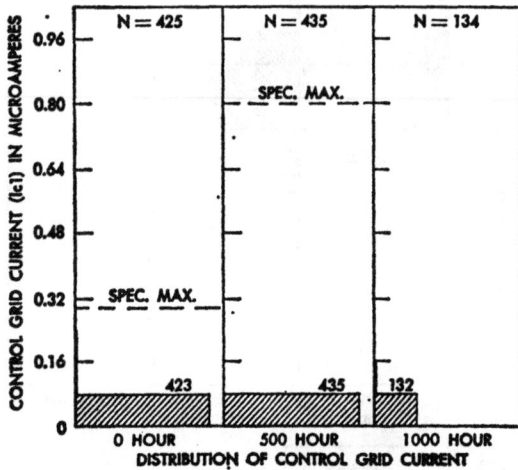

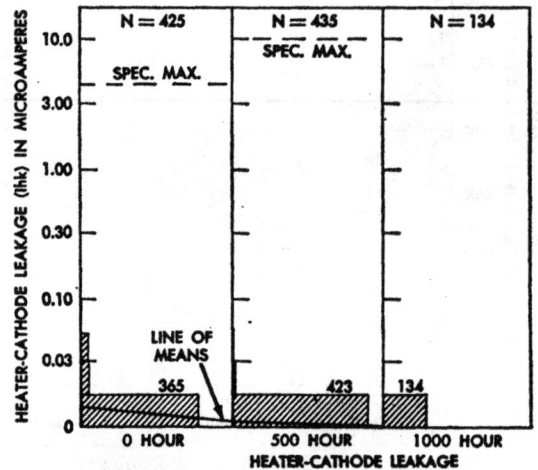

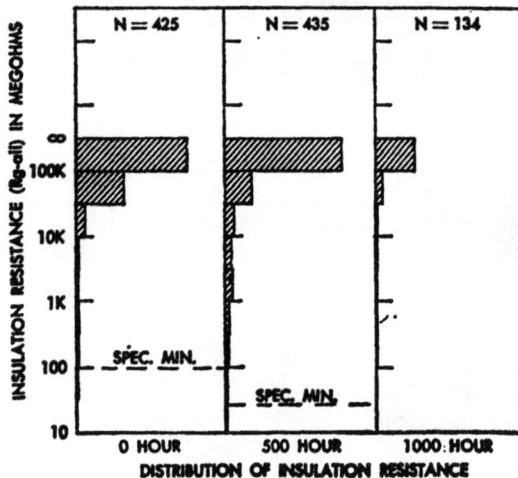

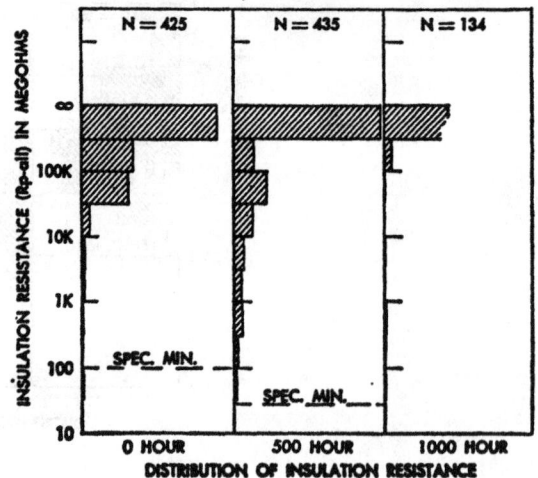

LIFE TEST PROPERTY BEHAVIOR
PROBABILITY OF SURVIVAL
MIL-E-1/97D   22 OCT. '57
PRODUCED IN 1955-'58 BY FOUR MANUFACTURERS

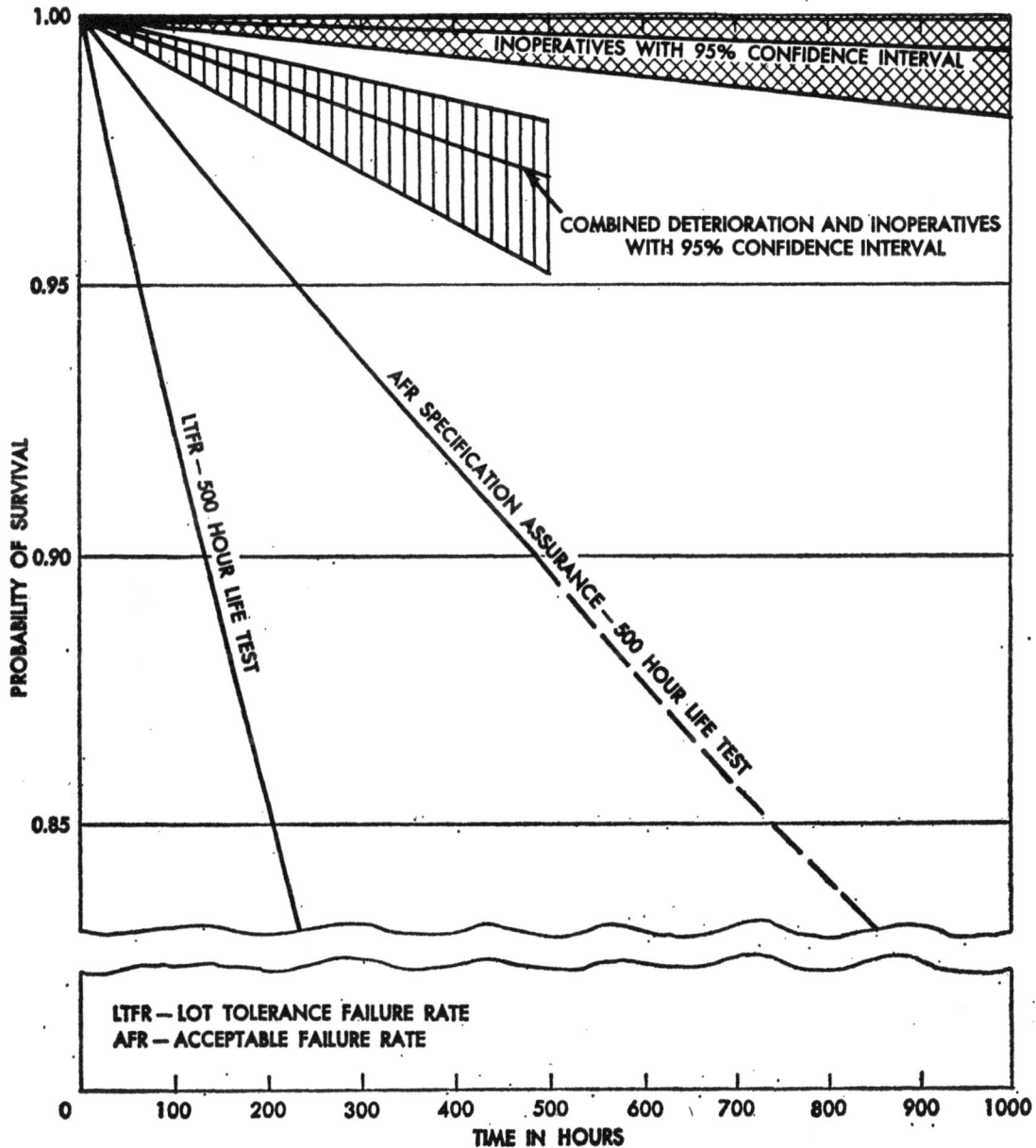

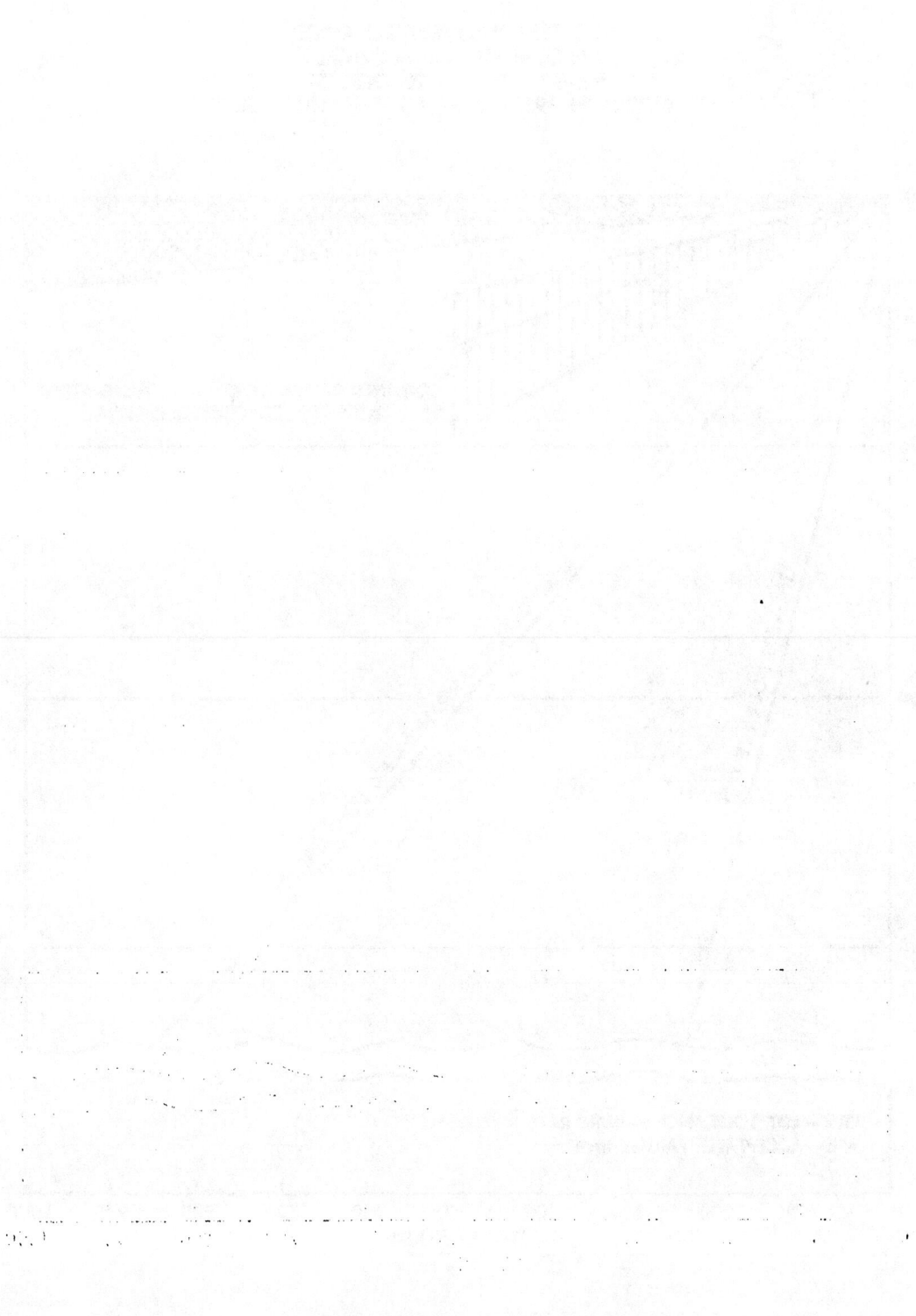

TUBE TYPE JAN-6021

DESCRIPTION:

The JAN-6021[1/] is an 8 lead, button base subminiature twin-triode having a Mu in the range of 30 to 40 and a transconductance in the range of 4450 to 6350. The JAN-6021 is similar in plate characteristics to the miniature type JAN-5670.

ELECTRICAL: The electrical characteristics are as follows:
Heater Voltage...6.3 V
Heater Current...300 mA
Cathode...Coated Unipotential

MOUNTING: Any type mounting is adequate.

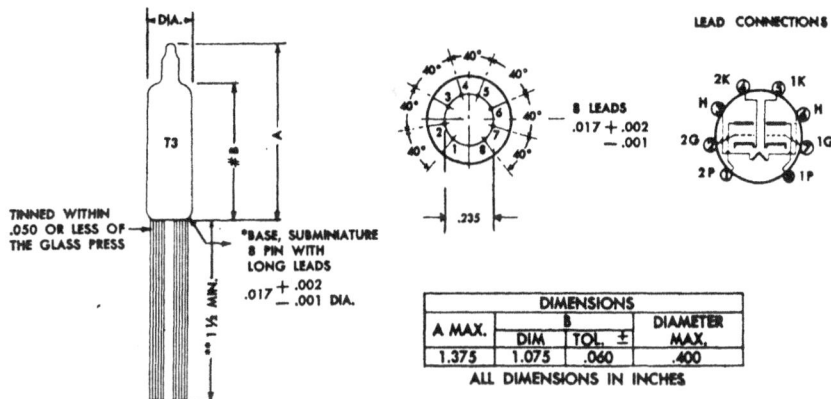

MEASURE FROM BASE SEAT TO BULB TOP-LINE AS DETERMINED BY RING GAGE OF .210 ± .001.

* LEAD DIAMETER TOLERANCE SHALL GOVERN BETWEEN .050 FROM THE GLASS TO .250 FROM THE GLASS.

** ALTERNATIVE LEAD LENGTH SHALL BE .200 ± .015 WHEN CUT LEADS ARE REQUIRED BY PROCUREMENT CONTRACT OR TSS. CUT LEADS SHALL BE ESSENTIALLY SQUARE CUT AND THE MAXIMUM BURR SHALL BE .003 INCREASE OVER THE ACTUAL LEAD DIAMETER.

RATINGS:	Ef	Eb	Ec	Ehk	Rk/k	Rg/g	Ib/b*	Ic/c*	Pp/p	T En-velope	Alt
Absolute	V	Vdc	Vdc	v	ohms	Meg	mAdc	mAdc	W	°C	ft
Maximum	6.6	165	0	200	---	1.1	22	5.5	0.7	+220	60,000
Minimum	6.0	---	-55	---	---	---	---	---	---	---	Note ---
Test Cond:	6.3	100	0	---	150	---	---	---	---	---	---

1/   The values and specification comments presented in this section are related to MIL-E-1/188B dated 23 Aug 1955.

*   Difficulty may be encountered if this tube is operated for long periods of time with very small values of cathode current. No specification assurance of life exists under conditions of cathode current approaching the maximum.

Note: If altitude rating is exceeded, reduction of instantaneous voltages (Ef excluded) may be required.

ACCEPTANCE TEST LIMITS SUMMARY

PROPERTY	MEASUREMENT CONDITIONS	INITIAL		500 HR LIFE TEST		1000 HR LIFE TEST		UNITS
		MIN	MAX	MIN	MAX	MIN	MAX	
Heater Current If		280	320	276	328	-	-	mA
Transconductance (1) Sm		4450	6350	-	-	-	-	umhos
Change in individual Δ Smt		-	-	-	25	-	-	%
Change in average Avg Δ Smt		-	-	-	15	-	-	%
Transconductance Change with Ef Δ SmEf	Ef = 5.7 V	-	15	-	15	-	-	%
Amplification Factor Mu		30	40	-	-	-	-	-
Plate Current (1) Ib		4.5	8.5	-	-	-	-	mAdc
Plate Current (2) Ib	Ec = -6.5Vdc; Rk = 0	-	100	-	-	-	-	uAdc
Plate Current (1) Ib difference between sections		-	1.6	-	-	-	-	mAdc
Pulse Emission is	Ef=6.0V; E pulse = 50V tp=25 u sec; prr = 200 pps	300	-	-	-	-	-	ma
Capacitance C gp	Ef = 0 No shield	1.2	1.8	-	-	-	-	uuf
C in		1.8	3.0	-	-	-	-	uuf
C out (1)		0.20	0.36	-	-	-	-	uuf
C out (2)		0.22	0.42	-	-	-	-	uuf
C gg		-	0.013	-	-	-	-	uuf
C pp		-	0.52	-	-	-	-	uuf
Control Grid Current Ic	Eb=150Vdc; Rk=300;Rg=1.0 Meg	0	-0.3	0	-0.9	-	-	uAdc
Control Grid Emission Ic	Ef=7.5V;Ec=-7.5 Vdc;Eb=150Vdc; Rk=0;Rg=1.0Meg Note:	0	-0.5	-	-	-	-	uAdc
Heater Cathode Leakage Ihk	Ehk=+100Vdc	-	5.0	-	10	-	-	uAdc
Ihk	Ehk=-100Vdc	-	-5.0	-	-10	-	-	uAdc
Insulation of Electrodes R(g-all)	Eg-all=-100Vdc	100	-	50	-	-	-	Meg
R(p-all)	Ep-all=-300Vdc	100	-	50	-	-	-	Meg

Measurement conditions are the same as stated under Test Conditions, unless otherwise indicated.

Note: The tube shall be preheated a minimum of five minutes at test conditions for this test (except Ec = 0; Rk = 500) prior to this test.

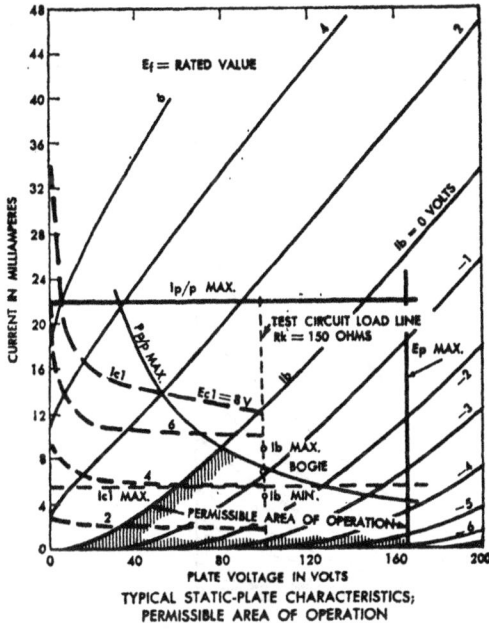

TYPICAL STATIC-PLATE CHARACTERISTICS;
PERMISSIBLE AREA OF OPERATION

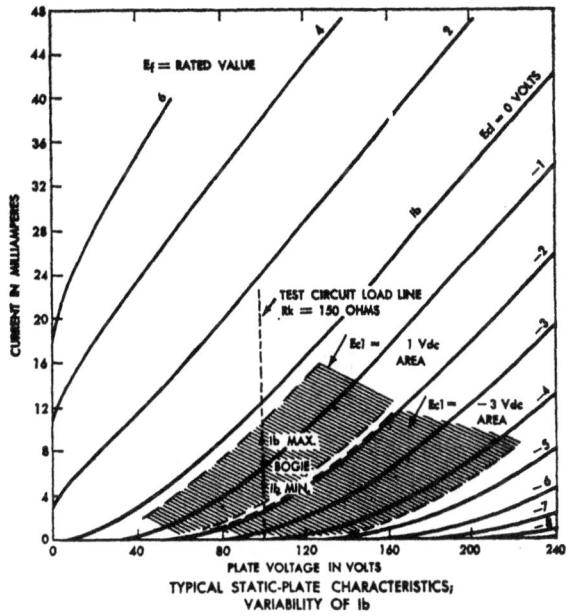

TYPICAL STATIC-PLATE CHARACTERISTICS;
VARIABILITY OF Ib

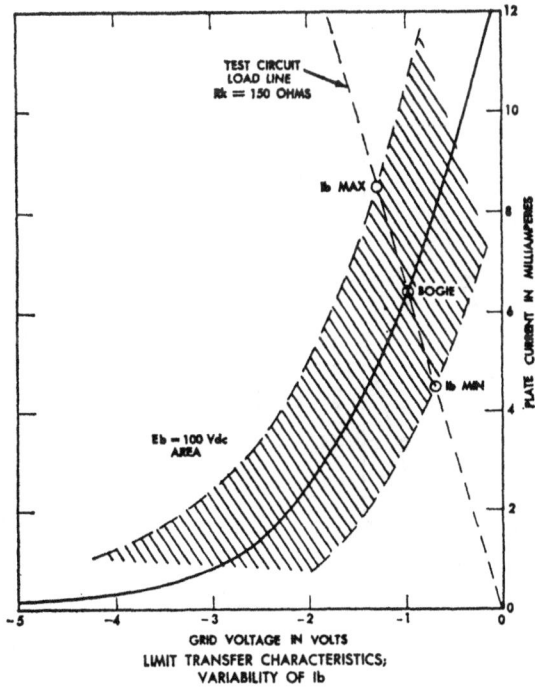

LIMIT TRANSFER CHARACTERISTICS;
VARIABILITY OF Ib

179

DESIGN CENTER CHARACTERISTICS
OBTAINED FROM DATA PUBLISHED BY ORIGINAL
RETMA REGISTRANT

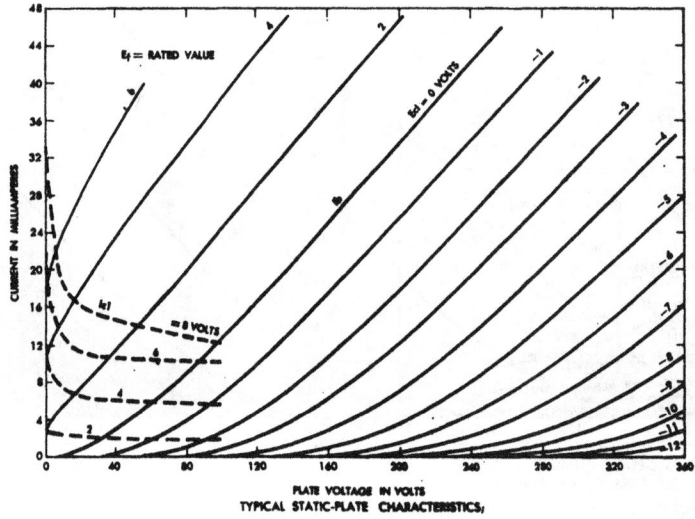

TYPICAL STATIC-PLATE CHARACTERISTICS;

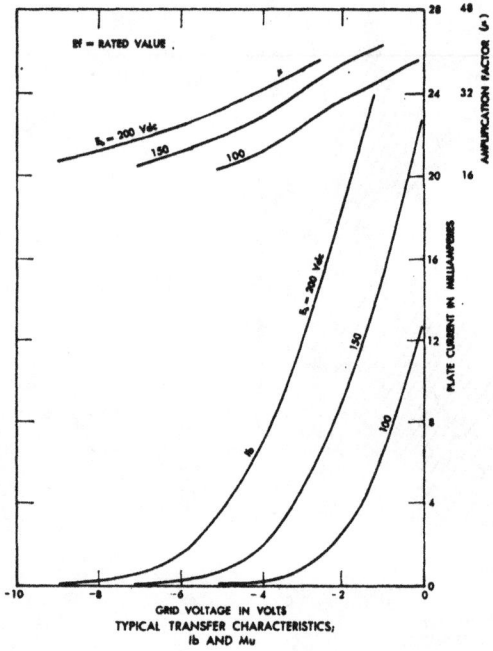

TYPICAL TRANSFER CHARACTERISTICS;
Ib AND Mu

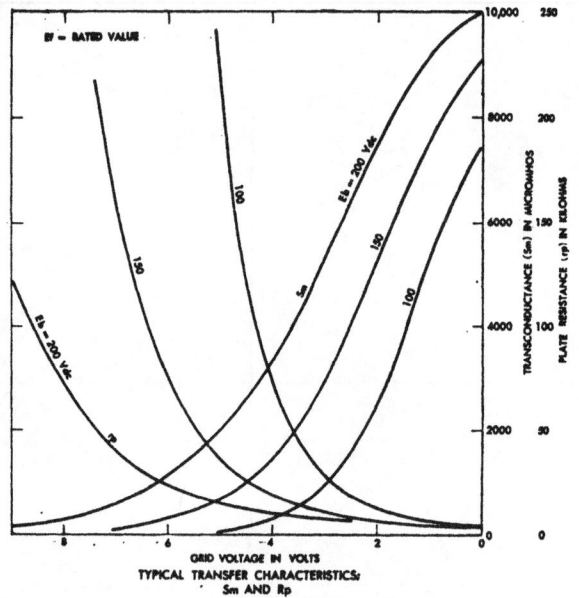

TYPICAL TRANSFER CHARACTERISTICS;
Sm AND Rp

180

LIFE TEST PROPERTY BEHAVIOR
MIL-E-1/188B   23 AUG. '55
PRODUCED IN 1956-'58 BY THREE MANUFACTURERS

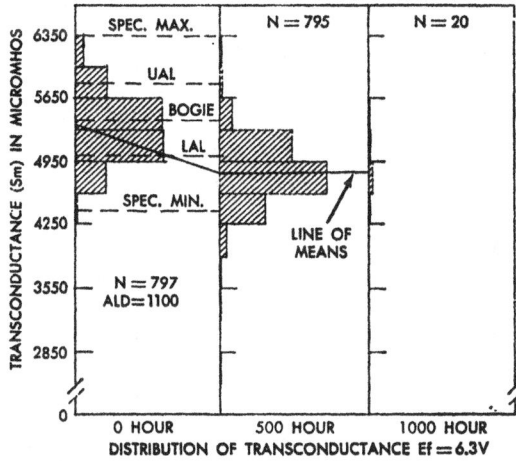

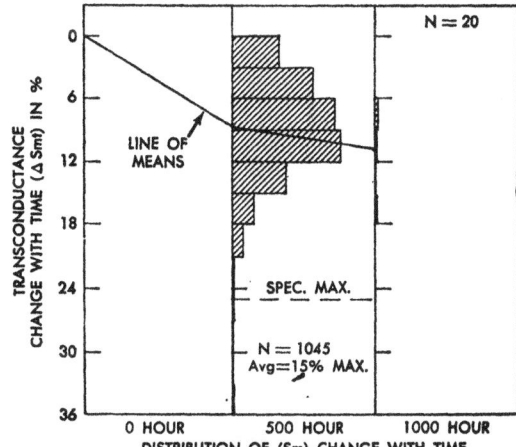

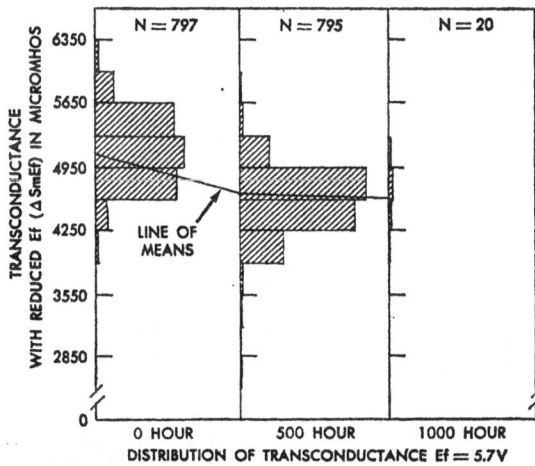

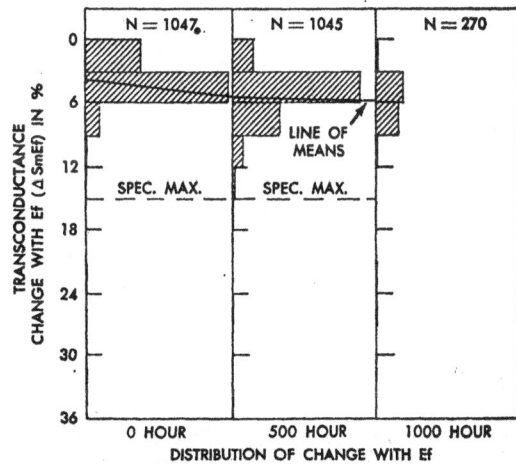

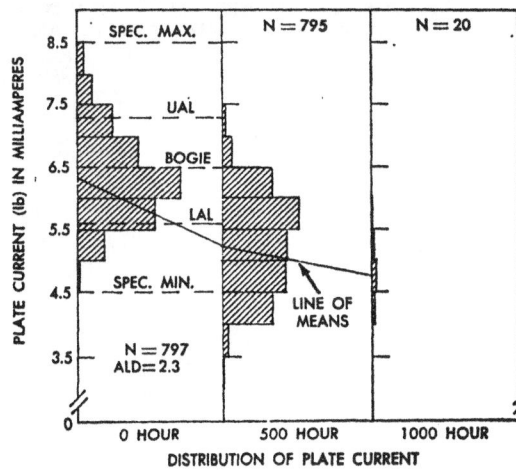

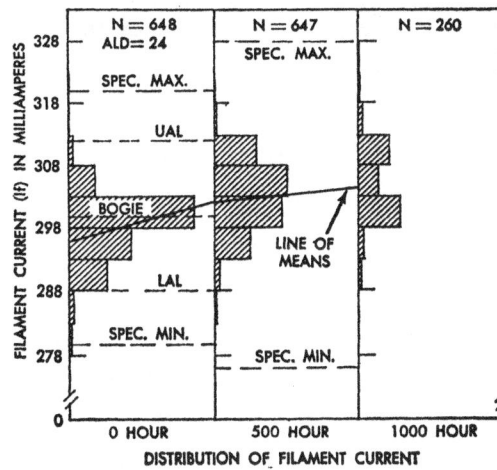

181

LIFE TEST PROPERTY BEHAVIOR
MIL-E-1/188B   23 AUG. '55
PRODUCED IN 1956-'58 BY THREE MANUFACTURERS

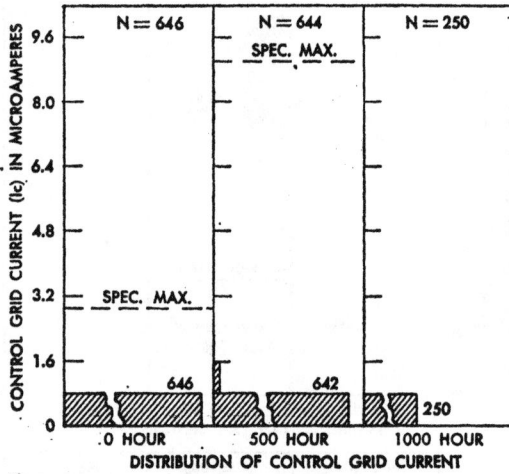

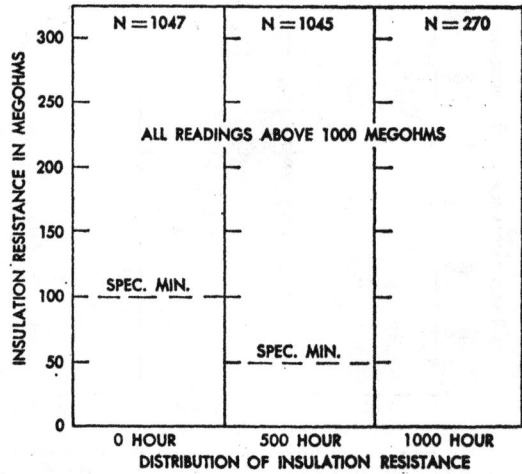

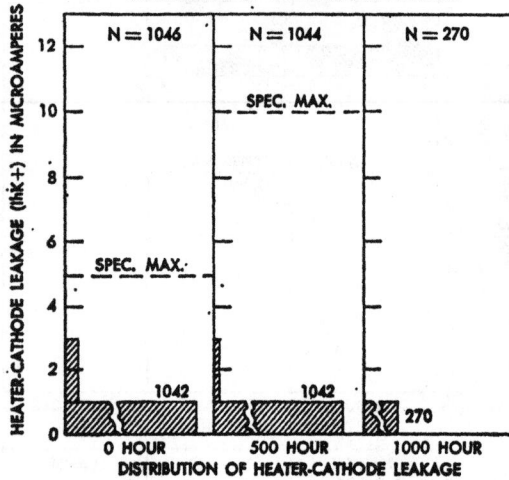

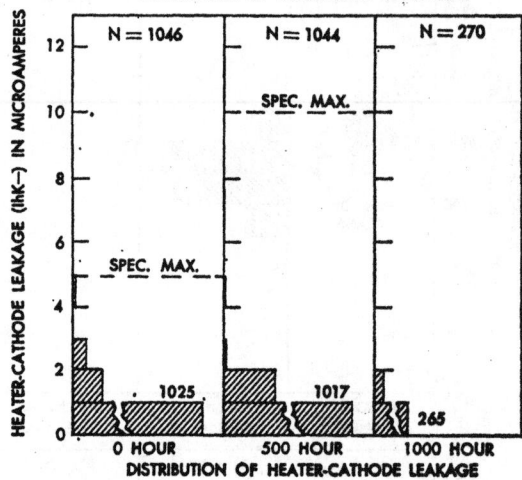

DISTRIBUTION OF CONTROL GRID CURRENT

DISTRIBUTION OF INSULATION RESISTANCE

DISTRIBUTION OF HEATER-CATHODE LEAKAGE

DISTRIBUTION OF HEATER-CATHODE LEAKAGE

LIFE TEST PROPERTY BEHAVIOR
PROBABILITY OF SURVIVAL
MIL-E-1/188B   23 AUG. '55
PRODUCED IN 1956-'58 BY THREE MANUFACTURERS

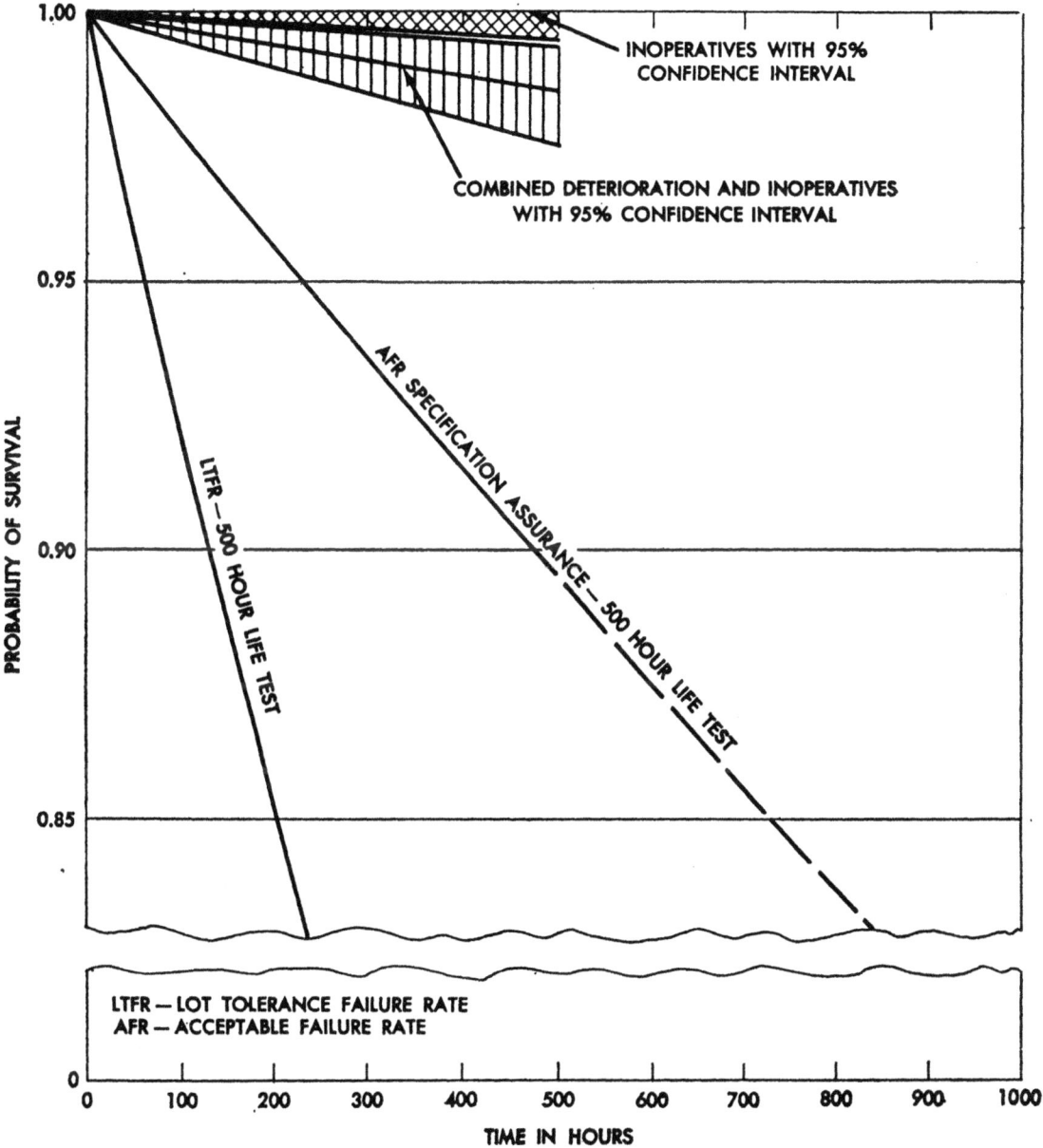

TUBE TYPE JAN-6110

DESCRIPTION:

The JAN-6110[1] is an 8 lead, button base, subminiature, double diode.

ELECTRICAL: The electrical characteristics are as follows:
Heater Voltage 6.3V
Heater Current, Design Center.....................................150 mA
**Cathode ....Coated Unipotential

MOUNTING: Any type of mounting is adequate.

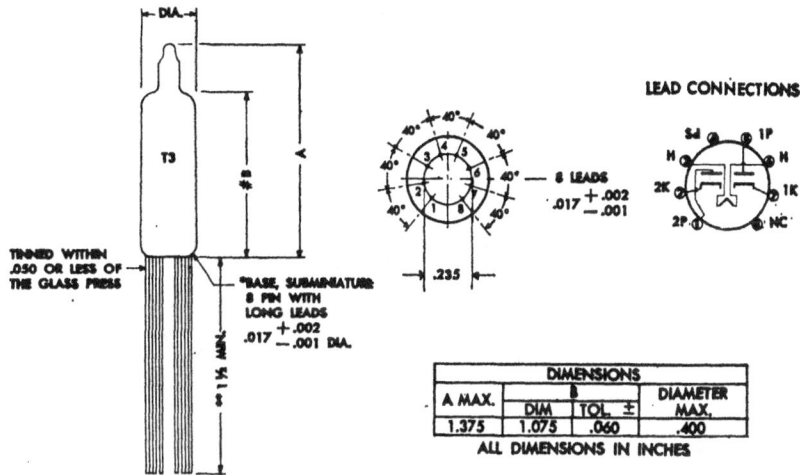

MEASURE FROM BASE SEAT TO BULB TOP-LINE AS DETERMINED BY RING GAGE OF .210 ± .001.

* LEAD DIAMETER TOLERANCE SHALL GOVERN BETWEEN .050 FROM THE GLASS TO .250 FROM THE GLASS.

** ALTERNATIVE LEAD LENGTH SHALL BE .200 ± .015 WHEN CUT LEADS ARE REQUIRED BY PROCUREMENT CONTRACT OR TSS. CUT LEADS SHALL BE ESSENTIALLY SQUARE CUT AND THE MAXIMUM BURR SHALL BE .003 INCREASE OVER THE ACTUAL LEAD DIAMETER.

RATINGS:	Ef	Epp/p	epx	Ehk	RL	CL	Io/p	ib/p	i surge/p	T Envelope	Alt
Absolute	V	Vac	v	v	ohms	uf	mAdc	ma	ma	°C	ft
Maximum	6.6	---	·460	3 60	---	--	4.4	26.5	160	≠220	60,000 Note
Minimum	6.0	---	---	---	---	--	---	---	---	---	---
TEST COND.:	6.3	165	---	---	20,000	8	---	---	---	---	---

** Difficulty may be encountered if this tube is operated for long periods of time with very small values of cathode current.

Note: If altitude rating is exceeded, reduction of instantaneous voltages (Ef excluded) may be required.

1/ The values and specification comments presented in this section are related to MIL-E-1/725C dated 26 Dec 1956.

ACCEPTANCE TEST LIMITS SUMMARY

PROPERTY		MEASUREMENT CONDITIONS	INITIAL		500 HR LIFE TEST		1000 HR LIFE TEST		UNITS
			MIN	MAX	MIN	MAX	MIN	MAX	
Heater Current	If		140	160	138	164	---	---	mA
Operation	Io	See note	7.8	---	6.8	---	---	---	mAdc
Plate Current	Ib	Ebb=0; Rp=40,000 ohms	2.0	22	---	---	---	---	uAdc
Difference between Sections	Ib'		---	5.0	---	---	---	---	uAdc
Emission	Is	Eb=10 Vdc	7.5	---	---	---	---	---	mAdc
Capacitance		0.405 in. dia. shield, Ef=0							
	C 1p to 2p		---	0.026	---	---	---	---	uuf
	C 1p to h+1k+sd		1.8	2.6	---	---	---	---	uuf
	C 2p to h+2K+sd		1.8	2.6	---	---	---	---	uuf
	C 1k to h+1p+sd		2.1	3.1	---	---	---	---	uuf
	C 2K to h+2p+sd		2.1	3.1	---	---	---	---	uuf
Heater-Cathode Leakage									
	Ihk	Ehk=+100 Vdc	---	10	---	20	---	---	uAdc
	Ihk	Ehk=-100 Vdc	---	-10	---	-20	---	---	uAdc
Insulation of Electrodes R(p-all)		Ep-all=-300 Vdc	100	---	25	---	---	---	Meg

Measurement conditions are the same as stated under Test Conditions, unless otherwise indicated.

Note: In a full wave circuit, adjust Zp (per plate) so that a bogie tube gives Io = 8.8 mAdc and ib not less than 24 ma per plate. A bogie tube has a tube drop Etd = 10 Vdc at Is = 15 mAdc per plate.

SIGNAL RECTIFIER APPLICATION: Chart "A" relates boundaries of permissible operation and the questionable area of operation to the plate characteristics. A further explanation of this curve is found in the basic section of Part III under Applications of Diodes, Permissible Operation Conditions.

SUPPLY VOLTAGE RECTIFIER APPLICATION: Rating charts I, II, and III represent areas of permissible operation. Requirements of all charts must be supplied simultaneously in capacitor-input filter applications. A further explanation of these curves is found in the basic section of Part III under Application of Diodes, Rating Charts.

TYPICAL PLATE CHARACTERISTICS: This chart was reproduced from data published by the original RETMA registrant of the type.

LIFE TEST PROPERTY BEHAVIOR: These charts are from data supplied by the manufacturer. A general discussion may be found in the basic section of this part.

TYPICAL PLATE CHARACTERISTICS, SINGLE SECTION

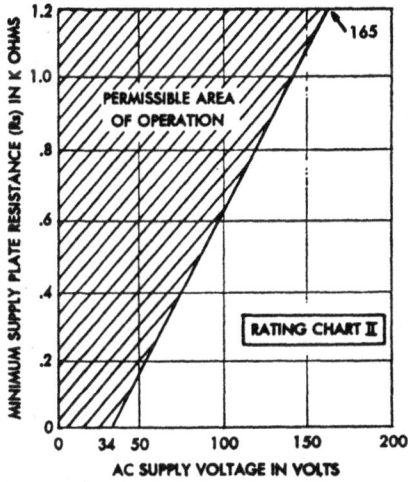

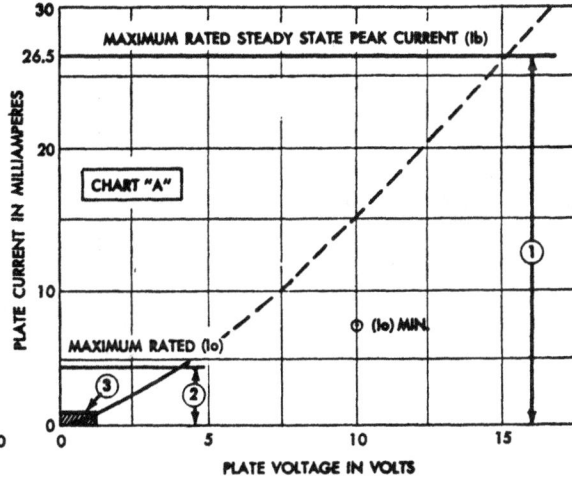

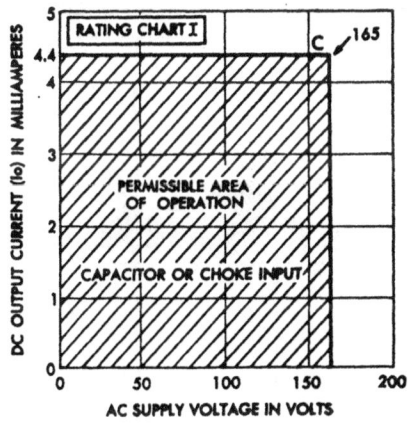

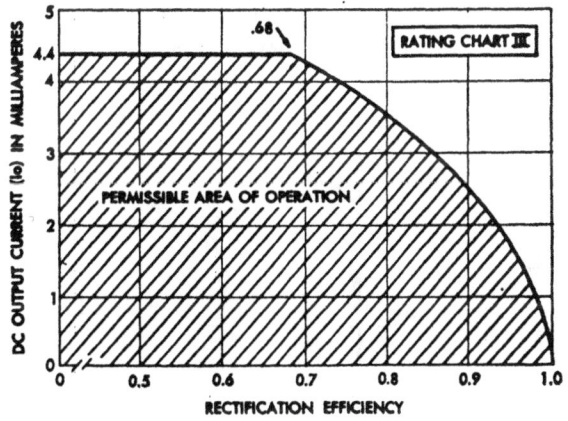

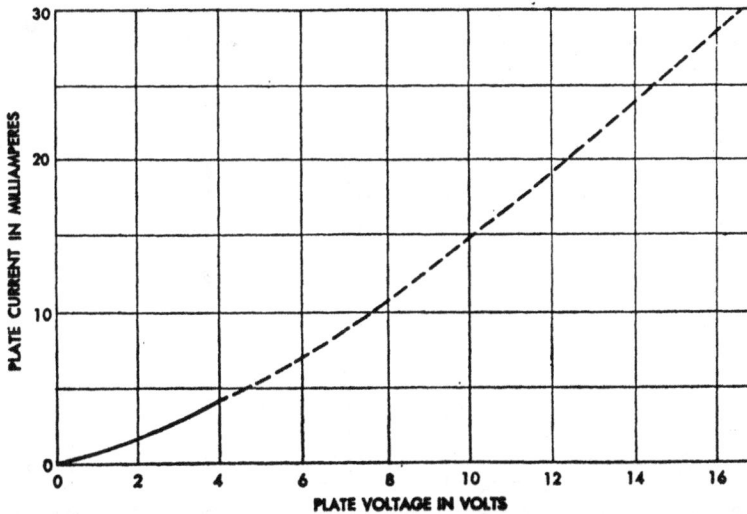

LIFE TEST PROPERTY BEHAVIOR
MIL-E-1/725 C   26 DEC 56
PRODUCED IN 1958 BY ONE MANUFACTURER

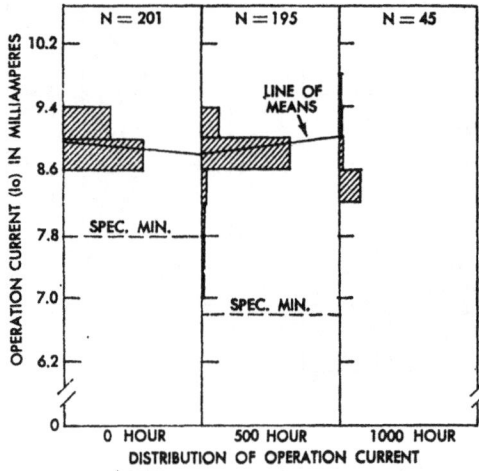

DISTRIBUTION OF OPERATION CURRENT

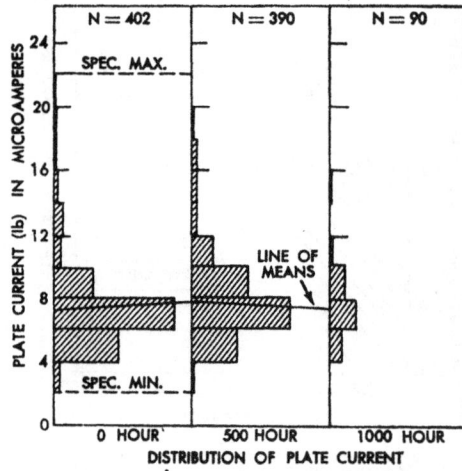

DISTRIBUTION OF PLATE CURRENT

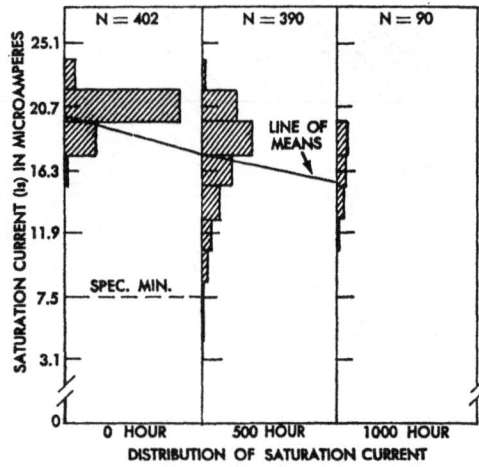

DISTRIBUTION OF SATURATION CURRENT

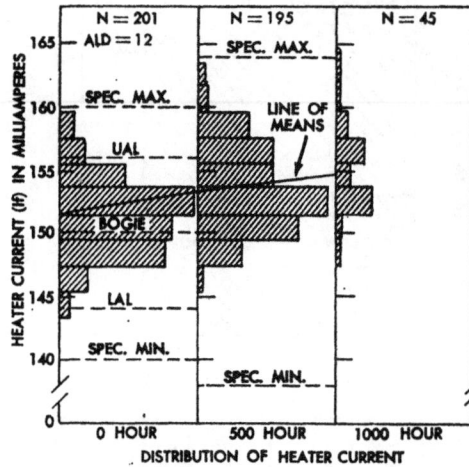

DISTRIBUTION OF HEATER CURRENT

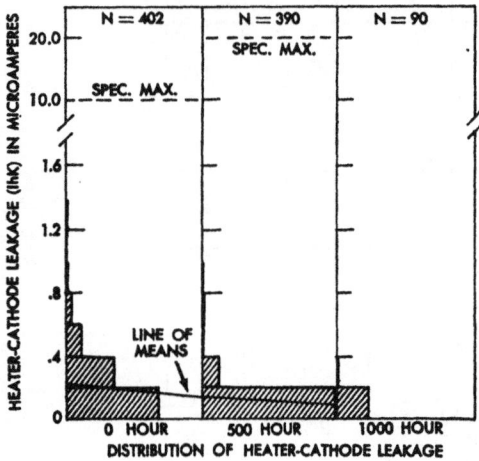

DISTRIBUTION OF HEATER-CATHODE LEAKAGE

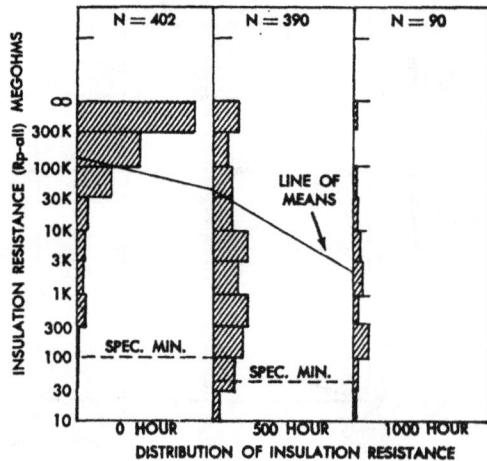

DISTRIBUTION OF INSULATION RESISTANCE

LIFE TEST PROPERTY BEHAVIOR
MIL-E-1/725 C   26 DEC 56
PRODUCED IN 1958 BY ONE MANUFACTURER

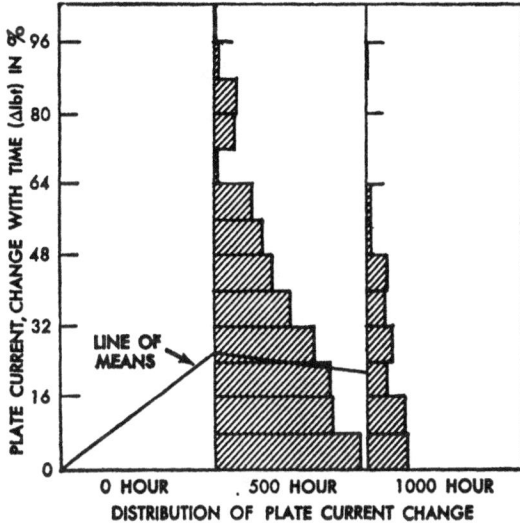

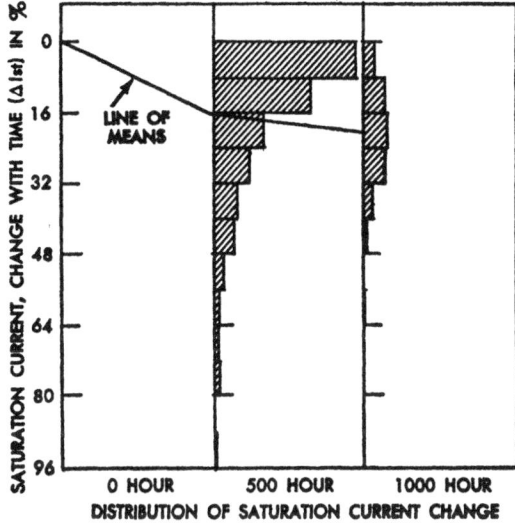

TUBE TYPE JAN-6205

DESCRIPTION:

The JAN-6205[1]/ is an 8 pin subminiature, sharp cutoff pentode having a transconductance in the range 4200 to 5800. The JAN-6205 is identical to JAN-5840 except having an external No. 3 grid connection and is similiar in plate characteristics to JAN-5702WA and the miniature type JAN-5654/6AK5W. Type 6205 has not been designed for control or gating purposes using the No. 3 grid.

ELECTRICAL: The electrical characteristics are as follows:
Heater Voltage...6.3 V
Heater Current...140-160 mA
Cathode...Coated Unipotential

MOUNTING: Any type mounting is adequate.

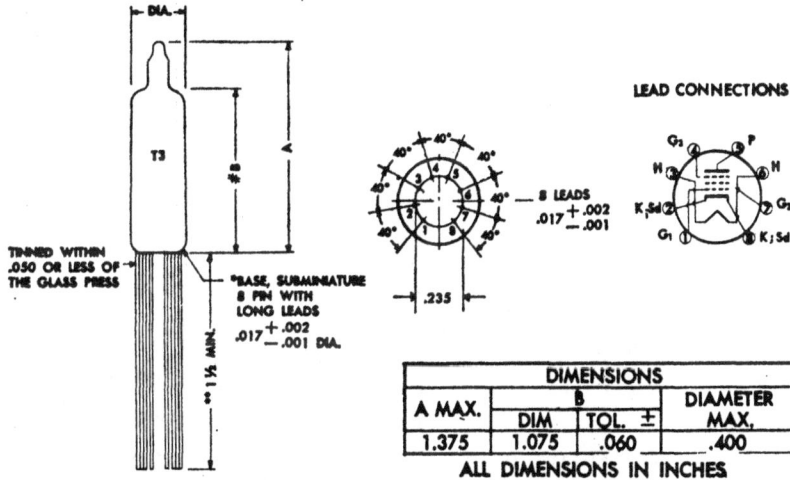

DIMENSIONS			
A MAX.	B		DIAMETER MAX.
	DIM	TOL. ±	
1.375	1.075	.060	.400

ALL DIMENSIONS IN INCHES

MEASURE FROM BASE SEAT TO BULB TOP-LINE AS DETERMINED BY RING GAGE OF .210 ± .001.

* LEAD DIAMETER TOLERANCE SHALL GOVERN BETWEEN .050 FROM THE GLASS TO .250 FROM THE GLASS.

** ALTERNATIVE LEAD LENGTH SHALL BE .200 ± .015 WHEN CUT LEADS ARE REQUIRED BY PROCUREMENT CONTRACT OR TSS. CUT LEADS SHALL BE ESSENTIALLY SQUARE CUT AND THE MAXIMUM BURR SHALL BE .003 INCREASE OVER THE ACTUAL LEAD DIAMETER.

RATINGS:	Ef	Eb	Ec1	Ec2	Ec3	Ehk	Rk	Rg1	Ik*	Pp	Pg2	T Envelope	Alt
Absolute	V	Vdc	Vdc	Vdc	Vdc	v	ohms	Meg	mAdc	W	W	°C	ft
Maximum	6.6	165	0	155	22	200	---	1.1	16.5	0.80	0.35	+220	60,000 Note.
Minimum	6.0	---	-55	---	---	---	---	---	---	---	---	---	---
Test Cond:	6.3	100	0	100	0	0	150	---	---	---	---	---	---

1/ The value and specification comments presented in this section are related to MIL-E-1/140B dated 5 Aug 1955.

* Difficulty may be encountered if this tube is operated for long periods of time with very small values of cathode current. No specification assurance of life exists under conditions of cathode current approaching the maximum.

Note: If altitude rating is exceeded, reduction of instantaneous voltages (Ef excluded may be required.

191

ACCEPTANCE TEST LIMITS SUMMARY

PROPERTY		MEASUREMENT CONDITIONS	INITIAL		500 HR LIFE TEST		1000 HR LIFE TEST		UNITS
			MIN	MAX	MIN	MAX	MIN	MAX	
Heater Current	If		140	160	138	164	–	–	mA
Transconductance	Sm		4200	5800	–	–	–	–	umhos
Change in individual	Δ Smt		–	–	–	20	–	–	%
Change in average	AvgΔSmt		–	–	–	15	–	–	%
Transconductance change with Ef	ΔSmEf	Ef=5.7V	–	10	–	15	–	–	%
Plate Resistance	rp		0.175	–	–	–	–	–	Meg
Plate Current (1)	Ib		5.5	9.5	–	–	–	–	mAdc
Plate Current (2)	Ib	Ec1=-9.0Vdc; Rk=0	–	50	–	–	–	–	uAdc
Screen Grid Current	Ic2		1.5	3.3	–	–	–	–	mAdc
Capacitance		0.405 in. dia shield, Ef=0							
	C g1p		–	0.015	–	–	–	–	uuf
	C in		3.5	4.9	–	–	–	–	uuf
	C out		2.9	3.9	–	–	–	–	uuf
Control Grid Current	Ic1	Rg1=1.0Meg	0	-0.3	0	-0.8	–	–	uAdc
Control Grid Emission	Ic1	Ef=7.5V;Ec1= -9.0Vdc;Rg1= 1.0Meg;Rk=0 Note	0	-0.5	–	–	–	–	uAdc
Heater Cathode Leakage	Ihk	Ehk=+100Vdc	–	5.0	–	10	–	–	uAdc
	Ihk	Ehk=-100Vdc	–	5.0	–	10	–	–	uAdc
Insulation of Electrodes	R(g1-all)	Eg1-all=-100V	100	–	50	–	–	–	Meg
	R(p-all)	Ep-all=-300V	100	–	50	–	–	–	Meg

Measurement conditions are the same as stated under Test Conditions, unless otherwise indicated.

Note: The tube shall be preheated a minimum of five minutes at test conditions for this test (except Ec1=0; Rk=150 ohms) prior to this test.

DESIGN CENTER CHARACTERISTICS
OBTAINED FROM DATA PUBLISHED BY ORIGINAL
RETMA REGISTRANT

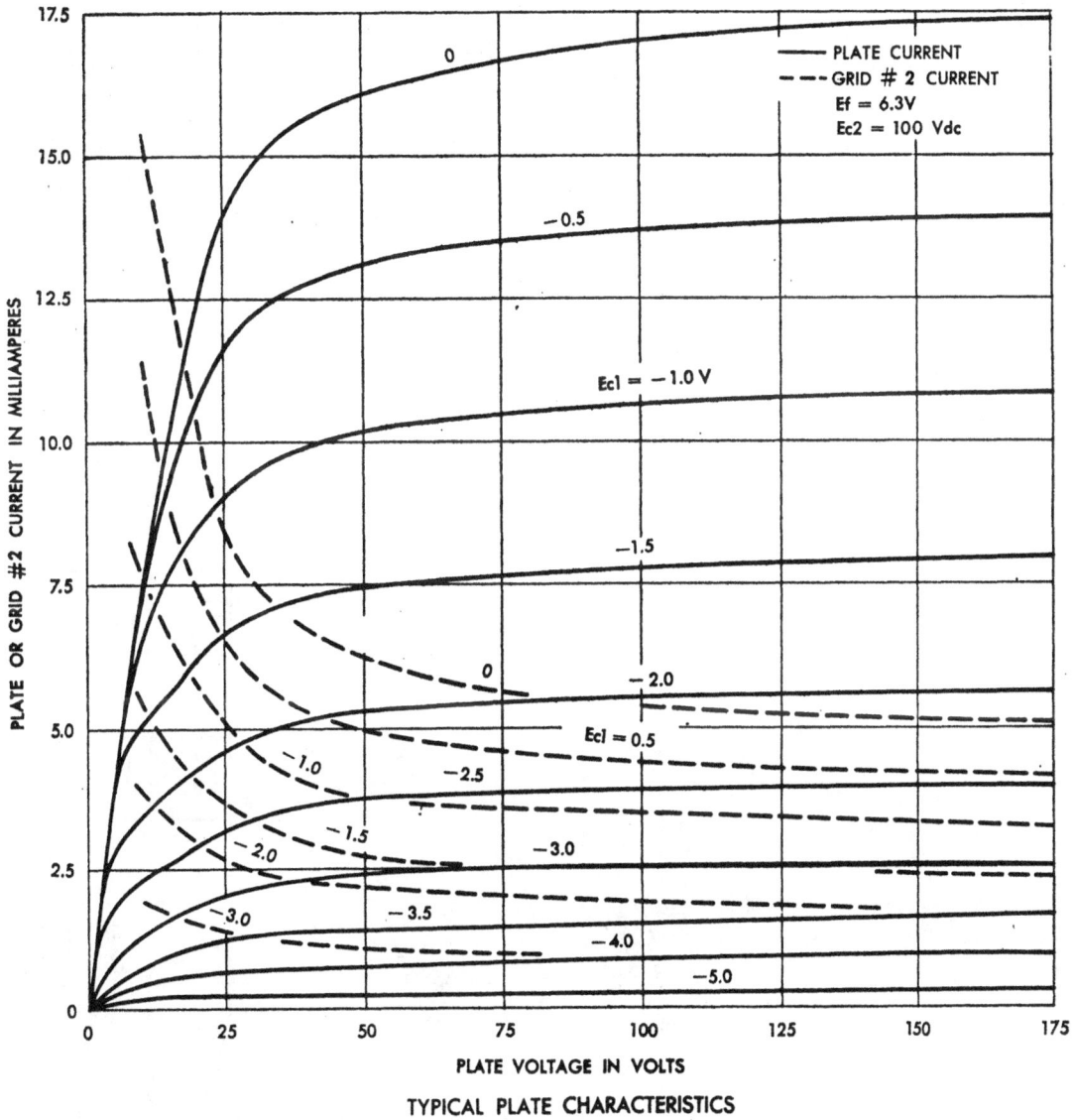

TYPICAL PLATE CHARACTERISTICS

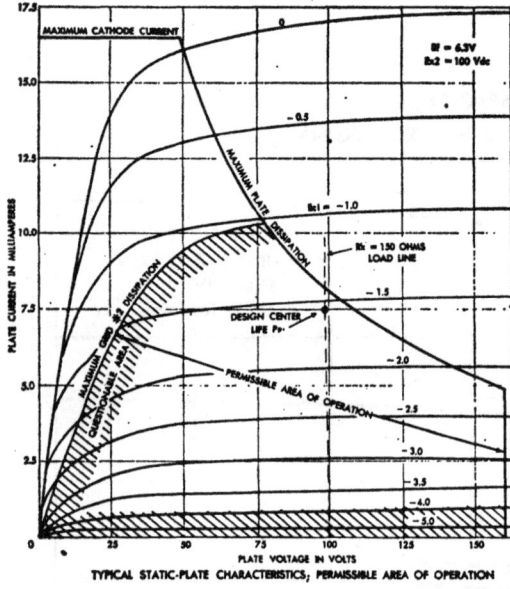

TYPICAL STATIC-PLATE CHARACTERISTICS; PERMISSIBLE AREA OF OPERATION

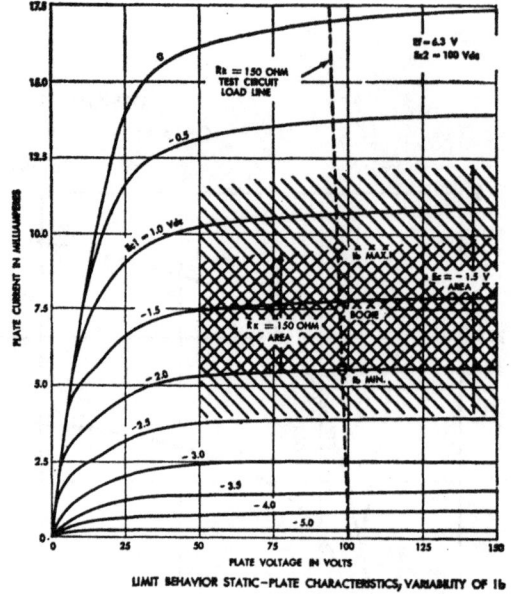

LIMIT BEHAVIOR STATIC-PLATE CHARACTERISTICS, VARIABILITY OF Ib

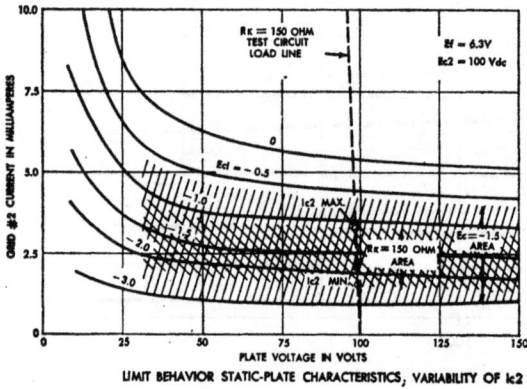

LIMIT BEHAVIOR STATIC-PLATE CHARACTERISTICS, VARIABILITY OF Ic2

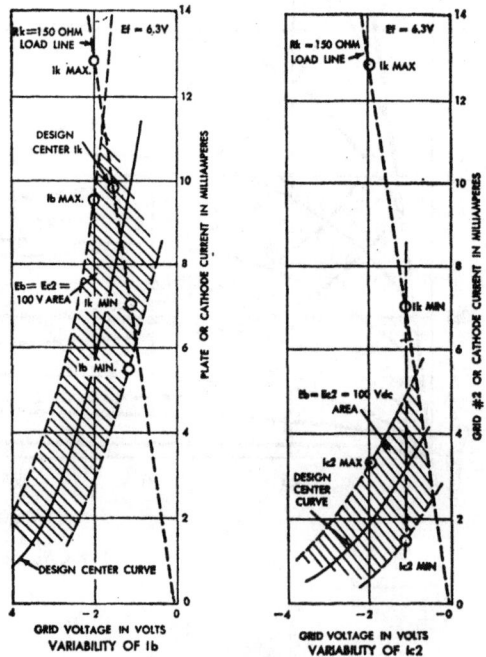

VARIABILITY OF Ib

VARIABILITY OF Ic2

LIMIT BEHAVIOR TRANSFER CHARACTERISTICS

194

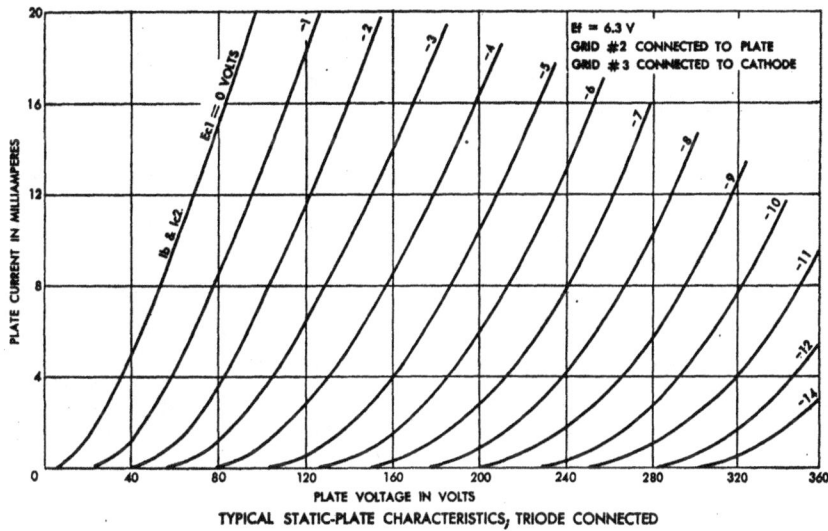

Ef = 6.3 V
GRID #2 CONNECTED TO PLATE
GRID #3 CONNECTED TO CATHODE

TYPICAL STATIC-PLATE CHARACTERISTICS, TRIODE CONNECTED

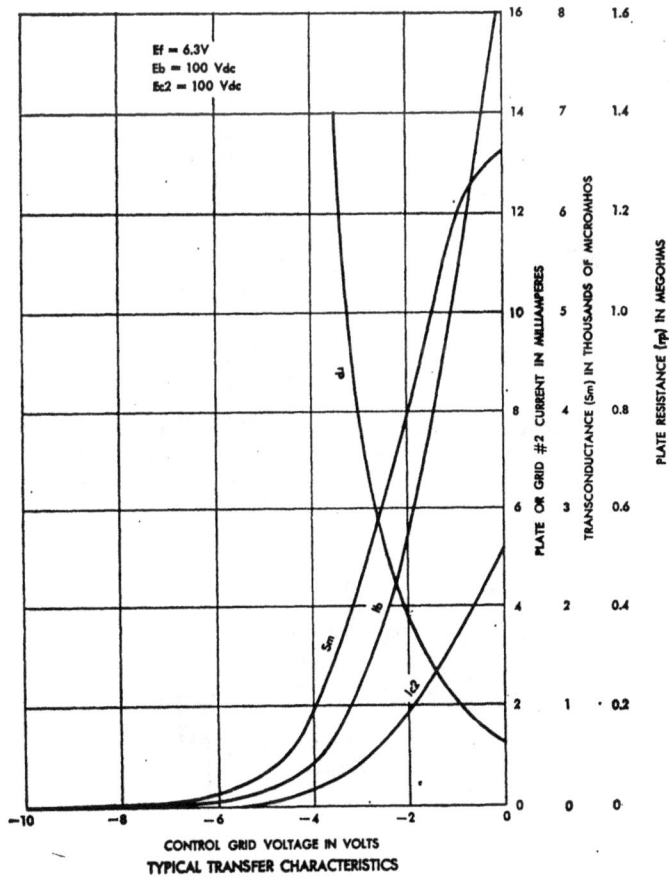

Ef = 6.3V
Eb = 100 Vdc
Ec2 = 100 Vdc

TYPICAL TRANSFER CHARACTERISTICS

195

LIFE TEST PROPERTY BEHAVIOR
MIL-E-1/140B   5 AUG. '55
PRODUCED IN 1957-'58 BY TWO MANUFACTURERS

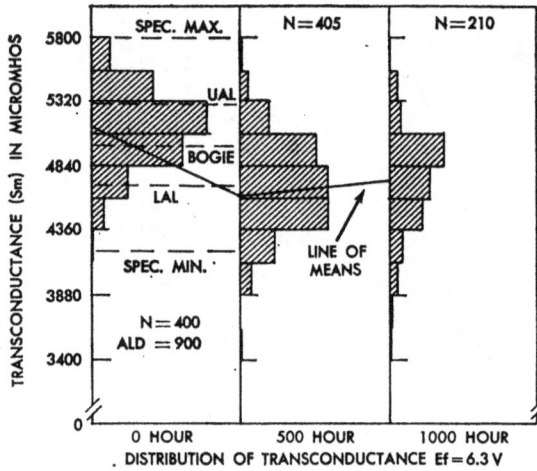

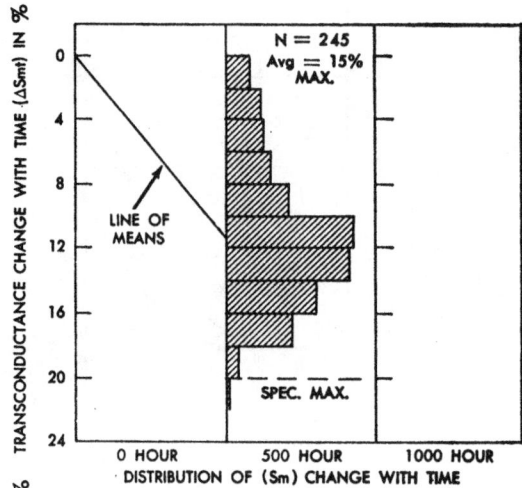

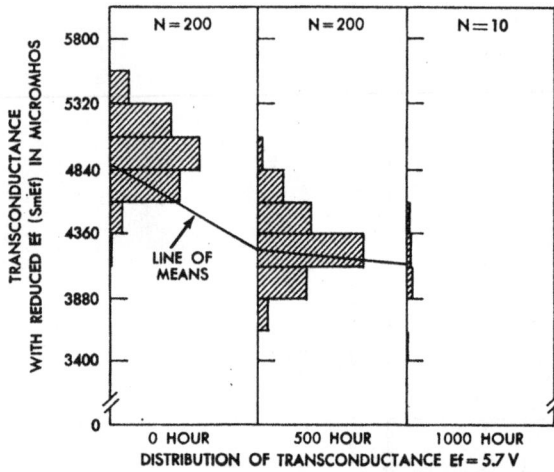

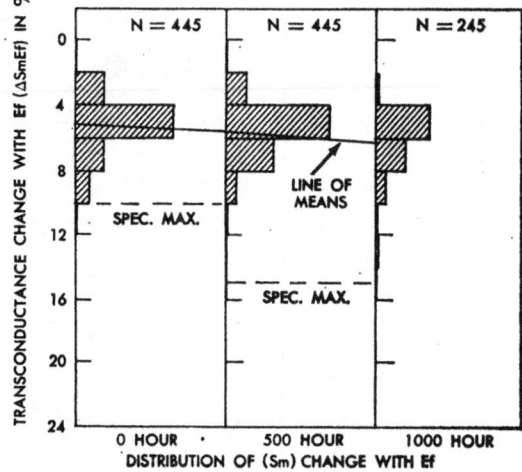

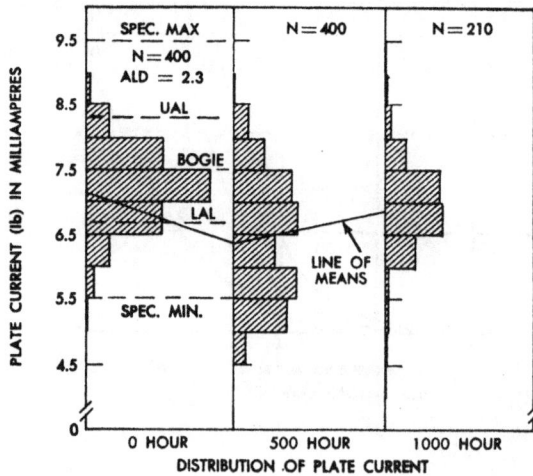

196

LIFE TEST PROPERTY BEHAVIOR
MIL-E-1/140B   5 AUG. '55
PRODUCED IN 1957-'58 BY TWO MANUFACTURERS

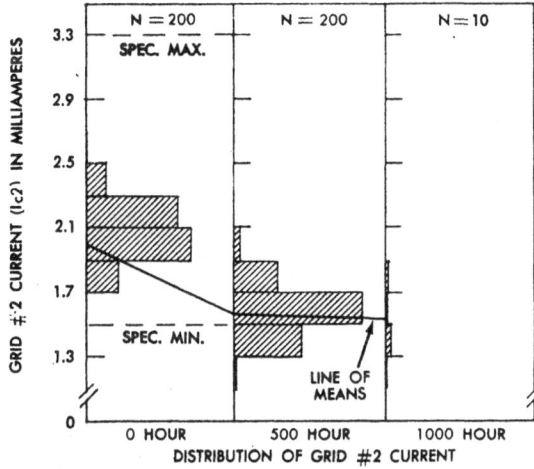

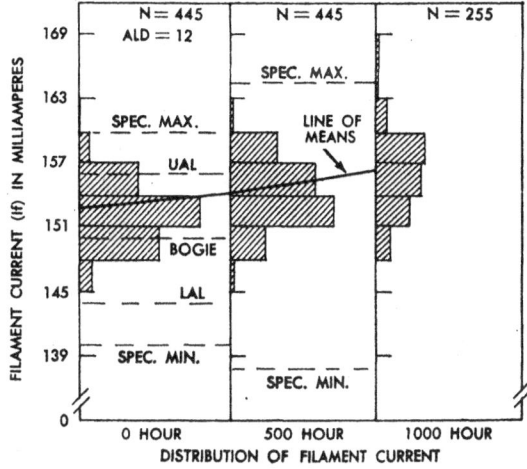

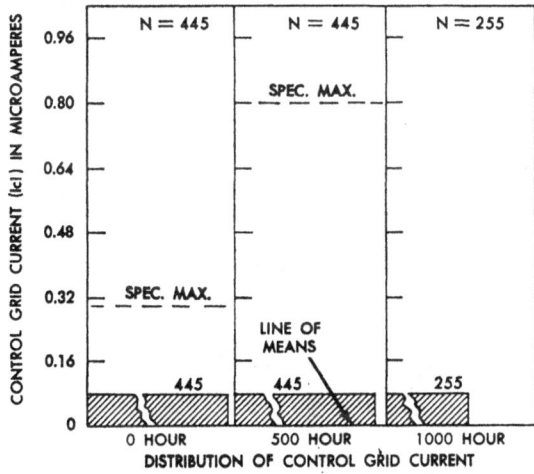

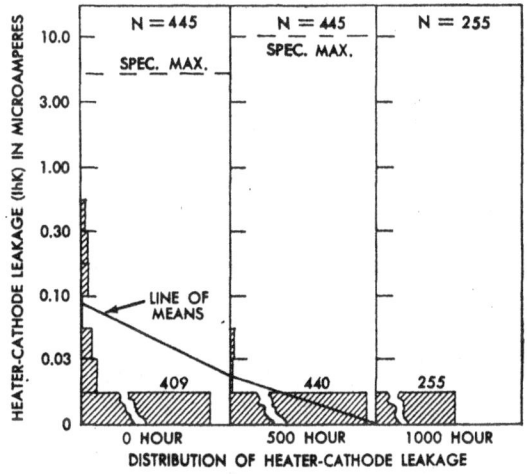

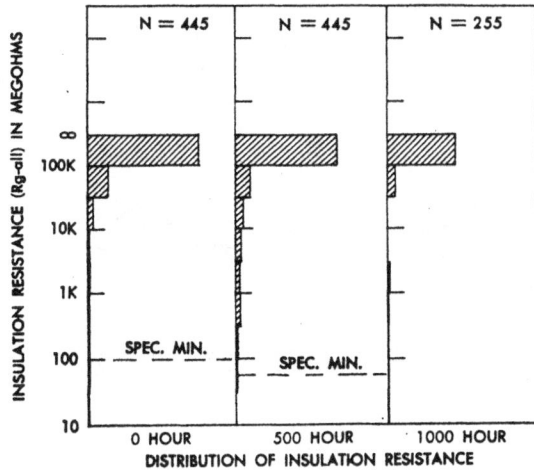

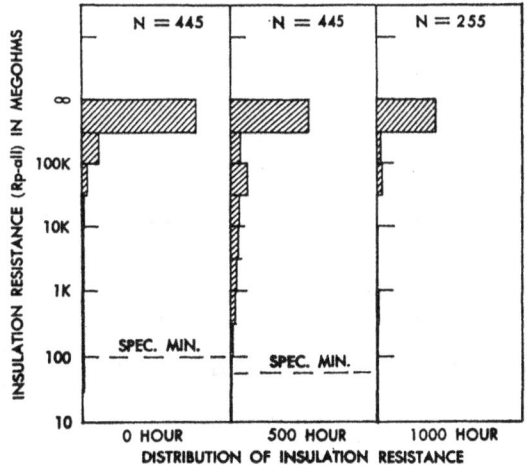

197

LIFE TEST PROPERTY BEHAVIOR
PROBABILITY OF SURVIVAL
MIL-E-1/140B   5 AUG. '55
PRODUCED IN 1957-'58 BY TWO MANUFACTURERS

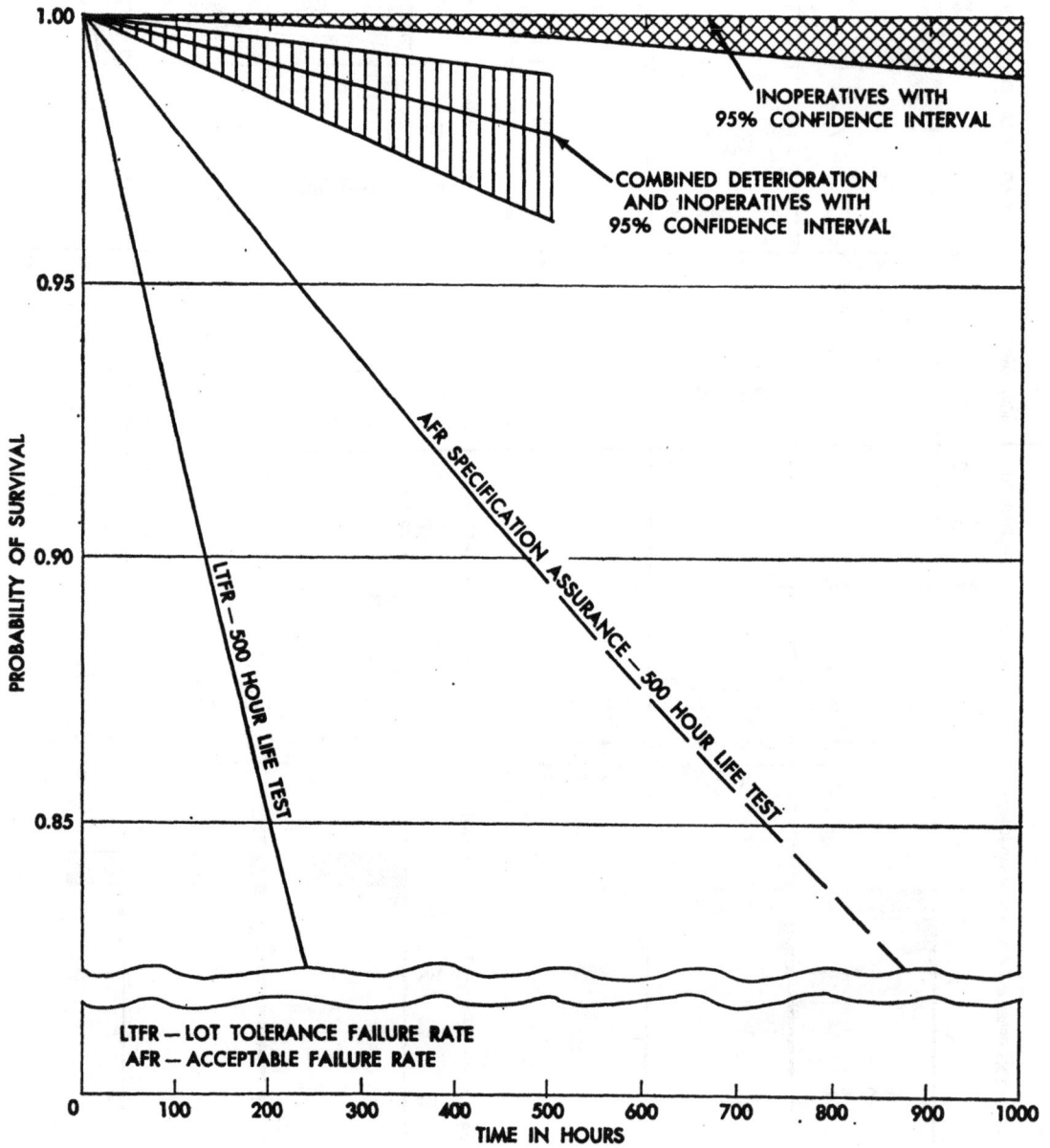

INOPERATIVES WITH
95% CONFIDENCE INTERVAL

COMBINED DETERIORATION
AND INOPERATIVES WITH
95% CONFIDENCE INTERVAL

AFR SPECIFICATION ASSURANCE—500 HOUR LIFE TEST

LTFR—500 HOUR LIFE TEST

LTFR — LOT TOLERANCE FAILURE RATE
AFR — ACCEPTABLE FAILURE RATE

PROBABILITY OF SURVIVAL

TIME IN HOURS

TUBE TYPE JAN-6206

DESCRIPTION:

The JAN-6206[1] is an 8 lead, button base, subminiature, semi-remote cut-off pentode having a transconductance in the range of 3800 to 5200. The JAN-6206 is identical to Type 5899 except for an external suppressor grid connection.

ELECTRICAL: The electrical characteristics are as follows:
Heater Voltage...6.3 V
Heater Current..140-160 A
Cathode...Coated Unipotential

MOUNTING: Any type mounting is adequate.

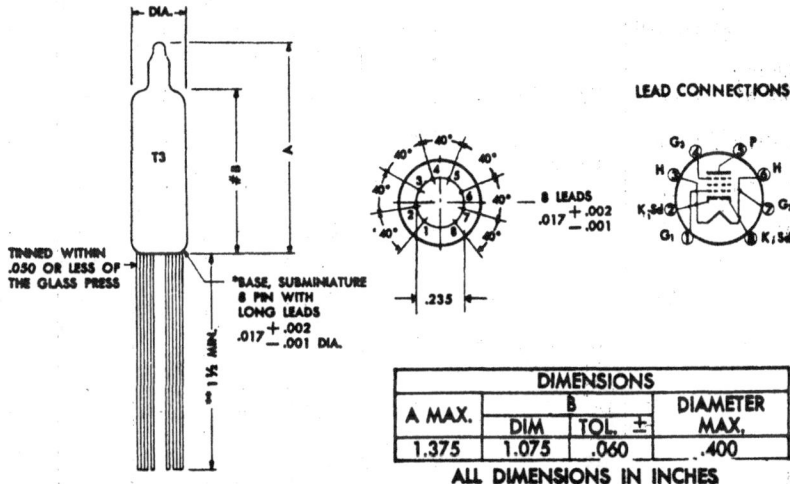

DIMENSIONS			
A MAX.	B		DIAMETER MAX.
	DIM	TOL. ±	
1.375	1.075	.060	.400

ALL DIMENSIONS IN INCHES

\# MEASURE FROM BASE SEAT TO BULB TOP-LINE AS DETERMINED BY RING GAGE OF .210 ± .001.

* LEAD DIAMETER TOLERANCE SHALL GOVERN BETWEEN .050 FROM THE GLASS TO .250 FROM THE GLASS.

** ALTERNATIVE LEAD LENGTH SHALL BE .200 ± .015 WHEN CUT LEADS ARE REQUIRED BY PROCUREMENT CONTRACT OR TSS. CUT LEADS SHALL BE ESSENTIALLY SQUARE CUT AND THE MAXIMUM BURR SHALL BE .003 INCREASE OVER THE ACTUAL LEAD DIAMETER.

RATINGS:	E_f	E_b	E_{c1}	E_{c2}	E_{c3}	E_{hk}	R_k	R_{g1}	I_k*	P_p	P_{g2}	T Envelope	Alt
Absolute	V	Vdc	Vdc	Vdc	Vdc	v	ohms	Meg	mAdc	W	W	°C	ft
Maximum Design	6.6	165	0	155	22	200	---	1.1	16.5	---	---	+220	60,000
Maximum	---	---	---	---	---	---	---	---	---	0.85	0.25	---	---
Minimum	6.0	---	-55	---	---	---	---	---	---	---	---	---	---
Test Cond:	6.3	100	0	100	0	0	120	---	---	---	---	---	---

[1] The values and specification comments presented in this section are related to MIL-E-1/97D dated 22 Oct 1957.

● Difficulty may be encountered if this tube is operated for long periods of time with very small values of cathode current. No specification assurance of life exists under conditions of cathode current approaching the maximum.

MIL-HDBK-211
APPENDIX-A
12 January 1960
JAN-6206

ACCEPTANCE TEST LIMITS SUMMARY

PROPERTY	MEASUREMENT CONDITIONS	INITIAL		500 HR LIFE TEST		1000 HR LIFE TEST		UNITS
		MIN	MAX	MIN	MAX	MIN	MAX	
Heater Current If		140	160	138	164	-	-	mA
Transconductance(1) Sm		3800	5200	-	-	-	-	umhos
Change in individual Δ Smt		-	-	-	20	-	-	%
Change in average Avg Δ Smt		-	-	-	15	-	-	%
Transconductance change with Ef Δ SmEf	Ef=5.7V	-	10	-	15	-	-	%
Transconductance(3) Sm	Ec1=-14Vdc; Rk=0	1.0	75	-	-	-	-	umhos
Plate Resistance rp		0.175	-	-	-	-	-	Meg
Plate Current Ib		5.2	9.2	-	-	-	-	mAdc
Screen Grid Current Ic2		1.0	3.0	-	-	-	-	mAdc
Capacitance	0.405 in.dia. shield,Ef=0							
C g1p		-	0.015	-	-	-	-	uuf
C in		3.5	4.5	-	-	-	-	uuf
C out		2.9	3.9	-	-	-	-	uuf
Control Grid Current Ic1	Rg1=1.0 Meg	0	-0.3	0	-0.8	-	-	uAdc
Control Grid Emission Ic1	Ef=7.5V;Ec1= -14Vdc;Rg1= 1.0Meg;Rk=0; NOTE	0	-0.5	-	-	-	-	uAdc
Heater Cathode Leakage Ihk	Ehk=+100Vdc	-	5.0	-	10	-	-	uAdc
Ihk	Ehk=-100Vdc	-	5.0	-	10	-	-	uAdc
Insulation of R(g1-all)	Eg1-all=-100V	100	-	50	-	-	-	Meg
Electrodes R(p-all)	Ep-all=-300V	100	-	50	-	-	-	Meg

Measurement conditions are the same as stated under Test Conditions, unless otherwise indicated.

NOTE: The tube shall be preheated a minimum of five minutes at test conditions for this test (except Ec1=0; Rk=120 ohms) prior to this test.

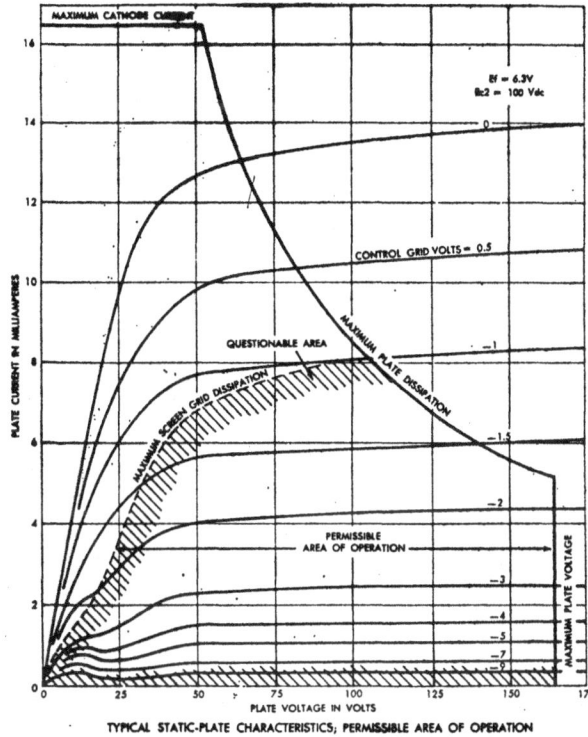

TYPICAL STATIC-PLATE CHARACTERISTICS; PERMISSIBLE AREA OF OPERATION

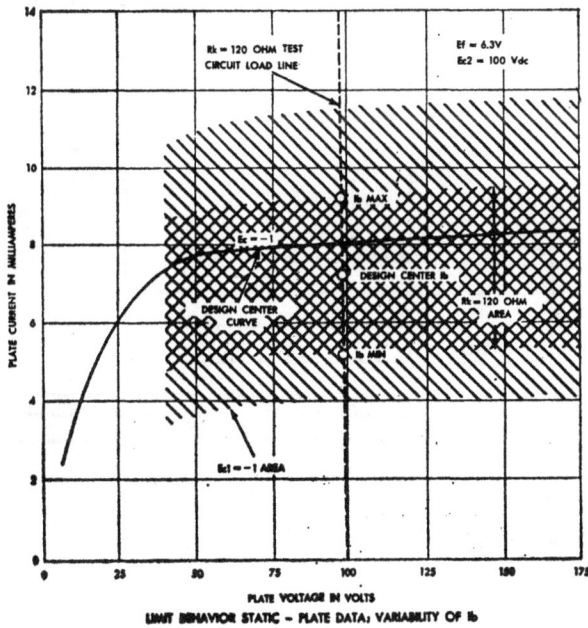

LIMIT BEHAVIOR STATIC - PLATE DATA; VARIABILITY OF Ib

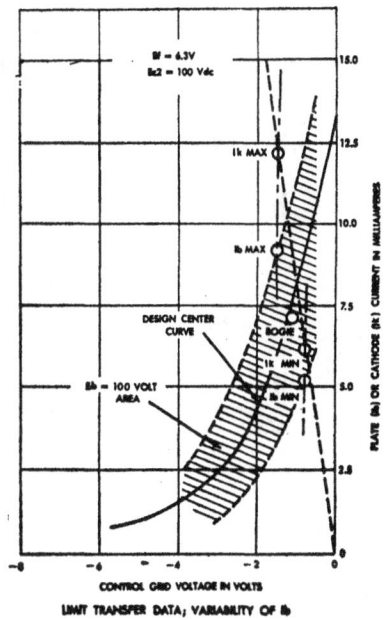

LIMIT TRANSFER DATA; VARIABILITY OF Ib

DESIGN CENTER CHARACTERISTICS
OBTAINED FROM DATA PUBLISHED BY ORIGINAL
RETMA REGISTRANT

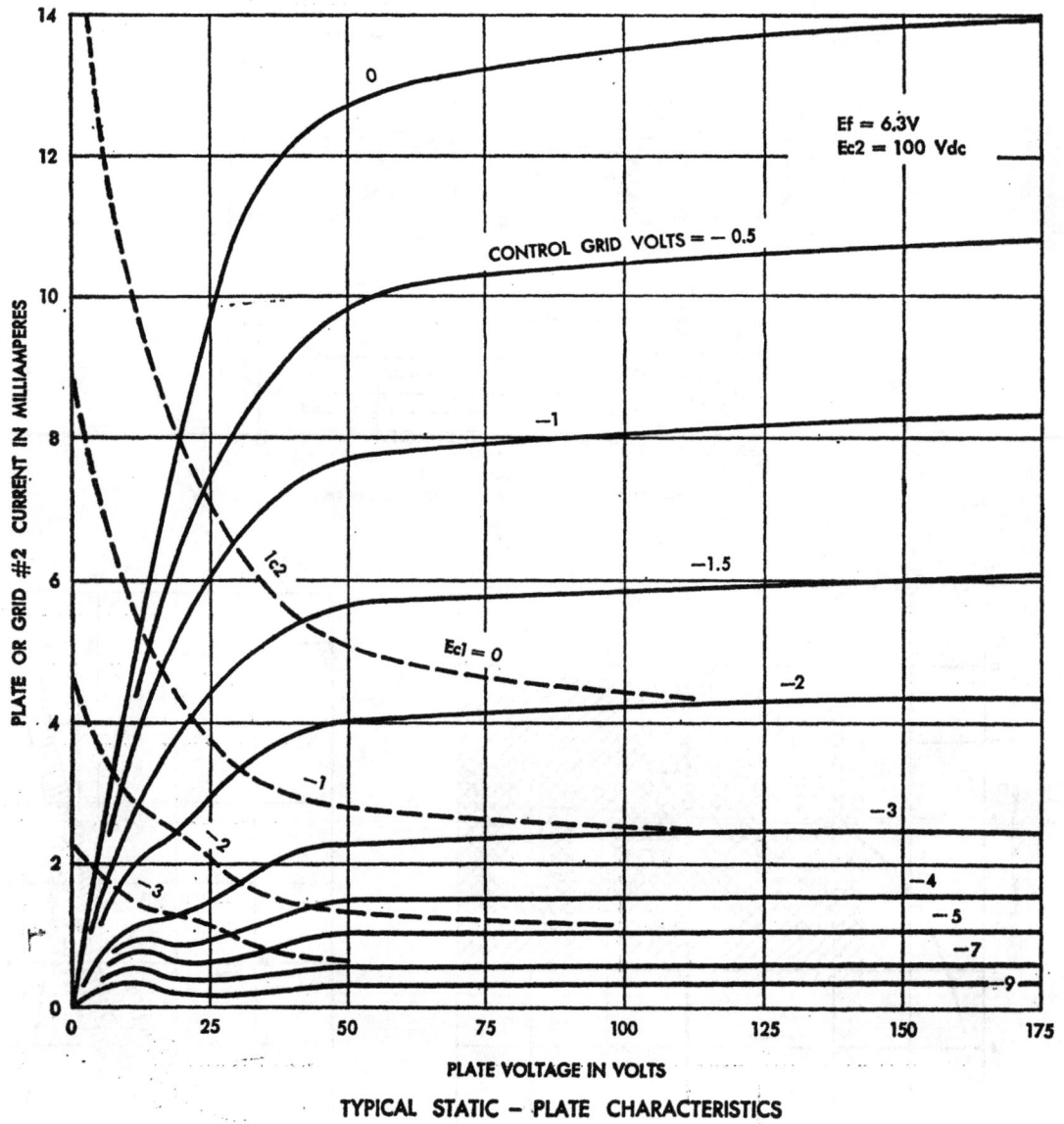

TYPICAL STATIC - PLATE CHARACTERISTICS

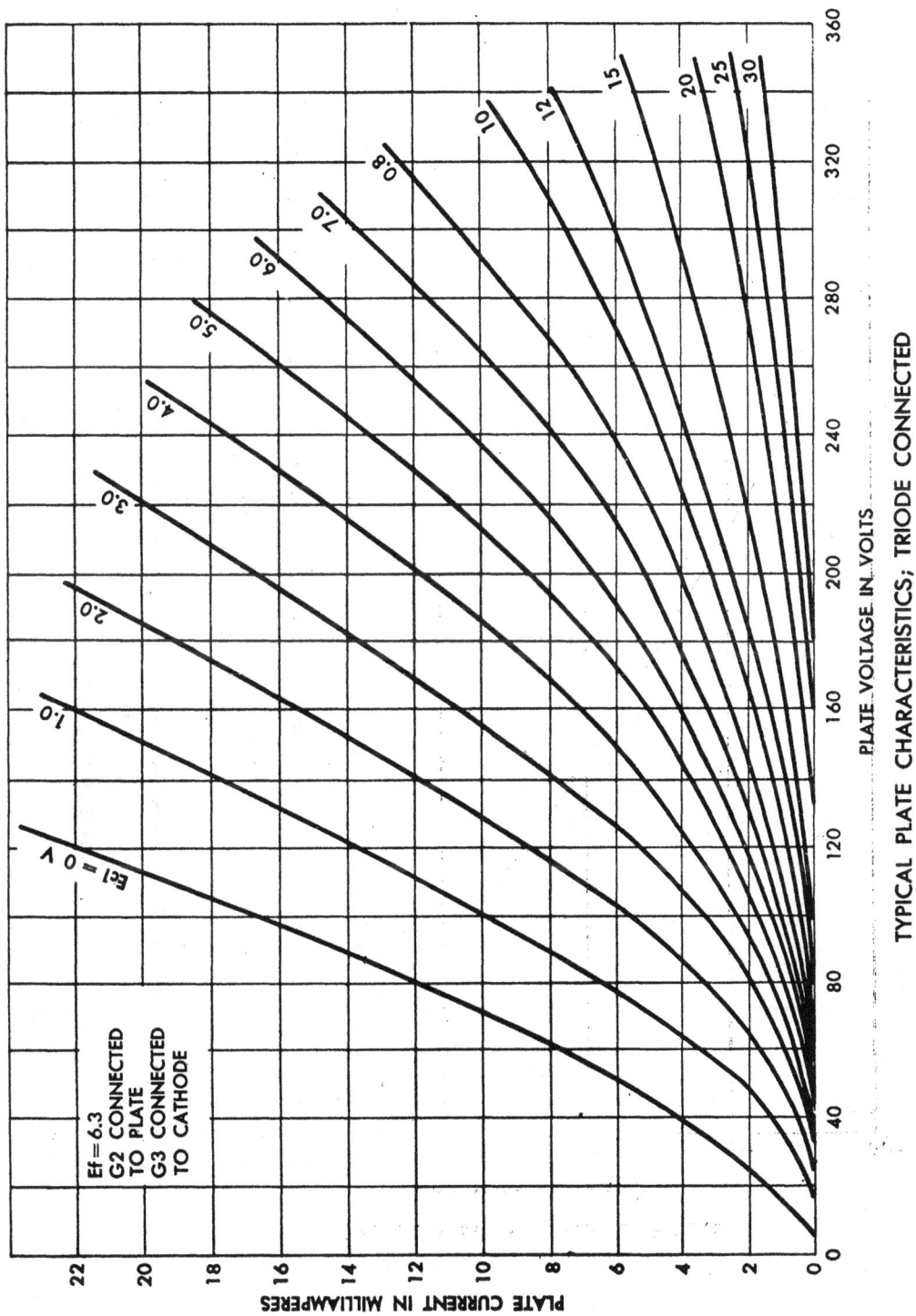

TYPICAL PLATE CHARACTERISTICS; TRIODE CONNECTED

Ef = 6.3V
Eb = 100 Vdc
Ec2 = 100 Vdc

CONTROL GRID VOLTS

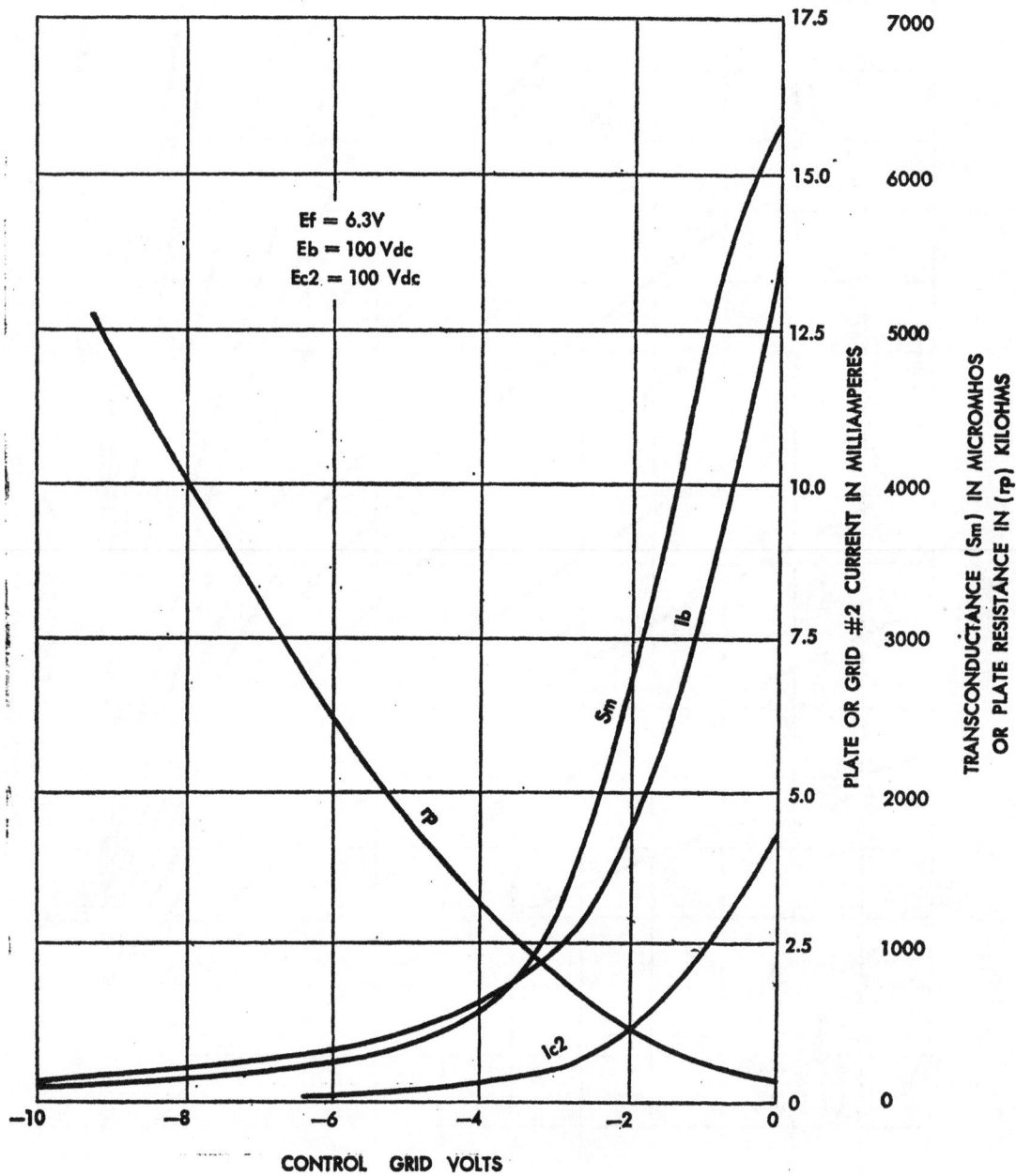

TYPICAL TRANSFER CHARACTERISTICS

PLATE OR GRID #2 CURRENT IN MILLIAMPERES

TRANSCONDUCTANCE (Sm) IN MICROMHOS
OR PLATE RESISTANCE IN (rp) KILOHMS

LIFE TEST PROPERTY BEHAVIOR
MIL-E-1/97D   22 OCT. '57
PRODUCED IN 1957-'58 BY ONE MANUFACTURER

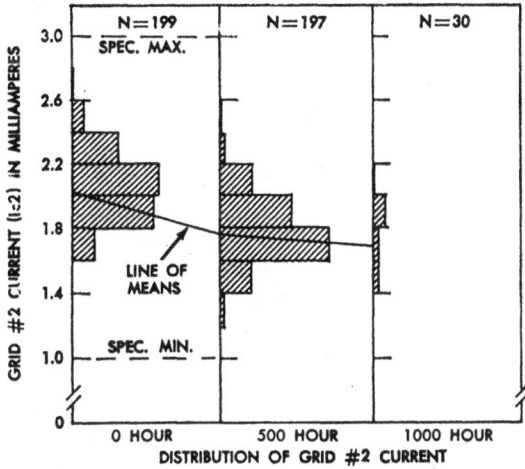

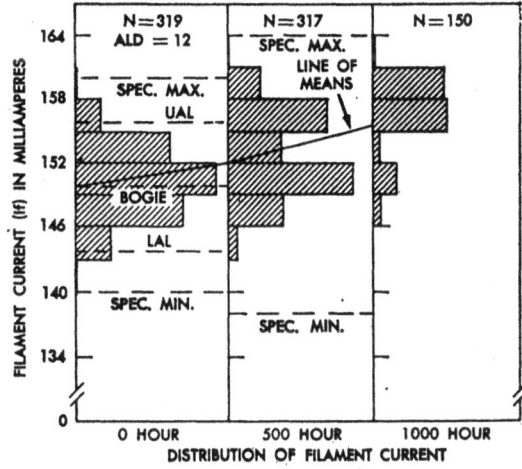

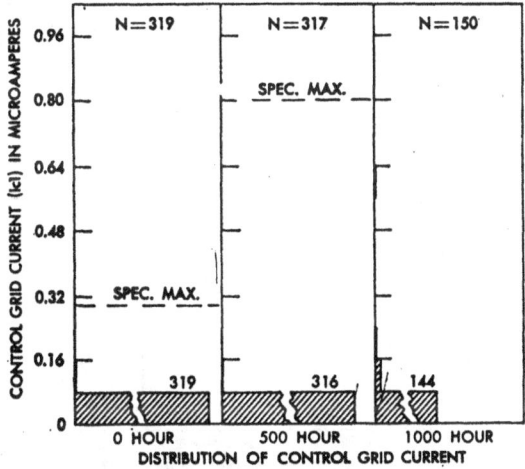

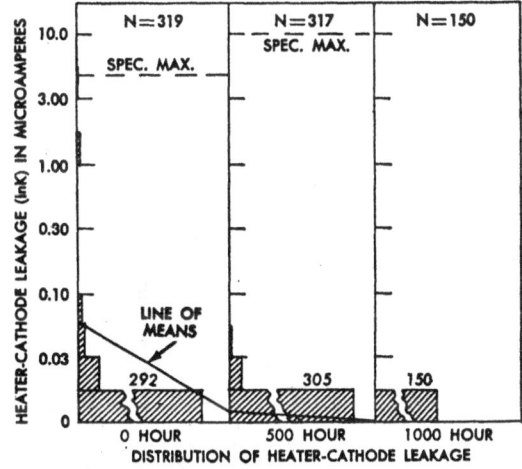

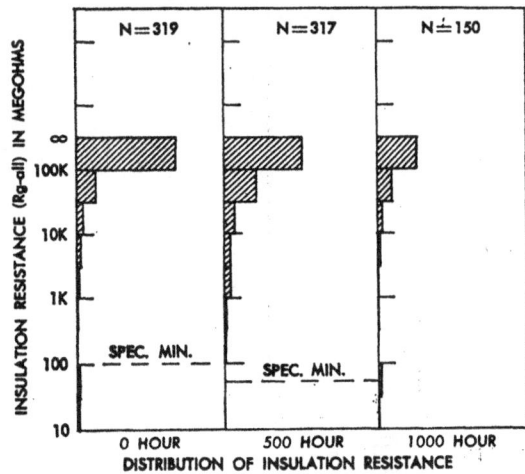

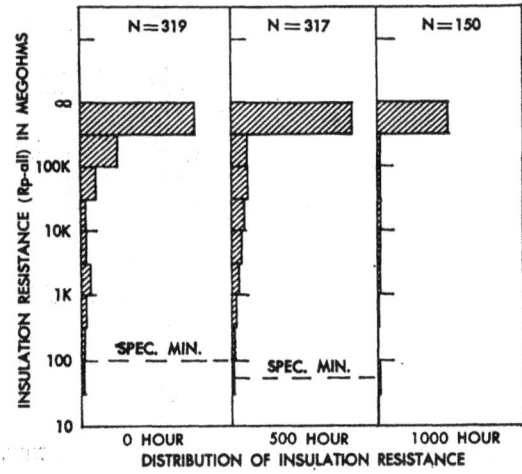

205

LIFE TEST PROPERTY BEHAVIOR
MIL-E-1/97D   22 OCT. '57
PRODUCED IN 1957-'58 BY TWO MANUFACTURERS

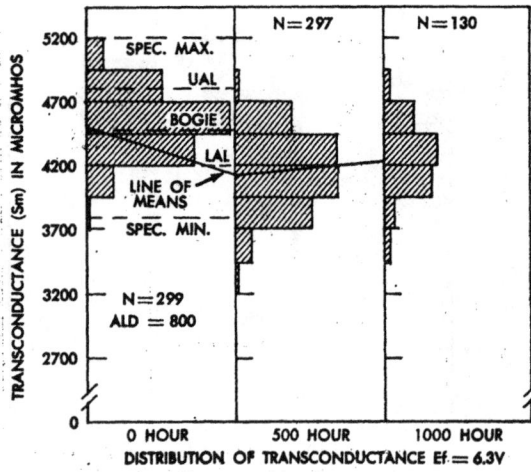

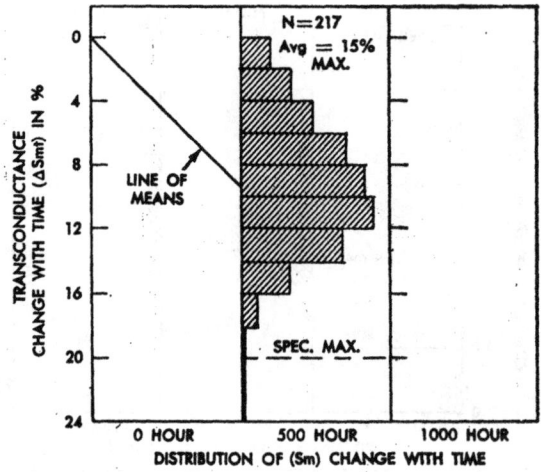

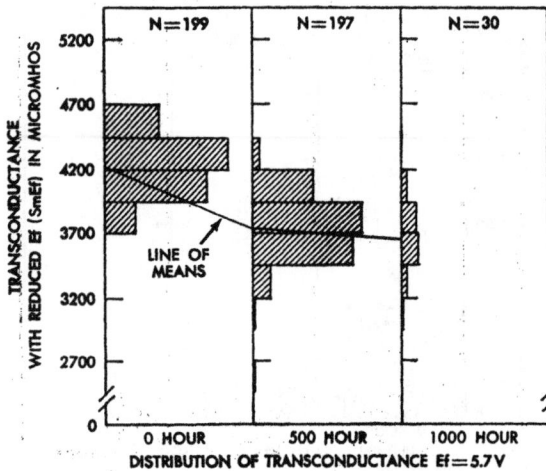

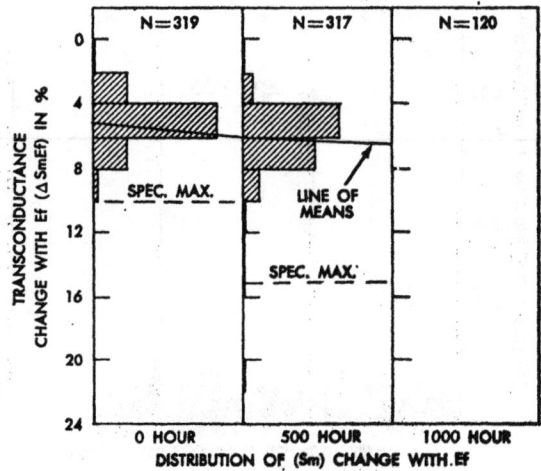

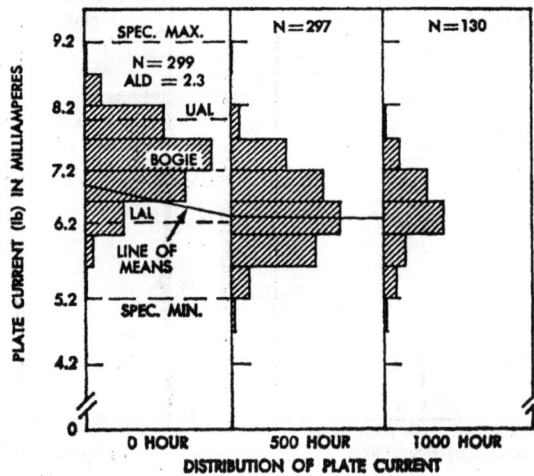

DISTRIBUTION OF TRANSCONDUCTANCE Ef = 6.3V

DISTRIBUTION OF (Sm) CHANGE WITH TIME

DISTRIBUTION OF TRANSCONDUCTANCE Ef = 5.7V

DISTRIBUTION OF (Sm) CHANGE WITH Ef

DISTRIBUTION OF PLATE CURRENT

206

LIFE TEST PROPERTY BEHAVIOR
PROBABILITY OF SURVIVAL
MIL-E-1/97D   22 OCT. '57
PRODUCED IN 1957-'58 BY TWO MANUFACTURERS

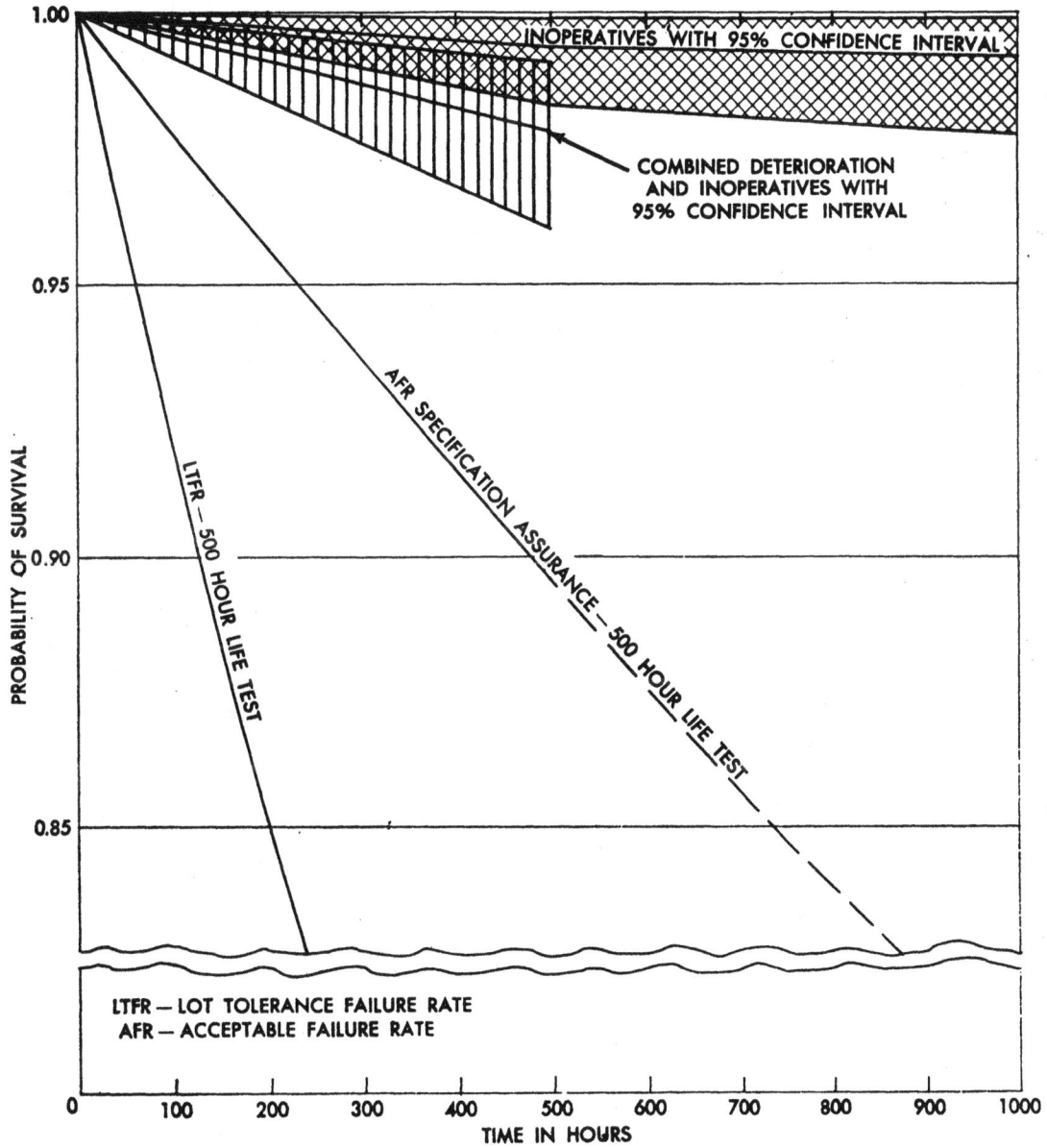

TUBE TYPE JAN-6222

DESCRIPTION:

The JAN-6222[1] is an 8 pin subminiature triode, with a Mu in the range of 60 to 80 and a transconductance ranging from 1400 to 2000.

ELECTRICAL: The electrical characteristics are as follows:
Heater Voltage...6.3 V
Heater Current...167-183 mA
Cathode..Coated Unipotential

MOUNTING: Any type mounting is adequate.

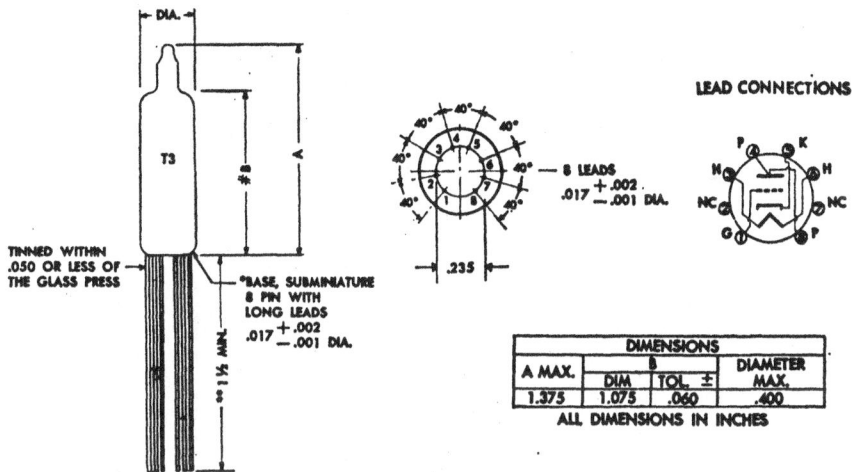

LEAD CONNECTIONS

8 LEADS .017 +.002 -.001 DIA.

.235

*BASE, SUBMINIATURE 8 PIN WITH LONG LEADS .017 +.002 -.001 DIA.

TINNED WITHIN .050 OR LESS OF THE GLASS PRESS

DIMENSIONS			
A MAX.	DIM	TOL ±	DIAMETER MAX.
1.375	1.075	.060	.400

ALL DIMENSIONS IN INCHES

MEASURE FROM BASE SEAT TO BULB TOP-LINE AS DETERMINED BY RING GAGE OF .210 ± .001.

* LEAD DIAMETER TOLERANCE SHALL GOVERN BETWEEN .050 FROM THE GLASS TO .250 FROM THE GLASS.

** ALTERNATIVE LEAD LENGTH SHALL BE .200 ± .015 WHEN CUT LEADS ARE REQUIRED BY PROCUREMENT CONTRACT OR TSS. CUT LEADS SHALL BE ESSENTIALLY SQUARE CUT AND THE MAXIMUM BURR SHALL BE .003 INCREASE OVER THE ACTUAL LEAD DIAMETER.

RATINGS:	Ef	Eb	Ec	Ehk	Rk	Rg	Ib	Ic	Pp*	T Envelope	Alt
Absolute	V	Vdc	Vdc	v	ohms	Meg	mAdc	mAdc	W	°C	ft
Maximum	6.6	165	0	+200	---	1.2	3.3	---	---	220	80,000
Design Maximum	---	---	---	---	---	---	---	---	0.30	---	---
Minimum	6.0	---	-55	-200	---	---	---	---	---	---	---
Test Cond:	6.3	100	0	0	1500	---	---	---	---	---	---

1/ The value and specification comments presented in this section are related to MIL-E-1/955A dated 22 Oct 1957.

*   No test at this rating exists in the specification.

** Difficulty may be encountered if this tube is operated for long periods of time with very small values of cathode current. No specification assurance of life exists under conditions of cathode approaching the maximum.

'TANCE TEST LIMITS SUMMARY

PROPERTY		MEASUREMENT CONDITIONS	INITIAL		500 HR LIFE TEST		1000 HR LIFE TEST		UNITS
			MIN	MAX	MIN	MAX	MIN	MAX	
Heater Current	If		167	183	165	185	160	190	mA
Transconductance	Sm		1400	2000	-	-	-	-	umhos
Change in individual	ΔSmt		-	-	-	15	-	20	%
Change in average	AvgΔSmt		-	-	-	15	-	-	%
Transconductance change with Ef	ΔSmEf	Ef=5.7V	-	5	-	15	-	-	%
Amplification Factor	Mu		60	80	-	-	-	-	-
Plate Current (1)	Ib		0.50	0.90	-	-	-	-	mAdc
Plate Current (2)	Ib	Ec=-2.5Vdc	-	50	-	-	-	-	uAdc
Plate Current (3)	Ib	Ec=-1.8Vdc	5	-	-	-	-	-	.uAdc
Capacitance	C gp	No shield,Ef=0	1.1	1.5	-	-	-	-	uuf
	C in		1.5	2.5	-	-	-	-	uuf
	C out		0.45	0.85	-	-	-	-	uuf
Control Grid Emission	Ic	Ef=7.5V;Ec=-2.5Vdc;Rg=1.0 Meg; Note	0	-0.3	-	-	-	-	uAdc
Control Grid Current	Ic	Eb=150Vdc;Rk=2700;Rg=1.0Meg	0	-0.3	0	-0.6	0	-0.8	uAdc
Heater Cathode Leakage	Ihk	Ehk=+100Vdc	-	5	-	10	-	15	uAdc
	Ihk	Ehk=-100Vdc	-	5	-	10	-	15	uAdc
Insulation of Electrodes	R(g1-all)	Eg1-all=-100V	100	-	50	-	-	-	Meg
	R(p-all)	Ep-all=-300V	100	-	50	-	-	-	Meg

Measurement conditions are the same as stated under Test Conditions, unless otherwise indicated.

Note: The tube shall be preheated a minimum of five minutes at test conditions for this test (except Ec=0) prior to this test.

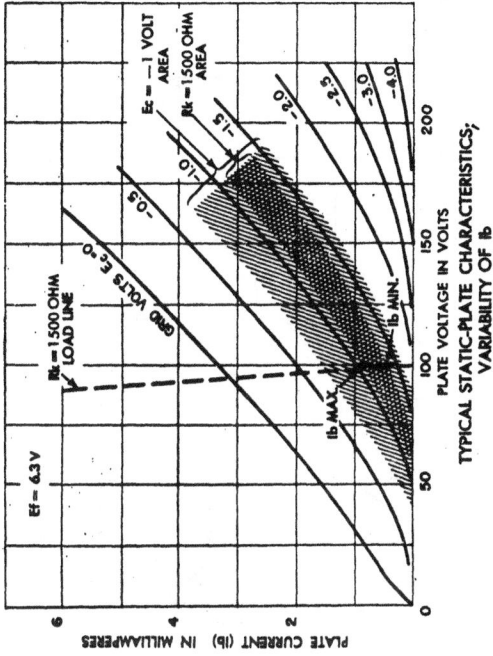

TYPICAL STATIC-PLATE CHARACTERISTICS;
VARIABILITY OF Ib

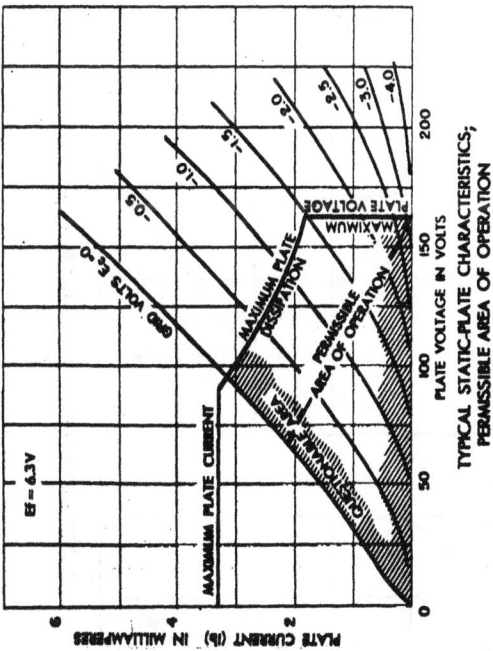

TYPICAL STATIC-PLATE CHARACTERISTICS;
PERMISSIBLE AREA OF OPERATION

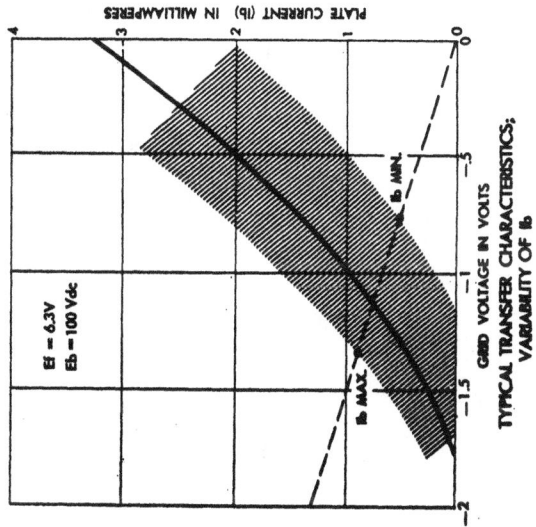

TYPICAL TRANSFER CHARACTERISTICS;
VARIABILITY OF Ib

DESIGN CENTER CHARACTERISTICS
OBTAINED FROM DATA PUBLISHED BY ORIGINAL
RETMA REGISTRANT

TYPICAL STATIC-PLATE CHARACTERISTICS

TYPICAL TRANSFER CHARACTERISTICS

TUBE TYPE JAN-6384

DESCRIPTION:

The JAN-6384[1] is a 6 pin octal base beam power pentode in a T-11 hard glass envelope. It is generic with JAN-6098 and JAN-6AR6.

ELECTRICAL: The electrical characteristics are as follows:
Heater Voltage..6.3 V
Heater Current..1.14-1.26 A
Cathode..Coated Unipotential

MOUNTING: Any type mounting is adequate.

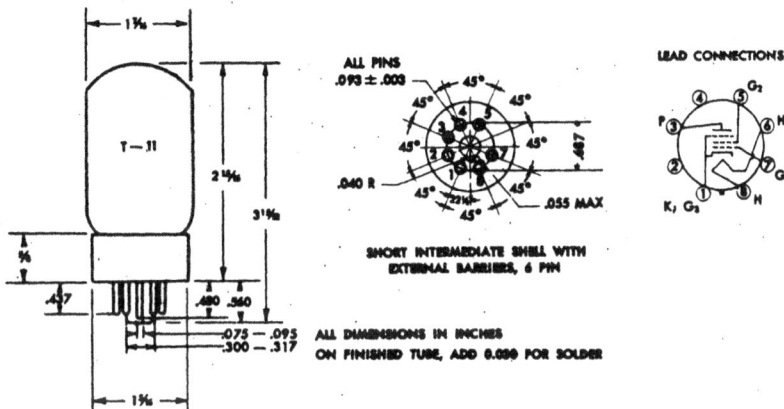

RATINGS:	Ef	Eb	Ec1	Ec2	Ehk	Rg1	Ik*	Ic1	Pp	Pg2	epy	ib	T Envelope	Alt
Absolute	V	Vdc	Vdc	Vdc	Vdc	Meg	mAdc	mAdc	W	W	v	a	°C	ft
Maximum	6.9	750	0	325	+450	0.1	125	+5.0	30	3.5	1500	1.0 Note	300	60,000
Minimum	5.7	---	-200	---	-450	---	---	---	---	---	---	---	---	---
Test Cond:	6.3	250	-22.5	250	0	---	---	---	---	---	---	---	---	---

1/ The values and specification comments presented in this section are related to MIL-E-1/1022 dated 28 Jun 1956.

* Difficulty may be encountered if this tube is operated for long periods of time with very small values of cathode current.

Note: See specification for duty factor limitations. Instantaneous plate potential shall not exceed 2500v. Screen potential shall not exceed 500V with a minimum series resistance of 20,000 ohms.

213

ACCEPTANCE TEST LIMITS SUMMARY

PROPERTY		MEASUREMENT CONDITIONS	INITIAL		500 HR LIFE TEST		1000 HR LIFE TEST		UNITS
			MIN	MAX	MIN	MAX	MIN	MAX	
Heater Current	If		1.14	1.26	1.12	1.28	1.12	1.28	A
Transconductance	Sm		4800	6000	4500	6000	4500	6000	umhos
Change in individual	Δ Smt		-	-	-	10	-	10	%
Transconductance change with Ef	Δ SmEf	Ef=5.7V	-	5	-	7	-	-	%
Plate Resistance	rp		0.015	-	-	-	-	-	Meg
Amplification Factor	Mu	TRI.Eb=Ec2=250	5.5	6.5	-	-	-	-	-
Plate Current (1)	Ib		66	88	-	-	-	-	mAdc
Plate Current (2)	Ib	Ec1=-60Vdc	-	0.5	-	-	-	-	mAdc
Plate Current (3)	Ib	Ec1=-40Vdc	5	-	-	-	-	-	mAdc
Pulsed Operation	ib	Ebb=750Vdc; Ecc1=-80Vdc; Ecc2=325Vdc; ec1=+100v	1.1	-	0.9	-	-	-	a
Screen Grid Current	Ic2		0.5	6.5	-	-	-	-	mAdc
Screen Grid Emission	Ic2	Eg2=150Vac; Eb=0 Note 2	-	-750	-	-	-	-	uAdc
Capacitance	C g1p	No shield,Ef=0	0.5	3.0	-	-	-	-	uuf
	C in		8.5	13.5	-	-	-	-	uuf
	C out		5.0	9.0	-	-	-	-	uuf
Control Grid Current (1)	Ic1		0	-0.2	0	-0.5	0	-0.5	uAdc
Control Grid Current (2)	Ic1	Ef=7.0V Note 1	0	-0.3	-	-	-	-	uAdc
Heater Cathode Leakage	Ihk	Ehk=+450Vdc	-	10	-	25	-	25	uAdc
	Ihk	Ehk=-450Vdc	-	10	-	25	-	25	uAdc
Insulation of R(g1-all) Electrodes R(p-all)		Eg1-all=-300V Ep-all=-500V	300 500	- -	150 250	- -	150 250	- -	Meg Meg
High Voltage Test	e	Ec1=-150Vdc; Ec2=300Vdc;Ebb =1250Vdc;Ic1= 1.0uAdc max; RL=5000	1150	-	-	-	-	-	v

Measurement conditions are the same as stated under Test Conditions, unless otherwise indicated.

Note 1: The tube shall be preheated a minimum of five minutes at test conditions for this test (except Ec1=0; Rk=250 ohms; Rg1=0.1 Meg) prior to this test.

Note 2: Operate the tube for five minutes at test conditions prior to the primary screen grid emission test. With the specified 60-cycle sinusoidal voltage applied to the screen, adjust Ec1 to provide Ic2=21 mAdc on the positive half-cycle. The primary emission shall be measured on the negative half cycles and shall not exceed the limits specified.

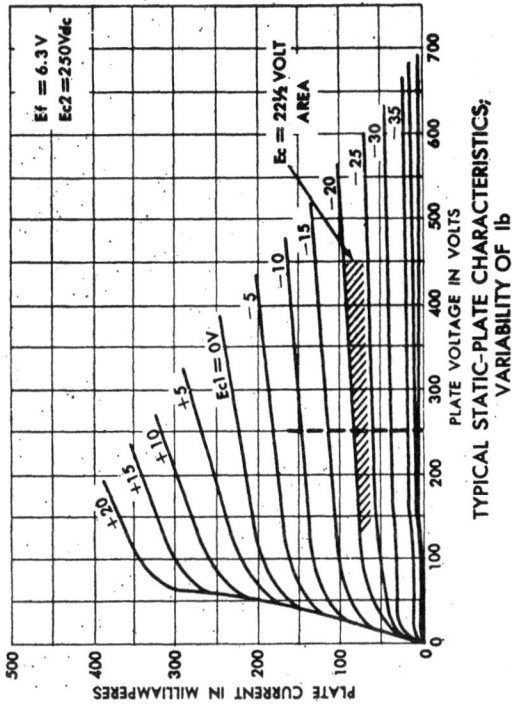

TYPICAL STATIC-PLATE CHARACTERISTICS;
VARIABILITY OF Ib

Ef = 6.3 V
Ec2 = 250 Vdc

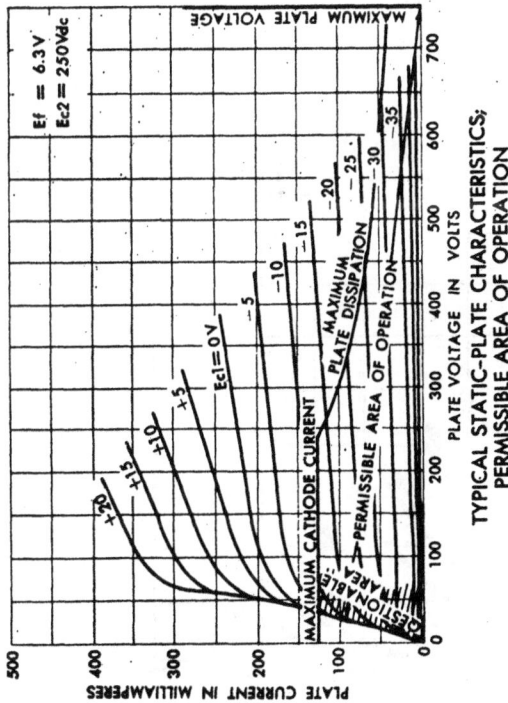

TYPICAL STATIC-PLATE CHARACTERISTICS;
PERMISSIBLE AREA OF OPERATION

Ef = 6.3 V
Ec2 = 250 Vdc

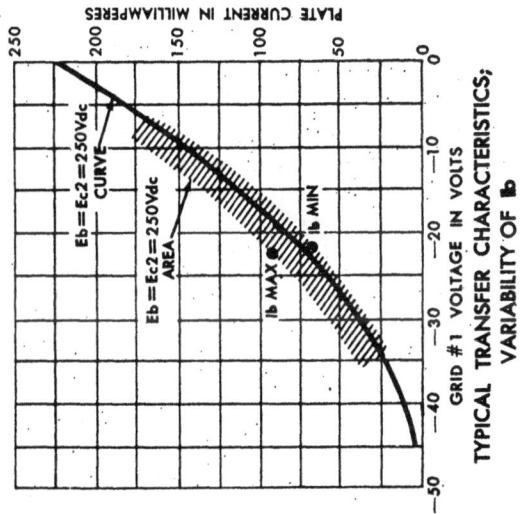

TYPICAL TRANSFER CHARACTERISTICS;
VARIABILITY OF Ib

215

DESIGN CENTER CHARACTERISTICS
OBTAINED FROM DATA PUBLISHED BY ORIGINAL
RETMA REGISTRANT

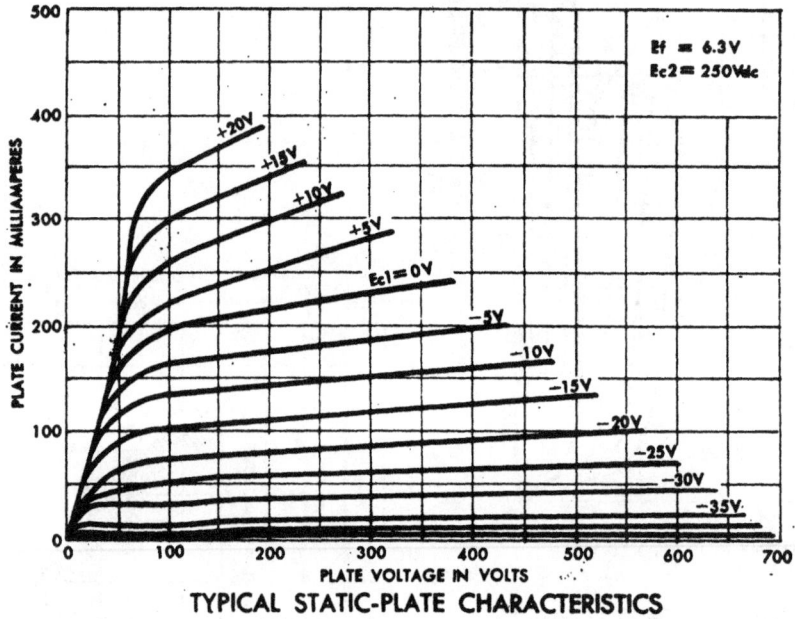

TYPICAL STATIC-PLATE CHARACTERISTICS

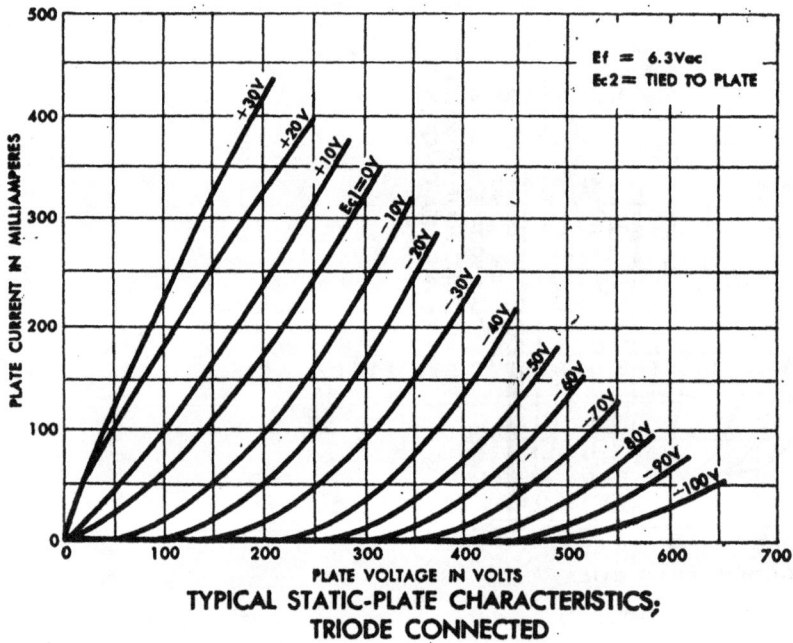

TYPICAL STATIC-PLATE CHARACTERISTICS;
TRIODE CONNECTED

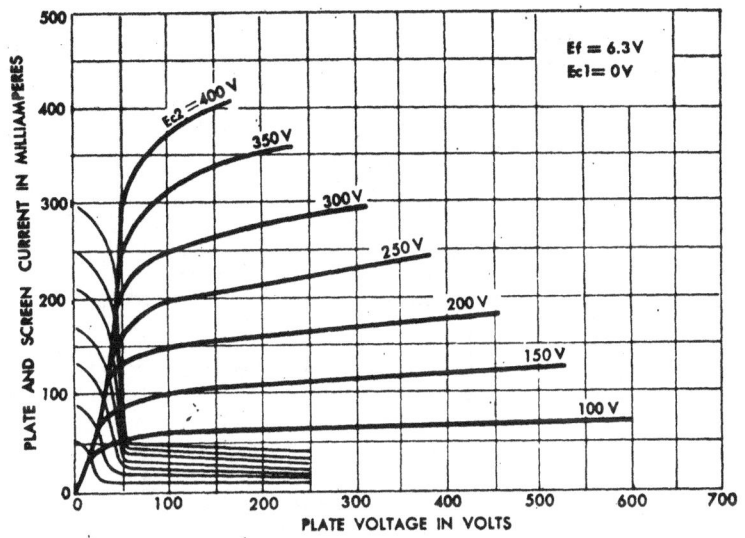

TYPICAL STATIC-PLATE CHARACTERISTICS

CHARACTERISTIC CURVES PRESENTED BELOW
ARE PRIMARILY FOR PULSE OPERATIONS FOR
WHICH THIS TUBE HAS RECEIVED USAGE

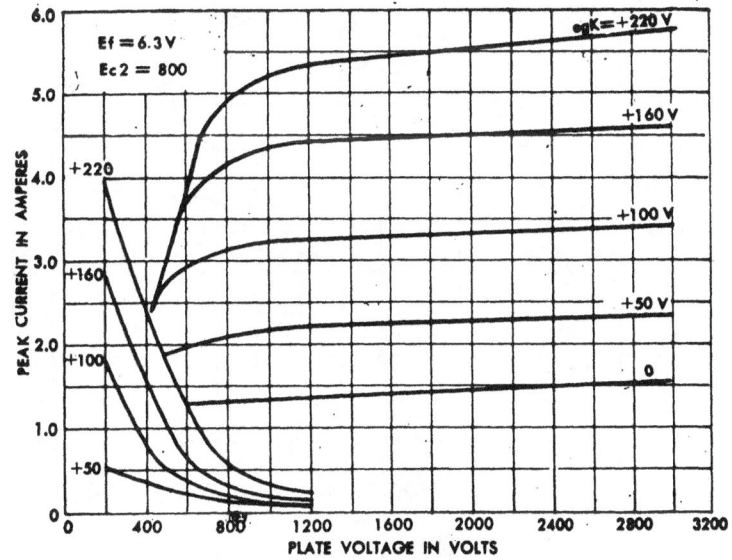

TYPICAL PULSE CHARACTERISTICS

217

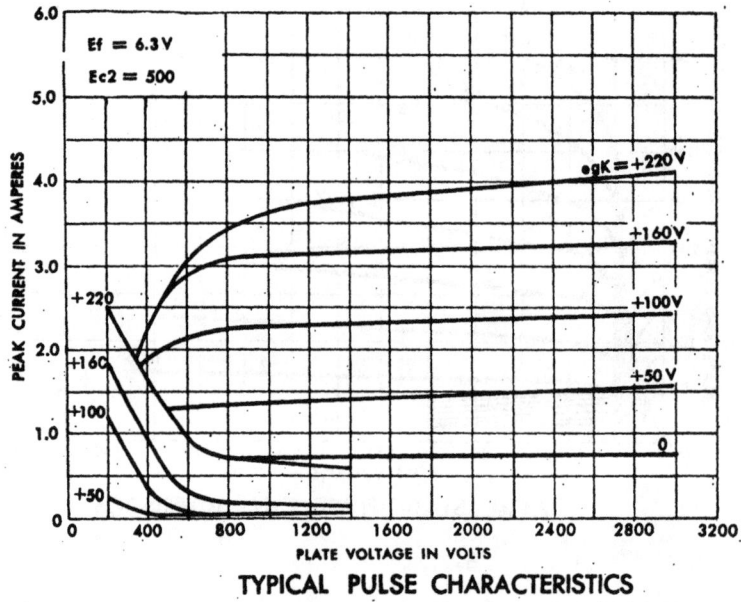

TYPICAL PULSE CHARACTERISTICS

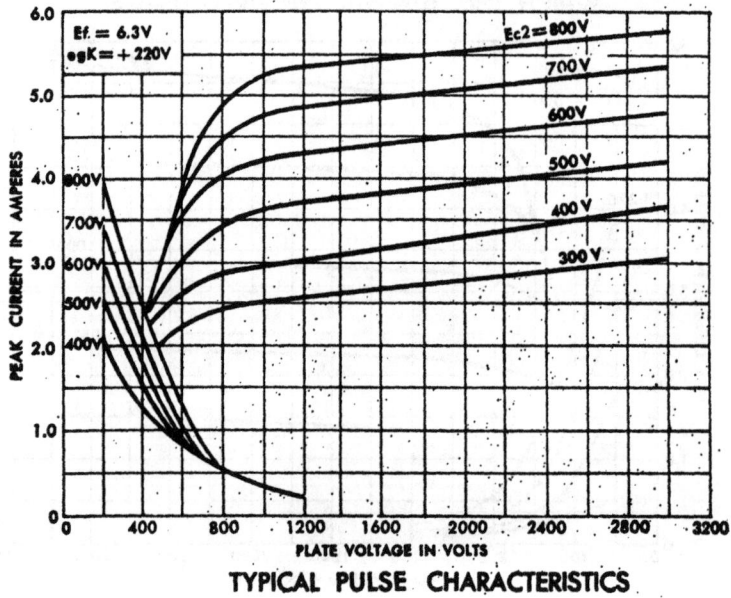

TYPICAL PULSE CHARACTERISTICS

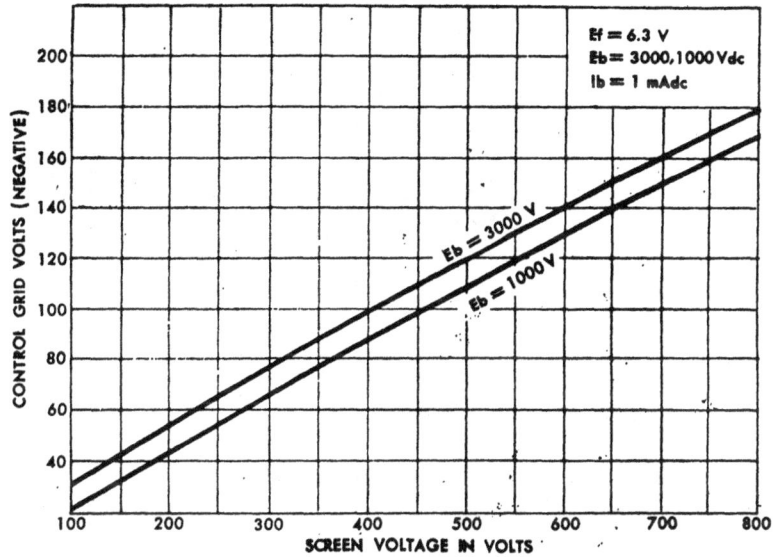

Ef = 6.3 V
Eb = 3000, 1000 Vdc
Ib = 1 mAdc

Eb = 3000 V
Eb = 1000 V

CONTROL GRID VOLTS (NEGATIVE)

SCREEN VOLTAGE IN VOLTS

TYPICAL CURRENT CUT-OFF CHARACTERISTICS

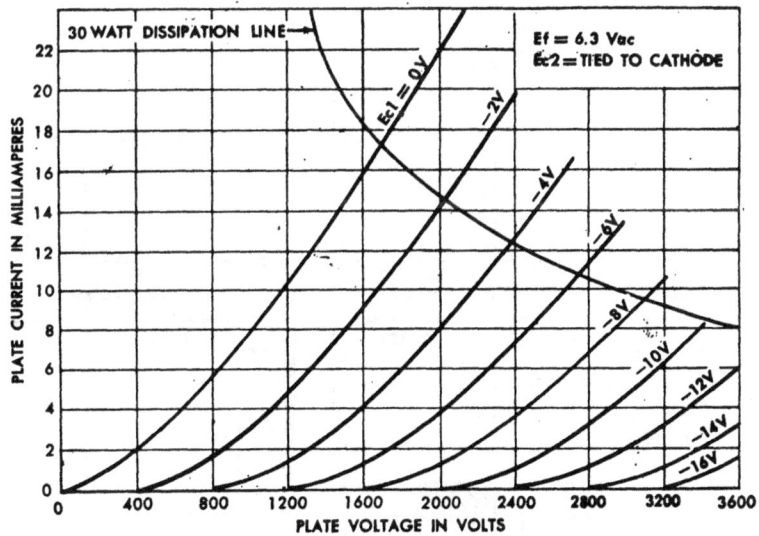

30 WATT DISSIPATION LINE →

Ef = 6.3 Vac
Ec2 = TIED TO CATHODE

Ec1 = 0V
-2V
-4V
-6V
-8V
-10V
-12V
-14V
-16V

PLATE CURRENT IN MILLIAMPERES

PLATE VOLTAGE IN VOLTS

TYPICAL PLATE CHARACTERISTICS,
SCREEN TIED TO CATHODE

LIFE TEST PROPERTY BEHAVIOR
MIL-E-1/1022   28 JUNE '56
PRODUCED IN 1957-'58 BY ONE MANUFACTURER

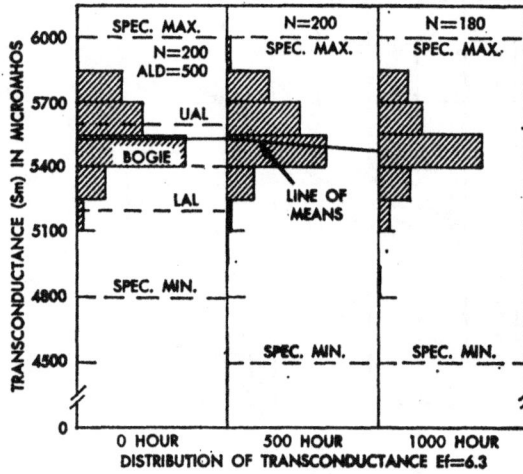

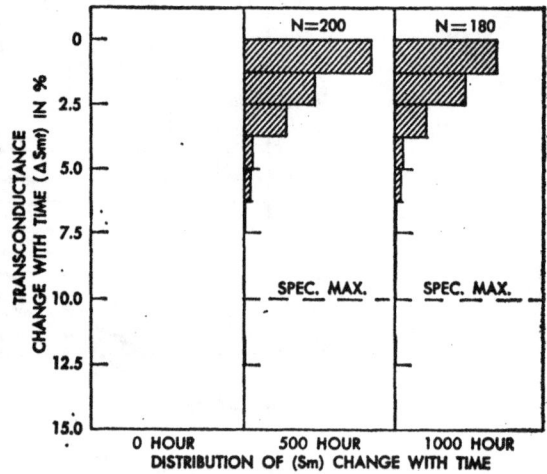

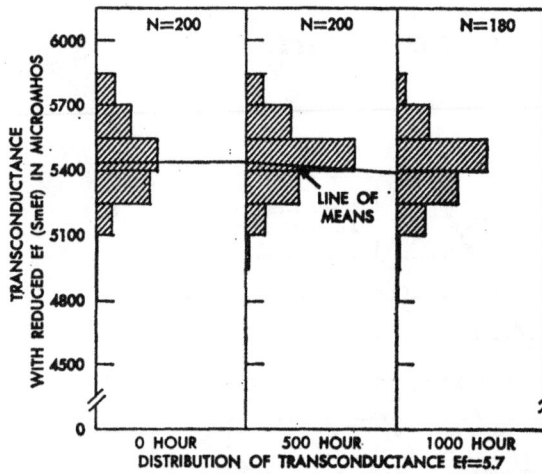

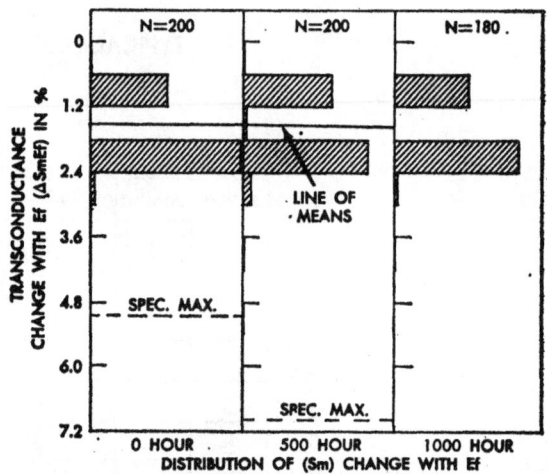

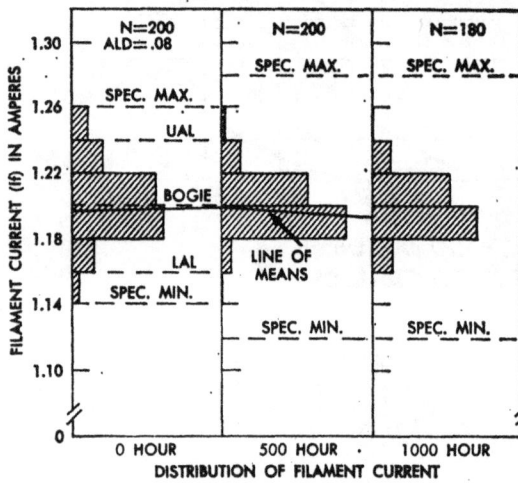

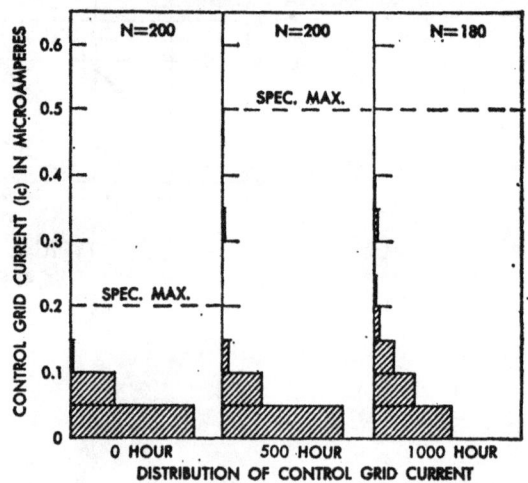

LIFE TEST PROPERTY BEHAVIOR
MIL-E-1/1022   28 JUNE '56
PRODUCED IN 1957-'58 BY ONE MANUFACTURER

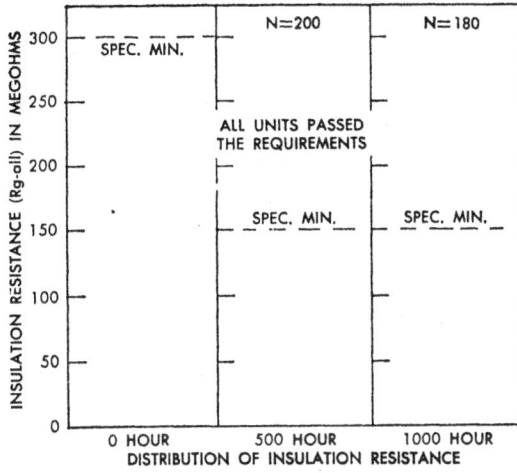

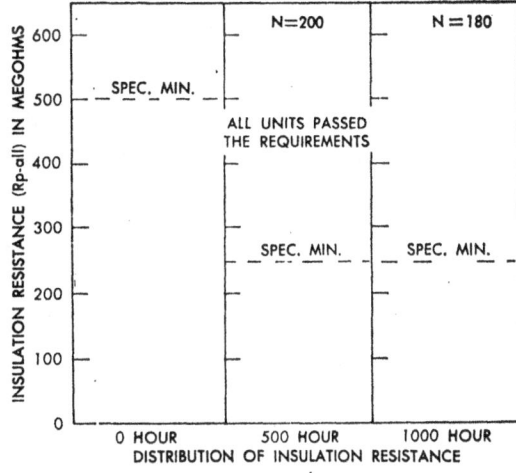

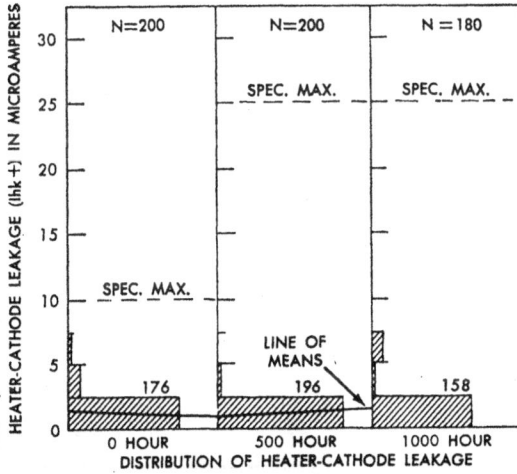

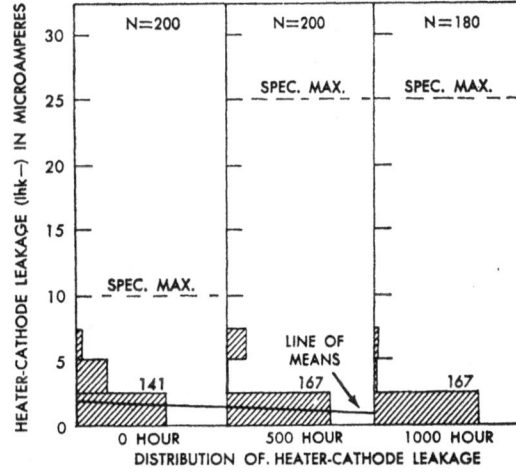

LIFE TEST PROPERTY BEHAVIOR
PROBABILITY OF SURVIVAL
MIL-E-1/1022   28 JUNE '56
PRODUCED IN 1957-'58 BY ONE MANUFACTURER

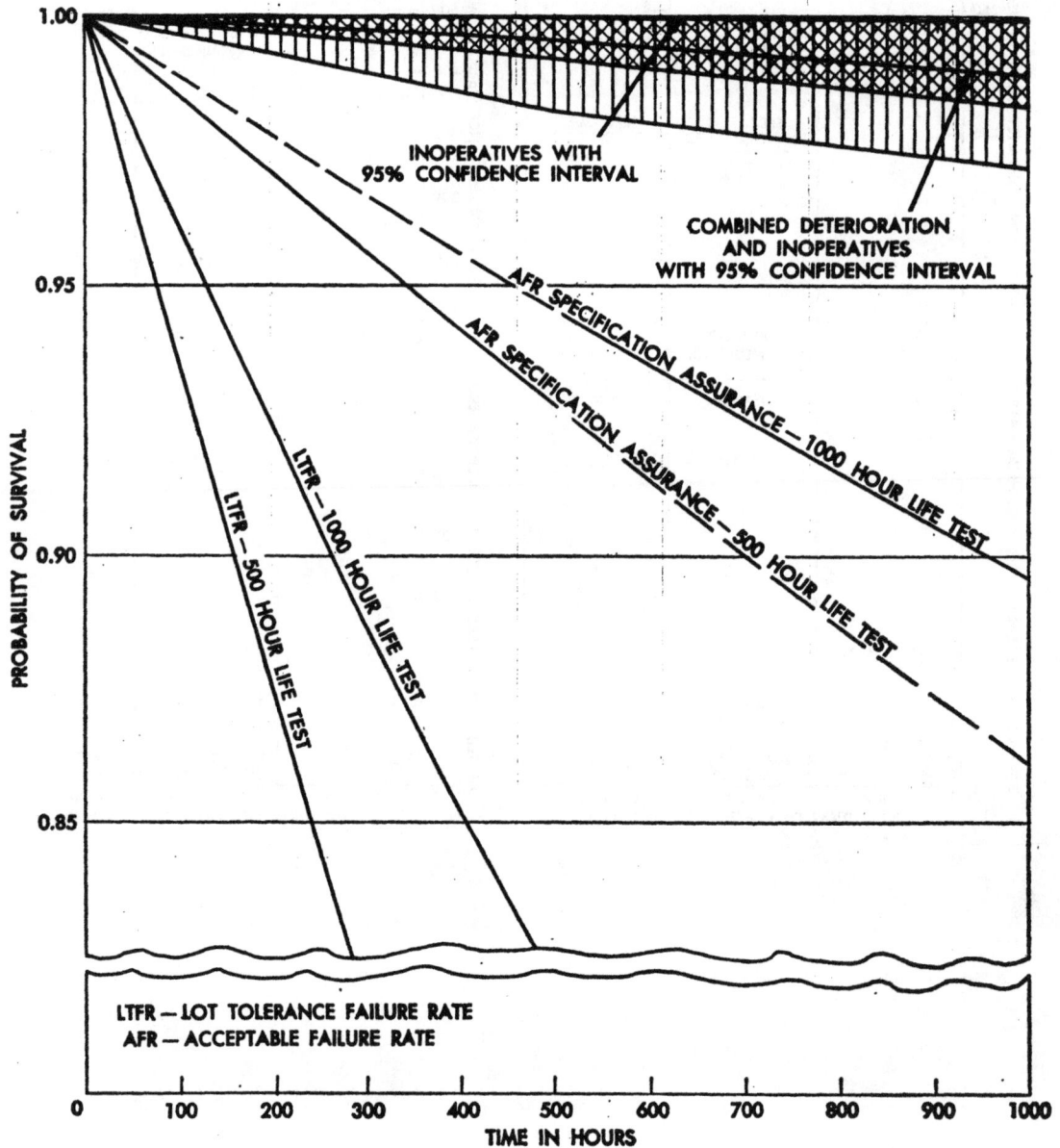

INOPERATIVES WITH
95% CONFIDENCE INTERVAL

COMBINED DETERIORATION
AND INOPERATIVES
WITH 95% CONFIDENCE INTERVAL

AFR SPECIFICATION ASSURANCE — 1000 HOUR LIFE TEST

AFR SPECIFICATION ASSURANCE — 500 HOUR LIFE TEST

LTFR — 1000 HOUR LIFE TEST

LTFR — 500 HOUR LIFE TEST

PROBABILITY OF SURVIVAL

LTFR — LOT TOLERANCE FAILURE RATE
AFR — ACCEPTABLE FAILURE RATE

TIME IN HOURS

☆ U. S. GOVERNMENT PRINTING OFFICE: 521545—1960—(V173)

www.ingramcontent.com/pod-product-compliance
Lightning Source LLC
Chambersburg PA
CBHW081501200326
41518CB00015B/2338